Springer Series in Materials Science

Volume 280

AF822776

The Springer Series in Materials Science covers the complete spectrum of materials physics, including fundamental principles, physical properties, materials theory and design. Recognizing the increasing importance of materials science in future device technologies, the book titles in this series reflect the state-of-the-art in understanding and controlling the structure and properties of all important classes of materials.

More information about this series at http://www.springer.com/series/856

Turab Lookman · Stephan Eidenbenz
Frank Alexander · Cris Barnes
Editors

Materials Discovery and Design

By Means of Data Science and Optimal Learning

Springer

Editors
Turab Lookman
Theoretical Division
Los Alamos National Laboratory
Los Alamos, NM, USA

Stephan Eidenbenz
Los Alamos National Laboratory
Los Alamos, NM, USA

Frank Alexander
Brookhaven National Laboratory
Brookhaven, NY, USA

Cris Barnes
Los Alamos National Laboratory
Los Alamos, NM, USA

ISSN 0933-033X ISSN 2196-2812 (electronic)
Springer Series in Materials Science
ISBN 978-3-030-07602-3 ISBN 978-3-319-99465-9 (eBook)
https://doi.org/10.1007/978-3-319-99465-9

This Springer imprint is published by the registered company Springer Nature Switzerland AG
The registered company address is: Gewerbestrasse 11, 6330 Cham, Switzerland

Preface

This book addresses aspects of data analysis and optimal learning as part of the co-design loop for future materials science innovation. The scientific process must cycle between theory and design of experiments and the conduct and analysis of them, in a loop that can be facilitated by more rapid execution. Computational and experimental facilities today generate vast amounts of data at an unprecedented rate. The role of visualization and inference and optimization methods, in distilling the data constrained by materials theory predictions, is key to achieving the desired goals of real-time analysis and control. The importance of this book lies in emphasizing that the full value of knowledge-driven discovery using data can only be realized by integrating statistical and information sciences with materials science, which itself is increasingly dependent on experimental data gathering efforts. This is especially the case as we enter a new era of big data in materials science with initiatives in exascale computation and with the planning and building of future coherent light source facilities such as the upgrade of the Linac Coherent Light Source at Stanford (LCLS-II), the European X-ray Free Electron Laser (EXFEL), and Matter Radiation in Extremes (MaRIE), the signature concept facility from Los Alamos National Laboratory. These experimental facilities, as well as present synchrotron light sources being upgraded and used in novel ways, are expected to generate hundreds of terabytes to several petabytes of in situ spatially and temporally resolved data per sample. The questions that then arise include how we can learn from this data to accelerate the processing and analysis of reconstructed microstructure, rapidly map spatially resolved properties from high throughput data, devise diagnostics for pattern detection, and guide experiments toward desired information and create materials with targeted properties or controlled functionality.

The book is an outgrowth of a conference held in Santa Fe, May 16–18, 2016 on "Data Science and Optimal Learning for Materials Discovery and Design". In addition, we invited a number of other authors active in these efforts, who did not participate in Santa Fe, to also contribute chapters. The authors are an interdisciplinary group of experts who include theorists surveying the open questions and future directions in the application of data science to materials problems, and experimentalists focusing on the challenges associated with obtaining, analyzing, and learning from data from large-scale user facilities, such as the Advanced Photon Source (APS) and LCLS. We have organized the chapters so that we start with a broad and fascinating perspective from **Lav Varshney** who discusses the relationship between accelerated materials discovery and problems in artificial intelligence, such as computational creativity, concept learning, and invention, as well as machine learning in other scientific domains. He shows how the connections lead to a number of common metrics including "dimension", information as measured in "bits" and Bayesian surprise, an entropy-related measure measured in "wows". With the thought-provoking title "Is Automated Materials Design and Discovery Possible?", **Mike McKerns** suggests that the tools traditionally used for finding materials with desired properties, which often make linear or quadratic approximations to handle the large dimensionality associated with the data, can be limiting as global optimization requires dealing with a highly nonlinear problem. He discusses the merits of the method of "Optimal Uncertainty Quantification" and the software tool *Mystic* as a possible route to handle such shortcomings. The importance of the choice and influence of material descriptors or features on the outcome of machine learning is the focus of the chapter by **Prasanna Balachandran et al.** They consider a number of materials data sets with different sets of features to independently track which of the sets finds most rapidly the compound with the largest target property. They emphasize that a relatively poor machine-learned model with large error but one that contains key features can be more efficient in accelerating the search process than a low-error model that lacks such features.

The bridge to the analysis of experimental data is provided by **Alisa Paterson et al.** who discuss the least squares and Bayesian inference approaches and show how they can be applied to X-ray diffraction data to study structure refinement. By considering single peak and full diffraction pattern fitting, they make the case that Bayesian inference provides a better model and generally affords the ability to escape from local minima and provide quantifiable uncertainties. They employ Markov Chain Monte Carlo algorithms to sample the distribution of parameters to construct the posterior probability distributions. The development of methods for extracting experimentally accessible spatially dependent information on structure and function from probes such as scanning transmission and scanning probe microscopies is the theme of the chapter by **Maxim Ziatdinov et al.** They emphasize the need to cross-correlate information from different experimental channels in physically and statistically meaningful ways and illustrate the use of machine learning and multivariate analysis to allow automated and accurate extraction and mapping of structural and functional material descriptors from experimental datasets. They consider a number of case studies, including strongly correlated materials.

The chapter by **Brian Patterson et al.** provides an excellent overview of the challenges associated with non-destructive 3D imaging and is a segue into the next three chapters also focused on imaging from incoherent and coherent light sources. This work features 3D data under dynamic time dependence at what is currently the most rapid strain rates available with present light sources. The chapter discusses issues and needs in the processing of large datasets of many terabytes in a matter of days from in situ experiments, and the developments required for automated reconstruction, filtering, segmentation, visualization, and animation, in addition to acquiring appropriate metrics and statistics characterizing the morphologies. **Reeju Pokharel** describes the technique and analysis tools associated with High Energy Diffraction Microscopy (HEDM) for characterizing polycrystalline microstructure under thermomechanical conditions. HEDM captures 3D views in a bulk sample at sub-grain resolution of about one micron. However, reconstruction from the diffraction signals is a computationally very intensive task. One of the challenges here is to develop tools based on machine learning and optimization to accelerate the reconstruction of images and decrease the time to analyze and use results to guide future experiments. The HEDM data can be utilized within a physics-based finite-element model of microstructure.

The final two chapters relate to aspects of light sources, in particular, advances in coherent diffraction imaging and the outstanding issues in the tuning and control of particle accelerators. In particular, **Edwin Fohtung et al**. discuss the recovery of the phase information from coherent diffraction data using iterative feedback algorithms to reconstruct the image of an object. They review recent developments including Bragg Coherent Diffraction Imaging (BCDI) for oxide nanostructures, as well as the big data challenges in BCDI. Finally, **Alex Sheinker** closes the loop by discussing the major challenges faced by future coherent light sources, such as fourth-generation Free Electron Lasers (FELs), in achieving extremely tight constraints on beam quality and in quickly tuning between various experimental setups under control. He emphasizes the need for feedback to achieve this control and outlines an extremum seeking method for automatic tuning and optimization.

The chapters in this book span aspects of optimal learning, from using information theoretic-based methods in the analysis of experimental data, to adaptive control and optimization applied to the accelerators that serve as light sources. Hence, the book is aimed at an interdisciplinary audience, with the subjects integrating aspects of statistics and mathematics, materials science, and computer science. It will be of timely appeal to those interested in learning about this emerging field. We are grateful to all the authors for their articles as well as their support of the editorial process.

Los Alamos, NM, USA — Turab Lookman
Los Alamos, NM, USA — Stephan Eidenbenz
Los Alamos, NM, USA — Cris Barnes
Brookhaven, NY, USA — Frank Alexander

Contents

Contributors

Prasanna V. Balachandran Los Alamos National Laboratory, Los Alamos, NM, USA; Department of Materials Science and Engineering, Department of Mechanical and Aerospace Engineering, University of Virginia, Charlottesville, VA, USA

Tilo Baumbach Institute for Photon Science and Synchrotron Radiation, Kasrlsruhe Institute of Technology, Eggenstein-Leopoldshafen, Germany

Nikhilesh Chawla 4D Materials Science Center, Arizona State University, Tempe, AZ, USA

Nikolaus L. Cordes Materials Science and Technology Division, Engineered Materials Group, Los Alamos National Laboratory, Los Alamos, NM, USA

Edwin Fohtung Department of Physics, New Mexico State University, Las Cruces, NM, USA; Los Alamos National Laboratory, Los Alamos, NM, USA

James E. Gubernatis Los Alamos National Laboratory, Los Alamos, NM, USA

Kevin Henderson Materials Science and Technology Division, Engineered Materials Group, Los Alamos National Laboratory, Los Alamos, NM, USA

John Hogden Los Alamos National Laboratory, Los Alamos, NM, USA

Jacob L. Jones Department of Materials Science and Engineering, North Carolina State University, Raleigh, NC, USA

Sergei V. Kalinin Oak Ridge National Laboratory, Institute for Functional Imaging of Materials, Oak Ridge, TN, USA; Oak Ridge National Laboratory, Center for Nanophase Materials Sciences, Oak Ridge, TN, USA

Dmitry Karpov Department of Physics, New Mexico State University, Las Cruces, NM, USA; Physical-Technical Institute, National Research Tomsk Polytechnic University, Tomsk, Russia

Turab Lookman Los Alamos National Laboratory, Los Alamos, NM, USA

Artem Maksov Oak Ridge National Laboratory, Institute for Functional Imaging of Materials, Oak Ridge, TN, USA; Oak Ridge National Laboratory, Center for Nanophase Materials Sciences, Oak Ridge, TN, USA; Bredesen Center for Interdisciplinary Research, University of Tennessee, Knoxville, TN, USA

Michael McKerns The Uncertainty Quantification Foundation, Wilmington, DE, USA

Alisa R. Paterson Department of Materials Science and Engineering, North Carolina State University, Raleigh, NC, USA

Brian M. Patterson Materials Science and Technology Division, Engineered Materials Group, Los Alamos National Laboratory, Los Alamos, NM, USA

Reeju Pokharel Los Alamos National Laboratory, Los Alamos, NM, USA

Brian J. Reich Department of Statistics, North Carolina State University, Raleigh, NC, USA

Alexander Scheinker Los Alamos National Laboratory, Los Alamos, NM, USA

Ralph C. Smith Department of Mathematics, North Carolina State University, Raleigh, NC, USA

James Theiler Los Alamos National Laboratory, Los Alamos, NM, USA

Lav R. Varshney Coordinated Science Laboratory and Department of Electrical and Computer Engineering, University of Illinois at Urbana-Champaign, Urbana, USA

Alyson G. Wilson Department of Statistics, North Carolina State University, Raleigh, NC, USA

Xianghui Xiao X-ray Photons Sciences, Argonne National Laboratory, Argonne, IL, USA

Dezhen Xue State Key Laboratory for Mechanical Behavior of Materials, Xi'an Jiaotong University, X'ian, China

Maxim Ziatdinov Oak Ridge National Laboratory, Institute for Functional Imaging of Materials, Oak Ridge, TN, USA; Oak Ridge National Laboratory, Center for Nanophase Materials Sciences, Oak Ridge, TN, USA

Chapter 1
Dimensions, Bits, and Wows in Accelerating Materials Discovery

Lav R. Varshney

Abstract In this book chapter, we discuss how the problem of accelerated materials discovery is related to other computational problems in artificial intelligence, such as computational creativity, concept learning, and invention, as well as to machine-aided discovery in other scientific domains. These connections lead, mathematically, to the emergence of three classes of algorithms that are inspired largely by the approximation-theoretic and machine learning problem of dimensionality reduction, by the information-theoretic problem of data compression, and by the psychology and mass communication problem of holding human attention. The possible utility of functionals including dimension, information [measured in bits], and Bayesian surprise [measured in wows], emerge as part of this description, in addition to measurement of quality in the domain.

1.1 Introduction

Finding new materials with targeted properties is of great importance to technological development in numerous fields including clean energy, national security, resilient infrastructure, and human welfare. Classical approaches to materials discovery rely mainly on trial-and-error, which requires numerous costly and time-intensive experiments. As such, there is growing interest in using techniques from the information sciences in accelerating the process of finding advanced materials such as new metal alloys or thermoelectric materials [1, 2]. Indeed the national *Materials Genome Initiative*—a large-scale collaboration to bring together new digital data, computational tools, and experimental tools—aims to quicken the design and deployment of advanced materials, cf. [3, 4]. In developing these computational tools, there is a

L. R. Varshney (✉)
Coordinated Science Laboratory and Department of Electrical and Computer Engineering, University of Illinois at Urbana-Champaign, Urbana 61801, USA
e-mail: varshney@illinois.edu

T. Lookman et al. (eds.), *Materials Discovery and Design*, Springer Series in Materials Science 280, https://doi.org/10.1007/978-3-319-99465-9_1

desire not only for supercomputing hardware infrastructure [5], but also advanced algorithms.

In most materials discovery settings of current interest, however, the algorithmic challenge is formidable. Due to the interplay between (macro- and micro-) structural and chemical degrees of freedom, computational prediction is difficult and inaccurate. Nevertheless, recent research has demonstrated that emerging statistical inference and machine learning algorithms may aid in accelerating the materials discovery process [1].

The basic process is as follows. Regression algorithms are first used to learn the functional relationship between features and properties from a corpus of some extant characterized materials. Next, an unseen material is tested experimentally and those results are used to enhance the functional relationship model; this unseen material should be chosen as *best* in some sense. Proceeding iteratively, more unseen materials are designed, fabricated, and tested and the model is further refined until a material that satisfies desired properties is obtained. This process is similar to the active learning framework (also called adaptive experimental design) [6], but unlike active learning, here the training set is typically very small: only tens or hundreds of samples as compared to the unexplored space that is combinatorial (in terms of constituent components) and continuous-valued (in terms of their proportions). It should be noted that the ultimate goal is not to learn the functional relationship accurately, but to discover the optimal material with the fewest trials, since experimentation is very costly.

What should be the notion of *best* in iteratively investigating new materials with particular desired properties? This is a *constructive machine learning* problem, where the goal of learning is not to find a good model of data but instead to find one or more particular instances of the domain which are likely to exhibit desired properties. Perhaps the criterion in picking the next sample should be to learn about a useful dimension in the feature space to get a sense of the entire space of possibilities rather than restricting to a small-dimensional manifold [7]. By placing attention on a new dimension of the space, new insights for discovery may be possible [8]. Perhaps the criterion for picking the next sample should be to choose the most informative, as in *infotaxis* in machine learning and descriptions of animal curiosity/behavior [9–13]. Perhaps the goal in driving materials discovery should be to be as surprising as possible, rather than to be as informative as possible, an algorithmic strategy for accelerated discovery one might call *surprise-taxis*. (As we will see, the Bayesian surprise functional is essentially the derivative of Shannon's mutual information [14], and so this can be thought of as a second-order method, cf. [15].)

In investigating these possibilities, we will embed our discussion in the larger framework of data-driven scientific discovery [16, 17] where theory and computation interact to direct further exploration. The overarching aim is to develop a viable *research tool* that is of relevance to materials scientists in a variety of industries, and perhaps even to researchers in further domains like drug cocktail discovery. The general idea is to provide researchers with cognitive support to augment their own intelligence [18], just like other technologies including pencil-and-paper [19, 20] or internet-based tools [21, 22] often lead to greater quality and efficiency of human thought.

When we think about human intelligence, we think about the kinds of abilities that people have, such as memory, deductive reasoning, association, perception, abductive reasoning, inductive reasoning, and problem solving. With technological advancement over the past century, computing technologies have progressed to the stage where they too have many of these abilities. The pinnacle of human intelligence is often said to be creativity and discovery, ensconced in such activities as music composition, scientific research, or culinary recipe design. One might wonder, then, can computational support help people to create and discover novel artifacts and ideas?

In addressing this question, we will take inspiration from related problems including computational creativity, concept learning, and invention, as well as from machine-aided discovery in other scientific domains. Connections to related problems lead, mathematically, to the emergence of three classes of accelerated discovery algorithms that are inspired largely by the approximation-theoretic [23] and machine learning problem of dimensionality reduction [24], by the information-theoretic problem of data compression [25, 26], and by the psychology and mass communication problem of holding human attention. The possible utility of functionals including dimension, information [measured in bits], and Bayesian surprise [measured in wows], emerge as part of this description, in addition to measurement of quality in the domain. It should be noted that although demonstrated in other creative and scientific domains, accelerated materials discovery approaches based on these approximation-theoretic and information-theoretic functionals remain speculative.

1.2 Creativity and Discovery

Whether considering literary manuscripts, musical compositions, culinary recipes, or scientific ideas, the basic argument framing this chapter is that it is indeed possible for computers to create novel, high-quality ideas or artifacts, whether operating autonomously or semi-autonomously by engaging with people. As one typical example, consider a culinary computational creativity system that uses repositories of existing recipes, data on the chemistry of food, and data on human hedonic perception of flavor to create new recipes that have never been cooked before, but that are flavorful [27–29]. As another example, consider a machine science system that takes the scientific literature in genomics, generates hypotheses, and tests them automatically to create new scientific knowledge [30]. Some classical examples of computational creativity include AARON, which creates original artistic images that have been exhibited in galleries around the world [31], and BRUTUS, which tells stories [32]. Several new applications, theories, and trends are now emerging in the field of computational creativity [33–35].

Although several specific algorithmic techniques have been developed in the literature, the basic structure of many computational creativity algorithms proceed by first taking existing artifacts from the domain of interest and intelligently performing a variety of transformations and modifications to generate new ideas; the design space has combinatorial complexity [36]. Next, these generated possibilities are assessed to predict if people would find them compelling as creative artifacts and the best are chosen. Some algorithmic techniques combine the generative and selective steps into a single optimization procedure.

A standard definition of creativity emerging in the psychology literature [37] is that: *Creativity is the generation of an idea or artifact that is judged to be novel and also to be appropriate, useful, or valuable by a suitably knowledgeable social group.* A critical aspect of any creativity algorithm is therefore determining a meaningful characterization of what constitutes a good artifact in the two distinct dimensions of novelty and utility. Note that each domain—whether literature or culinary art—has its own specific metrics for quality. However, independent of domain, people like to be surprised and there may be abstract information-theoretic measures for surprise [14, 38–40].

Can this basic approach to computational creativity be applied to accelerating discovery through machine science [41]? Most pertinently, one might wonder whether novelty and surprise are essential to problems like accelerating materials discovery, or is utility the only consideration. The wow factor of newly creative things or newly discovered facts is important in regimes with an excess of potential creative artifacts or growing scientific literature, not only for ensuring novelty but also for capturing people's attention. More importantly, however, it is important for pushing discovery into wholly different parts of the creative space than other computational/algorithmic techniques can. Designing for surprise is of utmost importance.

For machine science in particular, the following analogy to the three layers of communication put forth by Warren Weaver [42] seems rather apt.

Level A (The technical problem)
Communication: How accurately can the symbols of communication be transmitted? *Machine Science*: How accurately does gathered data represent the state of nature?
Level B (The semantic problem)
Communication: How precisely do the transmitted symbols convey the desired meaning? *Machine Science*: How precisely does the measured data provide explanation into the nature of the world?
Level C (The effectiveness problem)
Communication: How effectively does the received meaning affect conduct in the desired way? *Machine Science*: How surprising are the insights that are learned?

A key element of machine science is therefore not just producing accurate and explanatory data, but insights that are surprising as compared to current scientific understanding.

In the remainder of the chapter, we introduce three basic approaches to discovery algorithms, based on dimensions, information, and surprise.

1.3 Discovering Dimensions

One of the central problems in unsupervised machine learning for understanding, visualization, and further processing has been *manifold learning* or *dimensionality reduction*. The basic idea is to assume that a given set of data points that have some underlying low-dimensional structure are embedded in a high-dimensional Euclidean space, and the goal is to recover that low-dimensional structure. Note that the low-dimensional structure can be much more general than a classical smooth manifold [43, 44]. Such machine learning-based approaches generalize, in some sense, classical harmonic analysis and approximation theory where a fixed representation, say a truncated representation in the Fourier basis, is used as a low-dimensional representation [23].

The most classical approach, principal components analysis (PCA) [45, 46], is a linear transformation of data defined so the first principal component has the largest possible variance, accounting for as much of the data variability as possible. The second principal component has the highest variance possible under the constraint that it is orthogonal to the first principal component, and so on. This linear transformation method, accomplished by computing an eigenbasis, also turns possibly correlated variables into values of linearly uncorrelated variables. It can be extended to work with missing data [47]. One of the distinguishing features of PCA is that the learned transformation can be applied directly to data that was not used to train the transformation, so-called *out-of-sample extension*.

There are several nonlinear dimensionality reduction algorithms that first construct a sparsely-connected graph representation of local affinity among data points and then embed these points into a low-dimensional space, trying to preserve as much of the original affinity as possible. Examples include locally linear embedding [48], multidimensional scaling methods that try to preserve global information such as Isomap [49], spectral embeddings such as Laplacian eigenmaps [50], and stochastic neighbor embedding [51]. Direct out-of-sample extension is not possible with these techniques, and so further techniques such as the Nyström approximation are needed [52].

Another approach that supports direct out-of-sample extension is dimensionality reduction using an autoencoder. An autoencoder is a feedforward neural network that is trained to approximate the identity function, such that it maps a vector of values to itself. When used for dimensionality reduction, a hidden layer in the network is constrained to contain only a small number of neurons and so the network must learn to encode the vector into a small number of dimensions and then decode it back. Consequently, the first part of the network maps from high to low-dimensional space, and the second maps in the reverse manner.

With this background on dimensionality reduction, we can now present an accelerated discovery algorithm that essentially pursues dimensions in order to prioritize investigation of data. This Discovery through Eigenbasis Modeling of Uninteresting Data (DEMUD) algorithm, due to Wagstaff et al. [7], is essentially based on PCA and is meant not just to prioritize data for investigation but also provide domain-specific explanations for why a given item is potentially interesting. The reader will notice the fact that novel discovery algorithms could be developed using other dimensionality reduction techniques that can be updated and with direct out-of-sample extension in place of PCA, for example using autoencoders.

The basic idea of DEMUD is to use a notion of uninterestingness to judge what to select next. Data that has already been seen, data that is not of interest due to its category, or prior knowledge of uninterestingness are all used to iteratively model what should be ignored in selecting a new item of high interest. The specific technique used is to first compute a low-dimensional eigenbasis of uninteresting items using a singular value decomposition $U \Sigma V^T$ of the original dataset X and retaining the top k singular vectors (ranked by magnitude of the corresponding singular value). Data items are then ranked according to the reconstruction error in representing in this basis: items with largest error are said to have the most potential to be novel, as they are largely in an unmodeled dimension of the space. In order to initialize, we use the whole dataset, but then proceed iteratively in building up the eigenbasis. Specifically, the DEMUD algorithm takes the following three inputs: $X \in \mathbb{R}^{n \times d}$ as the input data, $X_U = \emptyset$ as the initial set of uninteresting items, and k as the number of principal components to be used in X_U. Then it proceeds as follows.

Algorithm 1 DEMUD [7]

1: Let $U = SVD(X, k)$ be the initial model of X_U and let μ be the mean of the data
2: **while** discovery is to continue and $X \neq \emptyset$ **do**
3: Compute reconstructions $\hat{x} = UU^T(x - \mu) + \mu$ for all $x \in X$
4: Compute error in reconstructions $R(x) = \|x - \hat{x}\|_2 = \|x - (UU^T(x - \mu) + \mu)\|_2$ for all $x \in X$
5: Choose $x' = \text{argmax}_{x \in X} R(x)$ to investigate next
6: Remove this data item from the data set and add it to the model, i.e. $X = X \backslash \{x'\}$ and $X_U = X_U \cup \{x'\}$.
7: Update U and μ by using the incremental SVD algorithm [53] with inputs (U, x', k).
8: **end while**

The ordering of data to investigate that emerges from the DEMUD algorithm is meant to quickly identify rare items of scientific value, maintain diversity in its selections, and also provide explanations (in terms of dimensions/subspaces to explore) to aid in human understanding. The algorithm has been demonstrated using hyperspectral data for exploring rare minerals in planetary science [7].

1.4 Infotaxis

Having discussed how the pursuit of novel dimensions in the space of data may accelerate scientific discovery, we now discuss how pursuit of information may do likewise. In Shannon information theory, the *mutual information* functional emerges from the noisy channel coding theorem in characterizing the limits of reliable communication in the presence of noise [54] and from the rate-distortion theorem in characterizing the limits of data compression [55]. In particular, the notion of information rate (e.g. measured in bits) emerges as a universal interface for communication systems. For two continuous-valued random variables, $X \in \mathscr{X}$ and $Y \in \mathscr{Y}$ with corresponding joint density $f_{XY}(x, y)$ and marginals $f_X(x)$ and $f_Y(y)$, the mutual information is given as

$$I(X; Y) = \int_{\mathscr{Y}} \int_{\mathscr{X}} f_{XY}(x, y) \log \frac{f_{XY}(x, y)}{f_X(x) f_Y(y)} dx dy.$$

If the base of the logarithm is chosen as 2, then the units of mutual information are bits. The mutual information can also be expressed as the difference between an unconditional entropy and a conditional one.

There are several methods for estimating mutual information from data, ranging from plug-in estimators for discrete-valued data to much more involved minimax estimators [56] and ensemble methods [57]. For continuous-valued data, there are a variety of geometric and statistical techniques that can also be used [58, 59].

Mutual information is often used to measure informativeness even outside the communication settings where the theorems are proven, since it is a useful measure of mutual dependence that indicates how much knowing one variable reduces uncertainty about the other. Indeed, there is an axiomatic derivation of the mutual information measure, where it is shown that it is the unique (up to choice of logarithm base) function that satisfies certain properties such as continuity, strong additivity, and an increasing-in-alphabet-size property. In fact, there are several derivations with differing small sets of axioms [60].

Of particular interest here is the pursuit of information as a method of discovery, in an algorithm that is called infotaxis [9–13]. The infotaxis algorithm was first explicitly discussed in [9] who described it as a model for animal foraging behavior. The basic insight of the algorithm is that it is a principled way to essentially encode exploration-exploitation trade-offs in search/discovery within an uncertain environment, and therefore has strong connections to reinforcement learning. There is a given but unknown (to the algorithm) probability distribution for the location of the source being searched for and the rate of information acquisition is also the rate of entropy reduction. The basic issue in discovering the source is that the underlying probability distribution is not known to the algorithm but must be estimated from available data. Accumulation of information allows a tighter estimate of the source distribution. As such, the searcher must choose either to move to the most likely source location or to pause and gather more information to make a better estimate of the source. Infotaxis allows a balancing of these two concerns by choosing to move (or stay still) in the direction that maximizes the expected reduction in entropy.

As noted, this algorithmic idea has been used to explain a variety of human/animal curiosity behaviors and also been used in several engineering settings.

1.5 Pursuit of Bayesian Surprise

Rather than moving within a space to maximize expected gain of information (maximize expected reduction of entropy), would it ever make sense to consider maximizing surprise instead. In the common use of the term, pursuit of surprise seems to indicate a kind of curiosity that would be beneficial for accelerating discovery, but is there a formal view of surprise as there is for information? How can we compute whether something is likely to be perceived as surprising?

A particularly interesting definition is based on a psychological and information-theoretic measure termed *Bayesian surprise*, due originally to Itti and Baldi [38, 40]. The surprise of each location on a feature map is computed by comparing beliefs about what is likely to be in that location before and after seeing the information. Indeed, novel and surprising stimuli spontaneously attract attention [61].

An artifact that is surprising is novel, has a wow factor, and changes the observer's world view. This can be quantified by considering a prior probability distribution of existing ideas or artifacts and the change in that distribution after the new artifact is observed, i.e. the posterior probability distribution. The difference between these distributions reflects how much the observer's world view has changed. It is important to note that surprise and saliency depend heavily on the observer's existing world view, and thus the same artifact may be novel to one observer and not novel to another. That is why Bayesian surprise is measured as a change in the observer's specific prior probability distribution of known artifacts.

Mathematically, the cognitively-inspired Bayesian surprise measure is defined as follows. Let $\mathcal{M}$ be the set of artifacts known to the observer, with each artifact in this repository being $M \in \mathcal{M}$. Furthermore, a new artifact that is observed is denoted D. The probability of an existing artifact is denoted $p(M)$, the conditional probability of the new artifact given the existing artifacts is $p(D|M)$, and via Bayes' theorem the conditional probability of the existing artifacts given the new artifact is $p(M|D)$. The Bayesian surprise is defined as the following relative entropy (Kullback-Leibler divergence):

$$s = D(p(M|D)||p(M)) = \int_{\mathcal{M}} p(M|D) \log \frac{p(M|D)}{p(M)} dM$$

One might wonder if Bayesian surprise, $s(D)$, has anything to do with measures of information such as Shannon's mutual information given in the previous section. In fact, if there is a definable distribution on new artifacts $q(D)$, the expected value of Bayesian surprise is the Shannon mutual information.

$$E[s(D)] = \int q(D) D(p(M|D)||p(M)) dD = \int \int_{\mathcal{M}} p(M, D) \log \frac{p(M|D)}{p(M)} dM dD,$$

which by definition is the Shannon mutual information $I(M; D)$. The fact that the average of the Bayesian surprise equals the mutual information points to the notion that surprise is essentially the derivative of information.

Let us define the weak derivative, which arises in the weak-* topology [62], as follows.

Definition Let $\mathscr{A}$ be a vector space, and f a real-valued functional defined on domain $\Omega \subset \mathscr{A}$, where Ω is a convex set. Fix an $a_0 \in \Omega$ and let $\theta \in [0, 1]$. If there exists a map $f'_{a_0} : \Omega \to \mathbb{R}$ such that

$$f'_{a_0}(a) = \lim_{\theta \downarrow 0} \frac{f[(1-\theta)a_0 + \theta a] - f(a_0)}{\theta}$$

for all $a \in \Omega$, then f is said to be *weakly differentiable in* Ω *at* a_0 and f'_{a_0} is the *weak derivative in* Ω *at* a_0.

If f is weakly differentiable in Ω at a_0 for all a_0 in Ω, then f is said to be *weakly differentiable*.

The precise relationship can be formalized as follows. For a fixed reference distribution $F_0 = q(D)$, the weak derivative of mutual information is:

$$I'_{F_0}(F) = \lim_{\theta \downarrow 0} \frac{(I((1-\theta)F_0 + \theta F_0) - I(F_0))}{\theta} = \int s(x)q(x)dx - I(F_0)$$

Indeed, even the Shannon capacity C of communication over a stochastic kernel $p(M|D)$ can be expressed in terms of the Bayesian surprise [63]:

$$C = \max_{q(D)} I(M; D) = \min_{p(M)} \max_{d \in \mathcal{M}} s(d),$$

therefore all communicated signals should be equally surprising when trying to maximize information rate of communication.

These formalisms are all well and good, but it is also important to have operational meaning for Bayesian surprise to go alongside. In fact, there are several kinds of operational meanings that have been established in a variety of fields.

- In defining Bayesian surprise, Itti and Baldi also performed several psychology experiments that demonstrated its connection to attraction of human attention across different spatiotemporal scales, modalities, and levels of abstraction [39, 40]. As a typical example of a such an experiment, human subjects were tasked with looking at a video of a soccer game while being measured using eye-tracking. The Bayesian surprise for the video was also computed. The places where the Bayesian surprise was large was also where the human subjects were looking. These classes of experiments have been further studied by several other research groups in psychology, e.g. [64–67].
- Bayesian surprise has not just been observed at a behavioral level, but also at a neurobiological level [68–70], where various brain processes concerned with attention have been related to Bayesian surprise.

- In the engineering of computational creativity systems, it has empirically been found that Bayesian surprise is a useful optimization criterion for ideas or artifacts to be rated as highly creative [27–29, 71]. Likewise in marketing [72], Bayesian surprise has been found to be an effective criterion for designing promotion campaigns [73].
- In the Bayesian model comparison literature, Bayesian surprise is also called *complexity* [74] and in thermodynamic formulations of Bayesian inference [75], an increase in Bayesian surprise is necessarily associated with a decrease in free-energy due to a reduction in prediction error. It should ne noted, however, that Bayes-optimal inference schemes do not optimize for Bayesian surprise in itself [74].
- In information theory, Bayesian surprise is sometimes called the *marginal information density* [76]. When communicating in information overload regimes, it is necessary for messages to not only provide information but also to attract attention in the first place. In many communication settings, the flood of messages is not only immense but also monotonously similar. Some have argued that "it would be far more effective to send one very unusual message than a thousand typical ones" [77]. The Bayesian surprise therefore arises in information-theoretic studies of optimal communication systems. One example is in highly-asynchronous communication, where the receiver must monitor the channel for long stretches of time before a transmitted signal appears [78]. Moreover, we have shown that Bayesian surprise is the natural cost function for communication just like log-loss [79] is the natural fidelity criterion for compression [14] (as follows from KKT conditions [80]). One can further note that there is a basic tradeoff between messages being informative and being surprising [14].

Given that Bayesian surprise has operational significance in a variety of psychology, neurobiology, statistics, creativity, and communication settings, as well as formal derivative relationships to mutual information, one might wonder if an accelerated discovery algorithm that aims to maximize Bayesian surprise might be effective. In particular, could surprise-taxis be a kind of second-order version of infotaxis? This direction may be promising since recent algorithms in accelerated materials discovery [81] imitate the human discovery process, e.g. by using an adaptive scheme based on Support Vector Regression (SVR) and Efficient Global Optimization (EGO) [82] and demonstrating on a certain family of alloys, M_2AX phases [83].

In developing a surprise-taxis algorithm for materials discovery, however, one may need to explicitly take notions of quality into account, rather than just pure novelty concerns, since there may be large parts of the discovery space that have low-quality possibilities: a Lagrangian balance between differing objectives of surprise and quality.

1.6 Conclusion

Although mathematically distinct, various problems in machine learning and artificial intelligence such as computational creativity, concept learning [84], invention, and accelerated discovery are all quite closely related philosophically. In this chapter, we have suggested that there may be value in bringing algorithmic ideas from these other related problems into accelerated materials discovery, especially the conceptual ideas of using dimensions, information, and surprise as key metrics for algorithmic pursuit.

It is an open question whether any of these ideas will be effective, as they have been in their original domains that include exploring minerals on distant planets [7], modeling the exploratory behavior of organisms such as moths and worms [9, 11], and creating novel and flavorful culinary recipes [27–29]. The data and informatics resources that are emerging in materials science, however, provide a wonderful opportunity to test this algorithmic hypothesis.

Acknowledgements Discussions with Daewon Seo, Turab Lookman, and Prasanna V. Balachandran are appreciated. Further encouragement from Turab Lookman in preparing this book chapter, despite the preliminary status of the work itself, is acknowledged.

References

1. T. Lookman, F.J. Alexander, K. Rajan (eds.), *Information Science for Materials Discovery and Design* (Springer, New York, 2016)
2. T.D. Sparks, M.W. Gaultois, A. Oliynyk, J. Brgoch, B. Meredig, Data mining our way to the next generation of thermoelectrics. Scripta Materialia **111**, 10–15 (2016)
3. A. Jain, S.P. Ong, G. Hautier, W. Chen, W.D. Richards, S. Dacek, S. Cholia, D. Gunter, D. Skinner, G. Ceder, K.A. Persson, The materials project: a materials genome approach to accelerating materials innovation. APL Mater. **1**(1), 011002 (2013)
4. M.L. Green, C.L. Choi, J.R. Hattrick-Simpers, A.M. Joshi, I. Takeuchi, S.C. Barron, E. Campo, T. Chiang, S. Empedocles, J.M. Gregoire, A.G. Kusne, J. Martin, A. Mehta, K. Persson, Z. Trautt, J. Van Duren, A. Zakutayev, Fulfilling the promise of the materials genome initiative with high-throughput experimental methodologies. Appl. Phys. Rev. **4**(1), 011105 (2017)
5. S. Curtarolo, G.L.W. Hart, M.B. Nardelli, N. Mingo, S. Sanvito, O. Levy, The high-throughput highway to computational materials design. Nat. Mater. **12**(3), 191–201 (2013)
6. B. Settles, Active learning literature survey. University of Wisconsin–Madison, Computer Sciences Technical Report 1648, 2009
7. K.L. Wagstaff, N.L. Lanza, D.R. Thompson, T.G. Dietterich, M.S. Gilmore, Guiding scientific discovery with explanations using DEMUD, in *Proceedings of the Twenty-Seventh AAAI Conference on Artificial Intelligence*, July 2013, pp. 905–911
8. J. Schwartzstein, Selective attention and learning. J. Eur. Econ. Assoc. **12**(6), 1423–1452 (2014)
9. M. Vergassola, E. Villermaux, B.I. Shraiman, 'Infotaxis' as a strategy for searching without gradients. Nature **445**(7126), 406–409 (2007)
10. J.L. Williams, J.W. Fisher III, A.S. Willsky, Approximate dynamic programming for communication-constrained sensor network management. IEEE Trans. Signal Process. **55**(8), 4300–4311 (2007)
11. A.J. Calhoun, S.H. Chalasani, T.O. Sharpee, Maximally informative foraging by *Caenorhabditis elegans*. eLife **3**, e04220 (2014)

12. R. Aggarwal, M.J. Demkowicz, Y.M. Marzouk, Information-driven experimental design in materials science, in *Information Science for Materials Discovery and Design*, ed. by T. Lookman, F.J. Alexander, K. Rajan (Springer, New York, 2016), pp. 13–44
13. K.J. Friston, M. Lin, C.D. Frith, G. Pezzulo, Active inference, curiosity and insight. Neural Comput. **29**(10), 2633–2683 (2017)
14. L.R. Varshney, To surprise and inform, in *Proceedings of the 2013 IEEE International Symposium on Information Theory*, July 2013, pp. 3145–3149
15. N. Agarwal, B. Bullins, E. Hazan, Second-order stochastic optimization for machine learning in linear time. J. Mach. Learn. Res. **18**(116), 1–40 (2017)
16. A. Karpatne, G. Atluri, J.H. Faghmous, M. Steinbach, A. Banerjee, A. Ganguly, S. Shekhar, N. Samatova, V. Kumar, Theory-guided data science: a new paradigm for scientific discovery from data. IEEE Trans. Knowl. Data Eng. **29**(10), 2318–2331 (2017)
17. V. Pankratius, J. Li, M. Gowanlock, D.M. Blair, C. Rude, T. Herring, F. Lind, P.J. Erickson, C. Lonsdale, Computer-aided discovery: toward scientific insight generation with machine support. IEEE Intell. Syst. **31**(4), 3–10 (2016)
18. B.F. Jones, The burden of knowledge and the 'death of the renaissance man': Is innovation getting harder? Rev. Econ. Stud. **76**(1), 283–317 (2009)
19. R. Netz, *The Shaping of Deduction in Greek Mathematics: A Study in Cognitive History* (Cambridge University Press, Cambridge, 1999)
20. L.R. Varshney, Toward a comparative cognitive history: Archimedes and D.H.J. Polymath, in *Proceedings of the Collective Intelligence Conference 2012*, Apr 2012
21. W.W. Ding, S.G. Levin, P.E. Stephan, A.E. Winkler, The impact of information technology on academic scientists' productivity and collaboration patterns. Manag. Sci. **56**(9), 1439–1461 (2010)
22. L.R. Varshney, The Google effect in doctoral theses. Scientometrics **92**(3), 785–793 (2012)
23. G.G. Lorentz, M. Golitschek, Y. Makovoz, *Constructive Approximation: Advanced Problems* (Springer, Berlin, 2011)
24. J.A. Lee, M. Verleysen, *Nonlinear Dimensionality Reduction* (Springer, New York, 2007)
25. T. Berger, *Rate Distortion Theory: A Mathematical Basis for Data Compression* (Prentice-Hall, Englewood Cliffs, NJ, 1971)
26. D.L. Donoho, M. Vetterli, R.A. DeVore, I. Daubechies, Data compression and harmonic analysis. IEEE Trans. Inf. Theory **44**(6), 2435–2476 (1998)
27. L.R. Varshney, F. Pinel, K.R. Varshney, D. Bhattacharjya, A. Schörgendorfer, Y.-M. Chee, A big data approach to computational creativity (2013). arXiv:1311.1213v1 [cs.CY]
28. F. Pinel, L.R. Varshney, Computational creativity for culinary recipes, in *Proceedings of the SIGCHI Conference on Human Factors in Computing Systems (CHI 2014)*, Apr 2014, pp. 439–442
29. F. Pinel, L.R. Varshney, D. Bhattacharjya, A culinary computational creativity system, in *Computational Creativity Research: Towards Creative Machines*, ed. by T.R. Besold, M. Schorlemmer, A. Smaill (Springer, 2015), pp. 327–346
30. R.D. King, J. Rowland, S.G. Oliver, M. Young, W. Aubrey, E. Byrne, M. Liakata, M. Markham, P. Pir, L.N. Soldatova, A. Sparkes, K.E. Whelan, A. Clare, The automation of science. Science **324**(5923), 85–89 (2009)
31. H. Cohen, The further exploits of AARON, painter, in *Constructions of the Mind: Artificial Intelligence and the Humanities*, ser. Stanford Humanities Review, vol. 4, no. 2, ed. by S. Franchi, G. Güzeldere (1995), pp. 141–160
32. S. Bringsjord, D.A. Ferrucci, *Artificial Intelligence and Literary Creativity: Inside the Mind of BRUTUS, a Storytelling Machine* (Lawrence Erlbaum Associates, Mahwah, NJ, 2000)
33. M.A. Boden, *The Creative Mind: Myths and Mechanisms*, 2nd edn. (Routledge, London, 2004)
34. A. Cardoso, T. Veale, G.A. Wiggins, Converging on the divergent: the history (and future) of the international joint workshops in computational creativity. A. I. Mag. **30**(3), 15–22 (2009)
35. M.A. Boden, Foreword, in *Computational Creativity Research: Towards Creative Machines*, ed. by T.R. Besold, M. Schorlemmer, A. Smaill (Springer, 2015), pp. v–xiii

36. M. Guzdial, M.O. Riedl, Combinatorial creativity for procedural content generation via machine learning, in *Proceedings of the AAAI 2018 Workshop on Knowledge Extraction in Games*, Feb 2018 (to appear)
37. R.K. Sawyer, *Explaining Creativity: The Science of Human Innovation* (Oxford University Press, Oxford, 2012)
38. L. Itti, P. Baldi, Bayesian surprise attracts human attention, in *Advances in Neural Information Processing Systems 18*, ed. by Y. Weiss, B. Schölkopf, J. Platt (MIT Press, Cambridge, MA, 2006), pp. 547–554
39. L. Itti, P. Baldi, Bayesian surprise attracts human attention. Vis. Res. **49**(10), 1295–1306 (2009)
40. P. Baldi, L. Itti, Of bits and wows: a Bayesian theory of surprise with applications to attention. Neural Netw. **23**(5), 649–666 (2010)
41. J. Evans, A. Rzhetsky, Machine science. Science **329**(5990), 399–400 (2010)
42. C.E. Shannon, W. Weaver, *The Mathematical Theory of Communication* (University of Illinois Press, Urbana, 1949)
43. N. Verma, S. Kpotufe, S. Dasgupta, Which spatial partition trees are adaptive to intrinsic dimension?, in *Proceedings of the Twenty-Fifth Conference on Uncertainty in Artificial Intelligence (UAI '09)*, June 2009, pp. 565–574
44. M. Tepper, A.M. Sengupta, D.B. Chklovskii, Clustering is semidefinitely not that hard: nonnegative SDP for manifold disentangling (2018). arXiv:1706.06028v3 [cs.LG]
45. K. Pearson, On lines and planes of closest fit to systems of points in space. Lond. Edinb. Dublin Philos. Mag. J. Sci. **2**(11), 559–572 (1901)
46. H. Hotelling, Analysis of a complex of statistical variables into principal components. J. Educ. Psychol. **24**(6), 417–441 (1933)
47. S. Bailey, Principal component analysis with noisy and/or missing data. Publ. Astron. Soc. Pac. **124**(919), 1015–1023 (2012)
48. S.T. Roweis, L.K. Saul, Nonlinear dimensionality reduction by locally linear embedding. Science **290**(5500), 2323–2326 (2000)
49. J.B. Tenenbaum, V. de Silva, J.C. Langford, A global geometric framework for nonlinear dimensionality reduction. Science **290**(5500), 2319–2323 (2000)
50. M. Belkin, P. Niyogi, Laplacian eigenmaps for dimensionality reduction and data representation. Neural Comput. **15**(6), 1373–1396 (2003)
51. L. van der Maaten, G. Hinton, Visualizing data using t-SNE. J. Mach. Learn. Res. **9**, 2579–2605 (2008)
52. Y. Bengio, J.-F. Paiement, P. Vincent, O. Delalleau, N.L. Roux, M. Ouimet, Out-of-sample extensions for LLE, Isomap, MDS, eigenmaps, and spectral clustering, in *Advances in Neural Information Processing Systems 16*, ed. by S. Thrun, L.K. Saul, B. Sch (2003)
53. J. Lim, D.A. Ross, R. Lin, M.-H. Yang, Incremental learning for visual tracking, in *Advances in Neural Information Processing Systems 17*, ed. by L.K. Saul, Y. Weiss, L. Bottou (MIT Press, 2005), pp. 793–800
54. C.E. Shannon, A mathematical theory of communication. Bell Syst. Tech. J. **27**, 379–423, 623–656 (1948)
55. C.E. Shannon, Coding theorems for a discrete source with a fidelity criterion. IRE Natl. Conv. Rec. (Part 4), 142–163 (1959)
56. J. Jiao, K. Venkat, Y. Han, T. Weissman, Minimax estimation of functionals of discrete distributions. IEEE Trans. Inf. Theory **61**(5), 2835–2885 (2015)
57. K.R. Moon, A.O. Hero, III, Multivariate f-divergence estimation with confidence, in *Advances in Neural Information Processing Systems 27*, ed. by Z. Ghahramani, M. Welling, C. Cortes, N.D. Lawrence, K.Q. Weinberger (MIT Press, 2014), pp. 2420–2428
58. A.O. Hero III, B. Ma, O.J.J. Michel, J. Gorman, Applications of entropic spanning graphs. IEEE Signal Process. Mag. **19**(5), 85–95 (2002)
59. Q. Wang, S.R. Kulkarni, S. Verdú, Universal estimation of information measures for analog sources. Found. Trends Commun. Inf. Theory **5**(3), 265–353 (2009)
60. J. Aczél, Z. Daróczy, *On Measures of Information and Their Characterization* (Academic Press, New York, 1975)

61. D. Kahneman, *Attention and Effort* (Prentice-Hall, Englewood Cliffs, NJ, 1973)
62. D.G. Luenberger, *Optimization by Vector Space Methods* (Wiley, New York, 1969)
63. I. Csiszár, J. Körner, *Information Theory: Coding Theorems for Discrete Memoryless Systems*, 3rd edn. (Akadémiai Kiadó, Budapest, 1997)
64. E. Hasanbelliu, K. Kampa, J.C. Principe, J.T. Cobb, Online learning using a Bayesian surprise metric, in *Proceedings of the 2012 International Joint Conference on Neural Networks (IJCNN)*, June 2012
65. B. Schauerte, R. Stiefelhagen, "Wow!" Bayesian surprise for salient acoustic event detection, in *Proceedings of the IEEE International Conference on Acoustics, Speech, and Signal Processing (ICASSP 2013)*, May 2013, pp. 6402–6406
66. K. Takahashi, K. Watanabe, Persisting effect of prior experience of change blindness. Perception **37**(2), 324–327 (2008)
67. T.N. Mundhenk, W. Einhuser, L. Itti, Automatic computation of an image's statistical surprise predicts performance of human observers on a natural image detection task. Vis. Res. **49**(13), 1620–1637 (2009)
68. D. Ostwald, B. Spitzer, M. Guggenmos, T.T. Schmidt, S.J. Kiebel, F. Blankenburg, Evidence for neural encoding of Bayesian surprise in human somatosensation. NeuroImage **62**(1), 177–188 (2012)
69. T. Sharpee, N.C. Rust, W. Bialek, Analyzing neural responses to natural signals: maximally informative dimensions. Neural Comput. **16**(2), 223–250 (2004)
70. G. Horstmann, The surprise-attention link: a review. Ann. New York Acad. Sci. **1339**, 106–115 (2015)
71. C. França, L.F.W. Goes, Á. Amorim, R. Rocha, A. Ribeiro da Silva, Regent-dependent creativity: a domain independent metric for the assessment of creative artifacts, in *Proceedings of the International Conference on Computational Creativity (ICCC 2016)*, June 2016, pp. 68–75
72. J.P.L. Schoormans, H.S.J. Robben, The effect of new package design on product attention, categorization and evaluation. J. Econ. Psychol. **18**(2–3), 271–287 (1997)
73. W. Sun, P. Murali, A. Sheopuri, Y.-M. Chee, Designing promotions: consumers' surprise and perception of discounts. IBM J. Res. Dev. **58**(5/6), 2:1–2:10 (2014)
74. H. Feldman, K.J. Friston, Attention, uncertainty, and free-energy. Front. Hum. Neurosci. **4**, 215 (2010)
75. K. Friston, The free-energy principle: a rough guide to the brain? Trends Cogn. Sci. **13**(7), 293–301 (2009)
76. J.G. Smith, The information capacity of amplitude- and variance-constrained scalar Gaussian channels. Inf. Control **18**(3), 203–219 (1971)
77. T.H. Davenport, J.C. Beck, *The Attention Economy: Understanding the New Currency of Business* (Harvard Business School Press, Boston, 2001)
78. V. Chandar, A. Tchamkerten, D. Tse, Asynchronous capacity per unit cost. IEEE Trans. Inf. Theory **59**(3), 1213–1226 (2013)
79. T.A. Courtade, T. Weissman, Multiterminal source coding under logarithmic loss. IEEE Trans. Inf. Theory **60**(1), 740–761 (2014)
80. M. Gastpar, B. Rimoldi, M. Vetterli, To code, or not to code: lossy source-channel communication revisited. IEEE Trans. Inf. Theory **49**(5), 1147–1158 (2003)
81. P.V. Balachandra, D. Xue, J. Theiler, J. Hogden, T. Lookman, Adaptive strategies for materials design using uncertainties. Sci. Rep. **6**, 19660 (2016)
82. D.R. Jones, M. Schonlau, W.J. Welch, Efficient global optimization of expensive black-box functions. J. Glob. Optim. **13**(4), 455–492 (1998)
83. M.F. Cover, O. Warschkow, M.M.M. Bilek, D.R. McKenzie, A comprehensive survey of M_2AX phase elastic properties. J. Phys.: Condens. Matter **21**(30), 305403 (2009)
84. H. Yu and L.R. Varshney, Towards deep interpretability (MUS-ROVER II): learning hierarchical representations of tonal music, in *Proceedings of the 6th International Conference on Learning Representations (ICLR)*, Apr 2017

Chapter 2
Is Automated Materials Design and Discovery Possible?

Michael McKerns

Abstract In materials design, we typically want to answer questions such as "Can we optimize the probability that a structure will produce the desired properties within some tolerance?" or "Can we optimize the probability that a transition will occur between the desired initial and final states?" In the vast majority of cases, these problems are addressed indirectly, and with a reduced-dimensional model that approximates the actual system. Why? The tools and techniques traditionally used are not sufficient to provide a general rigorous algorithmic approach to determining and/or validating models of the system. Solving for the structure that maximizes some property very likely will be a global optimization over a nonlinear surface with several local minima and nonlinear constraints, while the tools generally used are linear (or at best quadratic) solvers. This approximation is made to handle the large dimensionality of the problem and be able to apply some basic constraints on the space of possible solutions. Unfortunately, constraints from data, measurements, theory, and other physical information are often only applied post-optimization as a binary form of model validation. Additionally, sampling techniques like Monte Carlo, as well as machine learning and Bayesian inference (which strongly rely on existing observed data to infer the form of the solution), will not perform well when, in terms of structural configurations, discovering the materials in the state that produces the desired property is a rare-event. This is unfortunately the rule, rather than the exception—and thus most searches either require the solution to already have been observed, or at least to be in the locality of the optimum. Fortunately, recent developments in applied mathematics and numerical optimization provide a new suite of tools that should overcome the existing limitations, and make rigorous automated materials discovery and design possible.

M. McKerns (✉)
The Uncertainty Quantification Foundation, 300 Delaware Ave. Ste. 210,
Wilmington, DE 19801, USA
e-mail: mmckerns@uqfoundation.org

T. Lookman et al. (eds.), *Materials Discovery and Design*, Springer Series
in Materials Science 280, https://doi.org/10.1007/978-3-319-99465-9_2

2.1 Model Determination in Materials Science

2.1.1 The Status Quo

One of the ultimate goals of the physical material sciences is the development of detailed models that allow us to understand the properties of matter. While models may be developed from ab initio theory or from empirical rules, often models are fit directly to experimental results. Crystallographic structural analysis has pioneered model fitting; direct fitting of crystal structure models to diffraction datasets has been used routinely since the middle of the last century. In the past two decades, direct model fitting has been applied to other scattering techniques such as PDF analysis and X-ray spectroscopies. Combinations of experiments and theory to derive a single physical model is a broad frontier for materials science.

Models that use physically meaningful parameters may not be well-conditioned (meaning that the minimum is narrow and easily missed). Likewise, using parameters that are physically meaningful may result in problems that are not well-posed—meaning that there may not be a unique solution, since the effect of changing one parameter may be offset by adjustment to another. Despite this, models with physical parameters are most valuable for interpreting experimental measurements. In some cases there may be many model descriptions that provide equivalent fits, within experimental uncertainty. It is then not sufficient to identify a single minimum, since this leads to the misapprehension that this single answer has been proven. Identification of all such minima allows for the design of new experiments, or calculations to differentiate between them.

2.1.2 The Goal

The fundamental scientific limitation that has prevented more widespread deployment of model fitting has been that, until recently, relatively few types of measurements could be simulated at the level where quantitative agreement with experiments can be obtained. When simulations can directly reproduce experimental results, then parameters in the model can be optimized to improve the fit. However, to obtain unique solutions that are not overly affected by statistical noise, one needs to have many more observations than varied parameters (the crystallographic rule-of-thumb is 10:1). While accurate simulation of many types of experiments is now possible, the experimental data may not offer a sufficient number of observations to allow fitting of a very complex model. This changes when different types of experiments are combined, since each experiment may be sensitive to different aspects of the model. In addition to the advances in computation, modern user facilities now offer a wide assortment of experimental probes. Theory too can be added to the mix. It is clear that the frontier over the next decade will be to develop codes that optimize a single model to fit all types of data for a material—rather than to develop a different model from each experiment.

The task of model determination from pair distribution function (PDF) data has gained considerable interest because it is one of few techniques giving detailed short-, medium-, and long- range structural information for materials without long-range order. However, the task of automated model derivation is exceedingly more difficult without the assumption of a periodic lattice [40]. One approach is to use a greater range of experimental techniques in modeling, combining measurements from different instruments to reduce the ratio of observations to degrees of freedom. The challenge is that a computational framework is needed that can handle the complexity in the constraining information in a nonlinear global optimization, that is both generally applicable to structure solution and extensible to the (most-likely) requisite large-scale parallel computing.

For example, in powder diffraction crystallography, indexing the lattice from an unknown material, potentially in the presence of peaks from multiple phases, is an ill-conditioned problem where a large volume of parameter space must be searched for solutions with extremely sharp minima. Additionally, structure solution often is an ill-posed problem; however, crystallographic methodology assumes that if a well-behaved and plausible solution is identified, this solution is unique. An unusual counter example is [84], where molecular modeling was used to identify all possible physical models to fit the neutron and X-ray diffraction and neutron spectrometry data. Such studies should be routine rather than heroic.

2.2 Identification of the Research and Issues

2.2.1 Reducing the Degrees of Freedom in Model Determination

X-ray diffraction enables us to pinpoint the coordinates of atoms in a crystal, with a precision of around 10–4 nm. Determining the structure and arrangement of atoms in a solid is fundamental to understanding its properties, and this has become common practice for X-ray crystallographers over the past many years. However, with the emergence of nanotechnology, it has become abundantly clear that diffraction data alone may not be enough to uniquely solve the structure of nanomaterials. As part of a growing effort to incorporate the results of other techniques to constrain X-ray refinements, it has recently been proposed that combining information from spectroscopy with diffraction data can enable the unique solution for the structure of amorphous and nanostructured materials [14].

The forward problem of predicting the diffraction intensity given a particular density distribution is trivial, but the inverse, unraveling from the intensity distribution the density that gives rise to it, is a highly nontrivial problem in global optimization. In crystallography, the diffraction pattern is a wave-interference pattern, but we measure only the intensities (the squares of the waves) not the wave amplitudes. To get the amplitude, you take the square root of the intensity; however, in so doing you lose any knowledge of the phase of the wave, and thus half the information needed to reconstruct the density is also lost. When solving such inverse problems, you hope you can start with a uniqueness theorem that reassures you that, under ideal conditions, there is only one solution: one density distribution that corresponds to the measured intensity. Then you have to establish that your data set contains sufficient information to constrain that unique solution. This is a problem from information theory that originated with Reverend Thomas Bayes' work in the 18th century, and the work of Nyquist and Shannon in the 20th century [59, 72], and describes the fact that the degrees of freedom in the model must not exceed the number of pieces of independent information in the data.

In crystallography, the information is in the form of Bragg peak intensities and the degrees of freedom are the atomic coordinates. We use crystal symmetry to connect the model to the contents of a unit cell, and thus greatly reduce the degrees of freedom needed to describe the problem. A single diffraction measurement yields a multitude of Bragg peak intensities, providing ample redundant intensity information to make up for the lost phases. Highly efficient search algorithms, such as the conjugate gradient method, typically can readily accept parameter constraints, and in many cases, can find a solution quickly even in a very large search space. The problem is often so overconstrained that we can disregard a lot of directional information—in particular, even though Bragg peaks are orientationally averaged to a 1D function in a powder diffraction measurement, we still can get a 3D structural solution [16].

Moving from solving crystal structures to solving nanostructures will require a new set of tools, with vastly increased capabilities. For nanostructures, the information content in the data is degraded while the complexity of the model is much greater. At the nanoscale, finite size effects broaden the sharp Bragg peaks to the point where the broadening is sufficient enough that the peaks begin to overlap. We also can no longer describe the structure with the coordinates of a few atoms in a unit cell—we need the arrangement of hundreds or thousands of atoms in a nanoparticle. There also can be complicated effects, like finite-size induced relaxations in the core and the surface. Moreover, the measured scattering intensity asymptotically approaches zero as the nanoparticle gets smaller and the weak scattering of X-rays becomes hard to discern from the noise. In general, we measure the intensity from a multitude of nanoparticles or nanoclusters, and then struggle with how to deal with the averaged data.

The use of total scattering and atomic-pair distribution function (PDF) measurements for nanostructure studies is a promising approach [22]. In these experiments, powders of identical particles are studied using X-ray powder diffraction, resulting in good signals, but highly averaged data. Short wavelength X-rays or neutrons are used for the experiments giving data with good real-space resolution, and the resulting data are fit with models of the nanoparticle structures. Uniqueness is a real issue, as is the availability of good nanostructure solution algorithms. Attempts to fit amorphous structures, which have local order on the subnanometer scale and lots of disorder, yield highly degenerate results: many structure models, some completely physically nonsensical, give equivalent fits to the data within errors [28]. Degenerate solutions imply that there is insufficient information in the data set to constrain a unique solution. At this point we would like to seek additional constraints coming from prior knowledge about the system, or additional data sets, such that these different information sources can be combined to constrain a unique solution. This can be done either by adding constraints on how model parameters can vary (for example, crystal symmetries), or by adding terms to the target (or cost) function that is being minimized in the global optimization process.

In crystallography, it is considered a major challenge to be able to incorporate disparate information sources into the global optimization scheme, and to figure out how to weight their contributions to the cost function. There have been a few advances, such as Cliffe et al., where the authors introduced a variance term into the cost function that adds a cost when atomic environments of equivalent atoms in the model deviate too much from one other [14]. In the systems they studied, this simple term was the difference between successful and unsuccessful nanostructure solutions. We see that a relatively simple but well-chosen constraint added to the cost function can make a big difference in determining the unique structure solution. The impact of the constraints chosen by Cliffe et al. was to vastly reduce the volume of the search space for the global optimization algorithm, thus enabling the optimizer to converge within the limitations imposed by the simulated annealing algorithm itself. A similar effect has been seen in the work of Juhas et al., where adding ionic radii to a structure solution enabled the solution of structures from total scattering data [40]. Again, applying a simple constraint, which at first sight contained a rather limited amount of information, was all that was needed for success. The constraints applied in both of the above studies, however common sense, placed enormous restrictions on the solution space and the efficiency and uniqueness of solutions, and ultimately enabled the structure to be determined.

2.2.2 *OUQ and* `mystic`

The desire to combine information from different measurements, legacy data, models, assumptions, or other pieces of information into a global optimization problem is not unique to the field of crystallography, but has numerical and applied mathematical underpinnings that transcend any particular field of science. For example, recent advances in mechanical and materials engineering use a paradigm of applying different models, measurements, datasets, and other sources of information as constraints in global optimization problems posed to quantify the uncertainty in engineering systems of interest [66, 82]. In general, these studies have focused on the rigorous certification of the safety of engineering structures under duress, such as the probability of failure of a metal panel under ballistic impact [3, 44, 66, 82, 83] or the probability of elastoplastic failure of a tower under seismic stimulation [66]. Owhadi et al. has developed a mathematical framework called "optimal uncertainty quantification" (or OUQ), for solving these types of certification and other engineering design problems [66]. OUQ should also be directly leverageable in the inverse modeling of nanostructured materials.

The potential application of OUQ in the modeling of nanostructures is both broad and also unexplored. For example, when degenerate solutions are found in nanostructure refinement problems, it implies that there is insufficient information to constrain a unique solution for the nanostructure; however with OUQ we can rigorously establish whether or not there actually is sufficient information available to determine a unique solution. Further, we could leverage OUQ to discover what critical pieces of information would enable a unique solution to be found, or give us the likelihood that each of the degenerate solutions found is the true unique solution. OUQ could be used to rigorously identify the number of pieces of independent information in the data. We could also utilize uncertainty quantification to discover which design parameters or other information encapsulated in the constraints has the largest impact on the nanostrucuture, to determine which regions of parameter space have the largest impact on the outcome of the inverse problem, or to help us target the next best experiments to perform so we can obtain a unique solution. We can use OUQ to identify the impact of parameters within a hierarchical set of models; to determine, for example, whether finite-size induced relaxations in the nanostrucure core or on the surface have critical impact on the bulk properties of the material. Since engineering design problems, with similar objectives as the examples given above, have already been solved using uncertainty quantification—it would appear that the blocker to solving the nanostructure problem may only be one of implementation.

A practical implementation issue for OUQ is that many OUQ problems are one to two orders of magnitude larger than the standard inverse problem (say to find a local minima on some design surface). OUQ problems are often highly-constrained and high-dimensional global optimizations, since all of the available information about the problem is encapsulated in the constraints. In an OUQ problem, there are often numerous nonlinear and statistical constraints. The largest OUQ problem solved to date had over one thousand input parameters and over one thousand constraints [66]; however, nanostructure simulations where an optimizer is managing the arrangement of hundreds or thousands of atoms may quickly exceed that size. Nanostructure inverse problems may also seek to use OUQ to refine model potentials, or other aspects of a molecular dynamics simulation used in modeling the structure. The computational infrastructure for problems of this size can easily require distributed or massively parallel resources, and may potentially require a level of robust resource management that is on the forefront of computational science.

McKerns et al. has developed a software framework for high-dimensional constrained global optimization (called "`mystic`") that is designed to utilize large-scale parallelism on heterogeneous resources [51, 52, 54]. The `mystic` software provides many of the uncertainty quantification functionalities mentioned above, a suite of highly configurable global optimizers, and a robust toolkit for applying constraints and dynamically reducing dimensionality. `mystic` is built so the user can apply restraints on the solution set and penalties on the cost function in a robust manner—in `mystic`, all constraints are applied in functional form, and are therefore also independent of the optimization algorithm. Since `mystic`'s constraints solvers are functional (i.e. `x' = c(x)`, where `c` is a coordinate transformation to the valid solution set), any piece of information can be directly encoded in the constraints, including trust radii on surrogate models, measurement uncertainty in data, and statistical constraints on measured or derived quantities [66, 82]. Adaptive constraints solvers can be formulated that seek to reduce the volume of search space, applying and removing constraints dynamically during an optimization with the goal of, for example, reducing the dimensionality of the optimization as constraints are discovered to be redundant or irrelevant [82, 83]. Direct optimization algorithms, such as conjugate gradient, have had a long history of use in structural refinement, primarily due to efficiency of the algorithm; however, with `mystic`, nanostrucure refinements can leverage massively parallel global optimizations with the same convergence dynamics as the fastest of available direct methods [3, 44].

We can extrapolate from the lesson learned from the studies of Cliffe et al. [14] and Juhas et al. [40]. If applying a simple penalty constraint to reduce outliers in atomic environments of equivalent atoms can vastly reduce search space so that select nanostructures can be uniquely solved—we can begin to imagine what is possible when we add all available information to the refinement problem as constraints. We will be able to pose problems that not only yield us the answer of "which" nanostructure, but with `mystic` and OUQ, we should be able to directly and rigorously address the deeper questions that ask "why".

2.3 Introduction to Uncertainty Quantification

2.3.1 The UQ Problem

We present here a rigorous and unified framework for the statement and solution of uncertainty quantification (UQ) problems centered on the notion of available information. In general, UQ refers to any attempt to quantitatively understand the relationships among uncertain parameters and processes in physical processes, or in mathematical and computational models for them; such understanding may be deterministic or probabilistic in nature. However, to make the discussion specific, we start the description of the OUQ framework as it applies to the certification problem; Sect. 2.4 gives a broader description of the purpose, motivation and applications of UQ in the OUQ framework and a comparison with current methods.

By *certification* we mean the problem of showing that, with probability at least $1 - \varepsilon$, the real-valued response function G of a given physical system will not exceed a given safety threshold a. That is, we wish to show that

$$\mathbb{P}[G(X) \geq a] \leq \varepsilon. \tag{2.1}$$

In practice, the event $[G(X) \geq a]$ may represent the crash of an aircraft, the failure of a weapons system, or the average surface temperature on the Earth being too high. The symbol $\mathbb{P}$ denotes the probability measure associated with the randomness of (some of) the input variables X of G (commonly referred to as "aleatoric uncertainty").

Specific examples of values of ε used in practice are: 10^{-9} in the aviation industry (for the maximum probability of a catastrophic event per flight hour, see [77, p. 581] and [12]), 0 in the seismic design of nuclear power plants [21, 26] and 0.05 for the collapse of soil embankments in surface mining [36, p. 358]. In structural engineering [31], the maximum permissible probability of failure (due to any cause) is $10^{-4} K_s n_d / n_r$ (this is an example of ε) where n_d is the design life (in years), n_r is the number of people at risk in the event of failure and K_s is given by the following values (with 1/year units): 0.005 for places of public safety (including dams); 0.05 for domestic, office or trade and industry structures; 0.5 for bridges; and 5 for towers, masts and offshore structures. In US environmental legislation, the maximum acceptable increased lifetime chance of developing cancer due to lifetime exposure to a substance is 10^{-6} [48] ([43] draws attention to the fact that "there is no sound scientific, social, economic, or other basis for the selection of the threshold 10^{-6} as a cleanup goal for hazardous waste sites").

One of the most challenging aspects of UQ lies in the fact that in practical applications, the measure $\mathbb{P}$ and the response function G are not known a priori. This lack of information, commonly referred to as "epistemic uncertainty", can be described precisely by introducing $\mathcal{A}$, the set of all *admissible scenarios* (f, μ) for the unknown—or partially known—reality $(G, \mathbb{P})$. More precisely, in those applications, the available information does not determine $(G, \mathbb{P})$ *uniquely* but instead determines a set $\mathcal{A}$ such that any $(f, \mu) \in \mathcal{A}$ could a priori be $(G, \mathbb{P})$. Hence, $\mathcal{A}$ is a (possibly infinite-dimensional) set of measures and functions defining explicitly *information on and assumptions about G and $\mathbb{P}$*. In practice, this set is obtained from physical laws, experimental data and expert judgment. It then follows from $(G, \mathbb{P}) \in \mathcal{A}$ that

$$\inf_{(f,\mu)\in\mathcal{A}} \mu[f(X) \geq a] \leq \mathbb{P}[G(X) \geq a] \leq \sup_{(f,\mu)\in\mathcal{A}} \mu[f(X) \geq a]. \tag{2.2}$$

Moreover, it is elementary to observe that

- The quantities on the right-hand and left-hand of (2.2) are extreme values of optimization problems and elements of [0, 1].
- Both the right-hand and left-hand inequalities are optimal in the sense that they are the sharpest bounds for $\mathbb{P}[G(X) \geq a]$ that are consistent with the information and assumptions $\mathcal{A}$.

More importantly, in Proposition 2.5.1, we show that these two inequalities provide sufficient information to produce an optimal solution to the certification problem.

Example 2.3.1 To give a very simple example of the effect of information and optimal bounds over a class $\mathcal{A}$, consider the certification problem (2.1) when $Y := G(X)$ is a real-valued random variable taking values in the interval [0, 1] and $a \in (0, 1)$; to further simplify the exposition, we consider only the upper bound problem, suppress dependence upon G and X and focus solely on the question of which probability

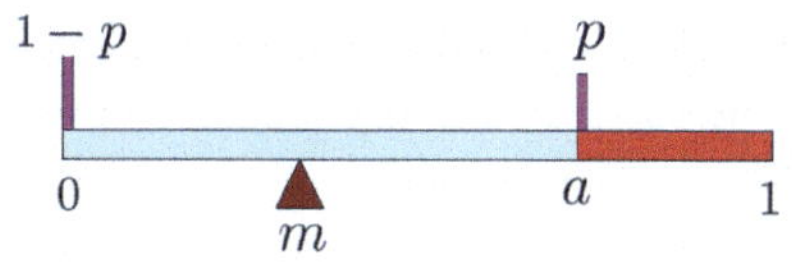

Fig. 2.1 You are given one pound of play-dough and a seesaw balanced around m. How much mass can you put on right hand side of a while keeping the seesaw balanced around m? The solution of this optimization problem can be achieved by placing any mass on the right hand side of a, exactly at a (to place mass on $[a, 1]$ with minimum leverage towards the right hand side of the seesaw) and any mass on the left hand side of a, exactly at 0 (for maximum leverage towards the left hand side of the seesaw)

measures ν on $\mathbb{R}$ are admissible scenarios for the probability distribution of Y. So far, *any* probability measure on $[0, 1]$ is admissible:

$$\mathscr{A} = \{\nu \mid \nu \text{ is a probability measure on } [0, 1]\}.$$

and so the optimal upper bound in (2.2) is simply

$$\mathbb{P}[Y \geq a] \leq \sup_{\nu \in \mathscr{A}} \nu[Y \geq a] = 1.$$

Now suppose that we are given an additional piece of information: the expected value of Y equals $m \in (0, a)$. These are, in fact, the assumptions corresponding to an elementary Markov inequality, and the corresponding admissible set is

$$\mathscr{A}_{\text{Mrkv}} = \left\{ \nu \,\middle|\, \begin{array}{c} \nu \text{ is a probability measure on } [0, 1], \\ \mathrm{e}_{\nu}[Y] = m \end{array} \right\}.$$

The least upper bound on $\mathbb{P}[Y \geq a]$ corresponding to the admissible set $\mathscr{A}_{\text{Mrkv}}$ is the solution of the infinite dimensional optimization problem

$$\sup_{\nu \in \mathscr{A}_{\text{Mrkv}}} \nu[Y \geq a] \tag{2.3}$$

Formulating (2.3) as a mechanical optimization problem (see Fig. 2.1), it is easy to observe that the extremum of (2.3) can be achieved only considering the situation where ν is the weighted sum of a Dirac delta mass at 0 (with weight $1 - p$) and a Dirac delta mass at a (with weight p). It follows that (2.3) can be reduced to the simple (one-dimensional) optimization problem: *Maximize p subject to $ap = m$*. It follows that Markov's inequality is the optimal bound for the admissible set $\mathscr{A}_{\text{Mrkv}}$.

$$\mathbb{P}[Y \geq a] \leq \sup_{\nu \in \mathscr{A}_{\text{Mrkv}}} \nu[Y \geq a] = \tfrac{m}{a}. \tag{2.4}$$

In some sense, the OUQ framework that we present here is the extension of this procedure to situations in which the admissible class $\mathscr{A}$ is complicated enough that a closed-form inequality such as Markov's inequality is unavailable, but optimal bounds can nevertheless be *computed* using reduction properties analogous to the one illustrated in Fig. 2.1.

2.4 Generalizations and Comparisons

2.4.1 Prediction, Extrapolation, Verification and Validation

In the previous section, the OUQ framework was described as it applies to the certification problem (2.1). We will now show that many important UQ problems, such as prediction, verification and validation, can be formulated as certification problems. This is similar to the point of view of [5], in which formulations of many problem objectives in reliability are shown to be representable in a unified framework.

A **prediction problem** can be formulated as, given ε and (possibly incomplete) information on $\mathbb{P}$ and G, finding a smallest $b - a$ such that

$$\mathbb{P}[a \leq G(X) \leq b] \geq 1 - \varepsilon, \tag{2.5}$$

which, given the admissible set $\mathscr{A}$, is equivalent to solving

$$\inf \left\{ b - a \,\middle|\, \inf_{(f,\mu)\in\mathscr{A}} \mu[a \leq f(X) \leq b] \geq 1 - \varepsilon \right\}. \tag{2.6}$$

Observe that $[a, b]$ can be interpreted as an optimal interval of confidence for $G(X)$ (although $b - a$ is minimal, $[a, b]$ may not be unique), in particular, with probability at least $1 - \varepsilon$, $G(X) \in [a, b]$.

In many applications the regime where experimental data can be taken is different than the deployment regime where prediction or certification is sought, and this is commonly referred to as the **extrapolation problem**. For example, in materials modeling, experimental tests are performed on materials, and the model run for comparison, but the desire is that these results tell us something where experimental tests are impossible, or extremely expensive to obtain.

In most applications, the response function G may be approximated via a (possibly numerical) model F. Information on the relation between the model F and the response function G that it is designed to represent (i.e. information on $(x, F(x), G(x))$) can be used to restrict (constrain) the set $\mathscr{A}$ of admissible scenarios $(G, \mathbb{P})$. This information may take the form of a bound on some distance between F and G or a bound on some complex functional of F and G [47, 71]. Observe that, in the context of the certification problem (2.1), the value of the model can be measured by changes induced on the optimal bounds $\mathscr{L}(\mathscr{A})$ and $\mathscr{U}(\mathscr{A})$. The problem of quantifying the relation (possibly the distance) between F and G is commonly referred to as the **validation problem**. In some situations F may be a numerical model involving millions of lines of code and (possibly) space-time discretization. The quantification of the uncertainty associated with the possible presence of bugs and discretization approximations is commonly referred to as the **verification problem**. Both, the validation and the verification problem, can be addressed in the OUQ framework by introducing information sets describing relations between G, F and the code.

2.4.2 Comparisons with Other UQ Methods

We will now compare OUQ with other widely used UQ methods and consider the certification problem (2.1) to be specific.

- Assume that n independent samples $Y_1, \dots, Y_n$ of the random variable $G(X)$ are available (i.e. n independent observations of the random variable $G(X)$, all distributed according to the measure of probability $\mathbb{P}$). If $\mathbb{1}[Y_i \geq a]$ denotes the random variable equal to one if $Y_i \geq a$ and equal to zero otherwise, then

$$p_n := \frac{\sum_{i=1}^{n} \mathbb{1}[Y_i \geq a]}{n} \tag{2.7}$$

  is an unbiased estimator of $\mathbb{P}[G(X) \geq a]$. Furthermore, as a result of Hoeffding's concentration inequality [34], the probability that p_n deviates from $\mathbb{P}[G(X) \geq n]$ (its mean) by at least $\varepsilon/2$ is bounded from above by $\exp(-\frac{n}{2}\varepsilon^2)$. It follows that if the number of samples n is large enough (of the order of $\frac{1}{\varepsilon^2} \log \frac{1}{\varepsilon}$), then the certification of (2.1) can be obtained through a Monte Carlo estimate (using p_n). As this example shows, **Monte Carlo strategies** [46] are simple to implement and do not necessitate prior information on the response function G and the measure $\mathbb{P}$ (other than the i.i.d. samples). However, they require a large number of (independent) samples of $G(X)$ which is a severe limitation for the certification of rare events (the $\varepsilon = 10^{-9}$ of the aviation industry would [12, 77] necessitate $O(10^{18})$ samples). Additional information on G and $\mathbb{P}$ can, in principle, be included (in a limited fashion) in Monte Carlo strategies via importance and weighted sampling [46] to reduce the number of required samples.
- The number of required samples can also be reduced to $\frac{1}{\varepsilon}(\ln \frac{1}{\varepsilon})^d$ using **Quasi-Monte Carlo Methods**. We refer in particular to the Koksma–Hlawka inequality [58], to [75] for multiple integration based on lattice rules and to [74] for a recent review. We observe that these methods require some regularity (differentiability) condition on the response function G and the possibility of sampling G at pre-determined points X. Furthermore, the number of required samples blows-up at an exponential rate with the dimension d of the input vector X.
- If G is regular enough and can be sampled at pre-determined points, and if X has a known distribution, then **stochastic expansion methods** [4, 20, 24, 29, 30, 91] can reduce the number of required samples even further (depending on the

regularity of G) provided that the dimension of X is not too high [11, 85]. However, in most applications, only incomplete information on $\mathbb{P}$ and G is available and the number of available samples on G is small or zero. X may be of high dimension, and may include uncontrollable variables and unknown unknowns (unknown input parameters of the response function G). G may not be the solution of a PDE, and may involve interactions between singular and complex processes such as (for instance) dislocation, fragmentation, phase transitions, physical phenomena in untested regimes, and even human decisions. We observe that in many applications of Stochastic Expansion methods, G and $\mathbb{P}$ are assumed to be perfectly known, and UQ reduces to computing the push forward of the measure $\mathbb{P}$ via the response (transfer) function $I_{\geq a} \circ G$ (to a measure on two points, in those situations $\mathscr{L}(\mathscr{A}) = \mathbb{P}[G \geq a] = \mathscr{U}(\mathscr{A})$).

- The investigation of variations of the response function G under variations of the input parameters X_i, commonly referred to as **sensitivity analysis** [69, 70], allows for the identification of critical input parameters. Although helpful in estimating the robustness of conclusions made based on specific assumptions on input parameters, sensitivity analysis, in its most general form, has not been targeted at obtaining rigorous upper bounds on probabilities of failures associated with certification problems (2.1). However, single parameter oscillations of the function G can be seen as a form of non-linear sensitivity analysis leading to bounds on $\mathbb{P}[G \geq a]$ via McDiarmid's concentration inequality [49, 50]. These bounds can be made sharp by partitioning the input parameter space along maximum oscillation directions and computing sub-diameters on sub-domains [83].
- If $\mathscr{A}$ is expressed probabilistically through a prior (an a priori measure of probability) on the set possible scenarios (f, μ) then **Bayesian inference** [7, 45] could in principle be used to estimate $\mathbb{P}[G \geq a]$ using the posterior measure of probability on (f, μ). This combination between OUQ and Bayesian methods avoids the necessity to solve the possibly large optimization problems (2.11) and it also greatly simplifies the incorporation of sampled data thanks to the Bayes rule. However, oftentimes, priors are not available or their choice involves some degree of arbitrariness that is incompatible with the certification of rare events. Priors may become asymptotically irrelevant (in the limit of large data sets) but, for small ε, the number of required samples can be of the same order as the number required by Monte-Carlo methods [73].

  When unknown parameters are estimated using priors and sampled data, it is important to observe that the convergence of the Bayesian method may fail if the underlying probability mechanism allows an infinite number of possible outcomes (e.g., estimation of an unknown probability on $\mathbb{N}$, the set of all natural numbers) [18]. In fact, in these infinite-dimensional situations, this lack of convergence (commonly referred to as inconsistency) is the rule rather than the exception [19]. As emphasized in [18], *as more data comes in, some Bayesian statisticians will become more and more convinced of the wrong answer*.

  We also observe that, for complex systems, the computation of posterior probabilities has been made possible thanks to advances in computer science. We refer to [81] for a (recent) general (Gaussian) framework for Bayesian inverse problems and [6] for a rigorous UQ framework based on probability logic with Bayesian updating. Just as Bayesian methods would have been considered computationally infeasible 50 years ago but are now common practice, OUQ methods are now becoming feasible and will only increase in feasibility with the passage of time and advances in computing.

The certification problem (2.1) exhibits one of the main difficulties that face UQ practitioners: many theoretical methods are available, but they require assumptions or conditions that, oftentimes, are not satisfied by the application. More precisely, the characteristic elements distinguishing these different methods are the assumptions upon which they are based, and some methods will be more efficient than others depending on the validity of those assumptions. UQ applications are also characterized by a set of assumptions/information on the response function G and measure $\mathbb{P}$, which varies from application to application. Hence, on the one hand, we have a list of theoretical methods that are applicable or efficient under very specific assumptions; on the other hand, most applications are characterized by an information set or assumptions that, in general, do not match those required by these theoretical methods. It is hence natural to pursue the development of a rigorous framework that does not add inappropriate assumptions or discard information.

We also observe that the effectiveness of different UQ methods cannot be compared without reference to the available information (some methods will be more efficient than others depending on those assumptions). Generally, none of the methods mentioned above can be used without adding (arbitrary) assumptions on probability densities or discarding information on the moments or independence of the input parameters. We also observe that it is by placing information at the center of UQ that the OUQ framework allows for the identification of best experiments. Without focus on the available information, UQ methods are faced with the risk of propagating inappropriate assumptions and producing a sophisticated answer to the wrong question. These distortions of the information set may be of limited impact on certification of common events but they are also of critical importance for the certification of rare events.

2.5 Optimal Uncertainty Quantification

In this section, we describe more formally the *Optimal Uncertainty Quantification* framework. In particular, we describe what it means to give optimal bounds on the probability of failure in (2.1) given information/assumptions about the system of interest, and hence how to rigorously certify or de-certify that system.

For the sake of clarity, we will start the description of OUQ with deterministic information and assumptions (when $\mathscr{A}$ is a deterministic set of functions and probability measures).

2.5.1 First Description

In the OUQ paradigm, information and assumptions lie at the core of UQ: the available information and assumptions describe sets of admissible scenarios over which optimizations will be performed. As noted by Hoeffding [35], assumptions about the system of interest play a central and sensitive role in any statistical decision problem, even though the assumptions are often only approximations of reality.

A simple example of an information/assumptions set is given by constraining the mean and range of the response function. For example, let $\mathscr{M}(\mathscr{X})$ be the set of probability measures on the set $\mathscr{X}$, and let $\mathscr{A}_1$ be the set of pairs of probability measures $\mu \in \mathscr{M}(\mathscr{X})$ and real-valued measurable functions f on $\mathscr{X}$ such that the mean value of f with respect to μ is b and the diameter of the range of f is at most D;

$$\mathscr{A}_1 := \left\{ (f, \mu) \,\middle|\, \begin{array}{c} f : \mathscr{X} \to \mathbb{R}, \\ \mu \in \mathscr{M}(\mathscr{X}), \\ \mathbb{E}_\mu[f] = b, \\ (\sup f - \inf f) \leq D \end{array} \right\}. \tag{2.8}$$

Let us assume that *all* that we know about the "reality" $(G, \mathbb{P})$ is that $(G, \mathbb{P}) \in \mathscr{A}_1$. Then any other pair $(f, \mu) \in \mathscr{A}_1$ constitutes an admissible scenario representing a valid possibility for the "reality" $(G, \mathbb{P})$. If asked to bound $\mathbb{P}[G(X) \geq a]$, should we apply different methods and obtain different bounds on $\mathbb{P}[G(X) \geq a]$? Since some methods will distort this information set and others are only using part of it, we instead view set $\mathscr{A}_1$ as a feasible set for an optimization problem.

The General OUQ Framework

In the general case, we regard the response function G as an unknown measurable function, with some possibly known characteristics, from one measurable space $\mathscr{X}$ of inputs to a second measurable space $\mathscr{Y}$ of values. The input variables are generated randomly according to an unknown random variable X with values in $\mathscr{X}$ according to a law $\mathbb{P} \in \mathscr{M}(\mathscr{X})$, also with some possibly known characteristics. We let a measurable subset $\mathscr{Y}_0 \subseteq \mathscr{Y}$ define the *failure region*; in the example given above, $\mathscr{Y} = \mathbb{R}$ and $\mathscr{Y}_0 = [a, +\infty)$. When there is no danger of confusion, we shall simply write [G fails] for the event $[G(X) \in \mathscr{Y}_0]$.

Let $\varepsilon \in [0, 1]$ denote the *greatest acceptable probability of failure*. We say that the system is *safe* if $\mathbb{P}[G \text{ fails}] \leq \varepsilon$ and the system is *unsafe* if $\mathbb{P}[G \text{ fails}] > \varepsilon$. By *information*, or a *set of assumptions*, we mean a subset

$$\mathscr{A} \subseteq \left\{ (f, \mu) \,\middle|\, \begin{array}{c} f : \mathscr{X} \to \mathscr{Y} \text{ is measurable,} \\ \mu \in \mathscr{M}(\mathscr{X}) \end{array} \right\} \tag{2.9}$$

that contains, at the least, $(G, \mathbb{P})$. The set $\mathscr{A}$ encodes all the information that we have about the real system $(G, \mathbb{P})$, information that may come from known physical laws, past experimental data, and expert opinion. In the example $\mathscr{A}_1$ above, the only information that we have is that the mean response of the system is b and that the diameter of its range is at most D; any pair (f, μ) that satisfies these two criteria is an *admissible scenario* for the unknown reality $(G, \mathbb{P})$. Since some admissible scenarios may be safe (i.e. have $\mu[f \text{ fails}] \leq \varepsilon$) whereas other admissible scenarios may be unsafe (i.e. have $\mu[f \text{ fails}] > \varepsilon$), we decompose $\mathscr{A}$ into the disjoint union $\mathscr{A} = \mathscr{A}_{\text{safe},\varepsilon} \uplus \mathscr{A}_{\text{unsafe},\varepsilon}$, where

$$\mathscr{A}_{\text{safe},\varepsilon} := \{(f, \mu) \in \mathscr{A} \mid \mu[f \text{ fails}] \leq \varepsilon\}, \tag{2.10a}$$

$$\mathscr{A}_{\text{unsafe},\varepsilon} := \{(f, \mu) \in \mathscr{A} \mid \mu[f \text{ fails}] > \varepsilon\}. \tag{2.10b}$$

Now observe that, given such an information/assumptions set $\mathscr{A}$, there exist upper and lower bounds on $\mathbb{P}[G(X) \geq a]$ corresponding to the scenarios compatible with assumptions, i.e. the values $\mathscr{L}(\mathscr{A})$ and $\mathscr{U}(\mathscr{A})$ of the optimization problems:

$$\mathscr{L}(\mathscr{A}) := \inf_{(f,\mu)\in\mathscr{A}} \mu[f \text{ fails}] \tag{2.11a}$$

$$\mathscr{U}(\mathscr{A}) := \sup_{(f,\mu)\in\mathscr{A}} \mu[f \text{ fails}]. \tag{2.11b}$$

Since $\mathscr{L}(\mathscr{A})$ and $\mathscr{U}(\mathscr{A})$ are well-defined in $[0, 1]$, and approximations are sufficient for most purposes and are necessary in general, the difference between sup and max should not be much of an issue. Of course, some of the work that follows is concerned with the attainment of maximizers, and whether those maximizers have any simple structure that can be exploited for the sake of computational efficiency. For the moment, however, simply assume that $\mathscr{L}(\mathscr{A})$ and $\mathscr{U}(\mathscr{A})$ can indeed be computed on demand. Now, since $(G, \mathbb{P}) \in \mathscr{A}$, it follows that

$$\mathscr{L}(\mathscr{A}) \leq \mathbb{P}[G \text{ fails}] \leq \mathscr{U}(\mathscr{A}).$$

Moreover, the upper bound $\mathscr{U}(\mathscr{A})$ is *optimal* in the sense that

$$\mu[f \text{ fails}] \leq \mathscr{U}(\mathscr{A}) \text{ for all } (f, \mu) \in \mathscr{A}$$

and, if $U' < \mathscr{U}(\mathscr{A})$, then there is an admissible scenario $(f, \mu) \in \mathscr{A}$ such that

$$U' < \mu[f \text{ fails}] \leq \mathscr{U}(\mathscr{A}).$$

That is, although $\mathbb{P}[G \text{ fails}]$ may be much smaller than $\mathscr{U}(\mathscr{A})$, there is a pair (f, μ) which satisfies the same assumptions as $(G, \mathbb{P})$ such that $\mu[f \text{ fails}]$ is approximately equal to $\mathscr{U}(\mathscr{A})$. Similar remarks apply for the lower bound $\mathscr{L}(\mathscr{A})$.

Moreover, the values $\mathscr{L}(\mathscr{A})$ and $\mathscr{U}(\mathscr{A})$, defined in (2.11) can be used to construct a solution to the certification problem. Let the certification problem be defined by an error function that gives an error whenever (1) the certification process produces "safe" and there exists an admissible scenario that is unsafe, (2) the certification process produces "unsafe" and there exists an admissible scenario that is safe, or (3) the certification process produces "cannot decide" and all admissible scenarios are safe or all admissible points are unsafe; otherwise, the certification process produces no error. The following proposition demonstrates that, except in the special case $\mathscr{L}(\mathscr{A}) = \varepsilon$, that these values determine an optimal solution to this certification problem.

Proposition 2.5.1 *If* $(G, \mathbb{P}) \in \mathscr{A}$ *and*

- $\mathscr{U}(\mathscr{A}) \leq \varepsilon$ *then* $\mathbb{P}[G \text{ fails}] \leq \varepsilon$.
- $\varepsilon < \mathscr{L}(\mathscr{A})$ *then* $\mathbb{P}[G \text{ fails}] > \varepsilon$.
- $\mathscr{L}(\mathscr{A}) < \varepsilon < \mathscr{U}(\mathscr{A})$ *the there exists* $(f_1, \mu_1) \in \mathscr{A}$ *and* $(f_2, \mu_2) \in \mathscr{A}$ *such that* $\mu_1[f_1 \text{ fails}] < \varepsilon < \mu_2[f_2 \text{ fails}]$.

In other words, provided that the information set $\mathscr{A}$ is valid (in the sense that $(G, \mathbb{P}) \in \mathscr{A}$) then if $\mathscr{U}(\mathscr{A}) \leq \varepsilon$, then, the system is provably safe; if $\varepsilon < \mathscr{L}(\mathscr{A})$, then the system is provably unsafe; and if $\mathscr{L}(\mathscr{A}) < \varepsilon < \mathscr{U}(\mathscr{A})$, then the safety of the system cannot be decided due to lack of information. The corresponding certification process and its optimality are represented in Table 2.1. Hence, solving the optimization problems (2.11) determines an optimal solution to the certification problem, under the condition that $\mathscr{L}(\mathscr{A}) \neq \varepsilon$. When $\mathscr{L}(\mathscr{A}) = \varepsilon$ we can still produce an optimal solution if we obtain further information. That is, when $\mathscr{L}(\mathscr{A}) = \varepsilon = \mathscr{U}(\mathscr{A})$, then the optimal process produces "safe". On the other hand, when $\mathscr{L}(\mathscr{A}) = \varepsilon < \mathscr{U}(\mathscr{A})$, the optimal solution depends on whether or not there exists a minimizer $(f, \mu) \in \mathscr{A}$ such that $\mu[f \text{ fails}] = \mathscr{L}(\mathscr{A})$; if so, the optimal process should declare "cannot decide", otherwise, the optimal process should declare "unsafe". Observe that, in Table 2.1, we have classified $\mathscr{L}(\mathscr{A}) = \varepsilon < \mathscr{U}(\mathscr{A})$ as "cannot decide". This "nearly optimal" solution appears natural and conservative without the knowledge of the existence or non-existence of optimizers.

Example 2.5.1 The bounds $\mathscr{L}(\mathscr{A})$ and $\mathscr{U}(\mathscr{A})$ can be computed exactly—and are non-trivial—in the case of the simple example $\mathscr{A}_1$ given in (2.8). Indeed, writing $x_+ := \max(x, 0)$, the optimal upper bound is given by

Table 2.1 The OUQ certification process provides a rigorous certification criterion whose outcomes are of three types: "Certify", "De-certify" and "Cannot decide"

	$\mathscr{L}(\mathscr{A}) := \inf_{(f,\mu)\in\mathscr{A}} \mu\big[f(X) \geq a\big]$	$\mathscr{U}(\mathscr{A}) := \sup_{(f,\mu)\in\mathscr{A}} \mu\big[f(X) \geq a\big]$
$\leq \varepsilon$	**Cannot decide** Insufficient Information	**Certify** **Safe** even in the Worst Case
$> \varepsilon$	**De-certify** **Unsafe** even in the Best Case	**Cannot decide** Insufficient Information

$$\mathscr{U}(\mathscr{A}_1) = p_{\max} := \left(1 - \frac{(a-b)_+}{D}\right)_+, \tag{2.12}$$

where the maximum is achieved by taking the measure of probability of the random variable $f(X)$ to be the weighted sum of two weighted Dirac delta masses[1]

$$p_{\max}\delta_a + (1 - p_{\max})\delta_{a-D}.$$

This simple example demonstrates an extremely important point: even if the function G is extremely expensive to evaluate, certification can be accomplished without recourse to the expensive evaluations of G.

2.6 The Optimal UQ Problem

2.6.1 From Theory to Computation

Rigorous quantification of the effects of epistemic and aleatoric uncertainty is an increasingly important component of research studies and policy decisions in science, engineering, and finance. In the presence of imperfect knowledge (sometimes called *epistemic uncertainty*) about the objects involved, and especially in a high-consequence decision-making context, it makes sense to adopt a posture of *healthy conservatism*, i.e. to determine the best and worst outcomes consistent with the available knowledge. This posture naturally leads to uncertainty quantification (UQ) being posed as an optimization problem. Such optimization problems are typically high-dimensional, and hence can be slow and expensive to solve computationally (depending on the nature of the constraining information).

In previous sections and [82], we outlined the theoretical framework for *optimal uncertainty quantification* (OUQ), namely the calculation of optimal lower and upper bounds on probabilistic output quantities of interest, given quantitative information about (underdetermined) input probability distributions and response functions. In their computational formulation [53, 54], OUQ problems require optimization over discrete (finite support) probability distributions of the form

$$\mu = \sum_{i=0}^{M} w_i \delta_{x_i},$$

[1] δ_x is the Dirac delta mass on x, i.e. the measure of probability on Borel subsets $A \subset \mathbb{R}$ such that $\delta_x(A) = 1$ if $x \in A$ and $\delta_x(A) = 0$ otherwise. The first Dirac delta mass is located at the minimum of the interval $[a, \infty]$ (since we are interested in maximizing the probability of the event $\mu[f(X) \geq a]$). The second Dirac delta mass is located at $x = a - D$ because we seek to maximize $p_{\max}$ under the constraints $p_{\max}a + (1 - p_{\max})x \leq b$ and $a - x \leq D$.

where $i = 0, \ldots, M$ is a finite range of indices, the w_i are non-negative weights that sum to 1, and the x_i are points in some input parameter space $\mathscr{X}$; δ_a denotes the Dirac measure (unit point mass) located at a point $a \in \mathscr{X}$, i.e., for $E \subseteq \mathscr{X}$,

$$\delta_a(E) := \begin{cases} 1, & \text{if } a \in E, \\ 0, & \text{if } a \notin E. \end{cases}$$

Many UQ problems such as certification, prediction, reliability estimation, risk analysis, etc. can be posed as the calculation or estimation of an expected value, i.e. an integral, although this expectation (integral) may depend in intricate ways upon various probability measures, parameters, and models. This point of view on UQ is similar to that of [5], in which formulations of many problem objectives in reliability are represented in a unified framework, and the decision-theoretic point of view of [76]. In the presentation below, an important distinction is made between the "real" values of objects of interest, which are decorated with daggers (e.g. $g^\dagger$ and $\mu^\dagger$), versus possible models or other representatives for those objects, which are not so decorated.

The system of interest is a measurable *response function* $g^\dagger : \mathscr{X} \to \mathscr{Y}$ that maps a measurable space $\mathscr{X}$ of *inputs* into a measurable space $\mathscr{Y}$ of outputs. The inputs of this response function are distributed according to a probability measure $\mu^\dagger$ on $\mathscr{X}$; $\mathscr{P}(\mathscr{X})$ denotes the set of all probability measures on $\mathscr{X}$. The UQ objective is to determine or estimate the expected value under $\mu^\dagger$ of some measurable *quantity of interest* $q : \mathscr{X} \times \mathscr{Y} \to \mathbb{R}$, i.e.

$$\mathrm{e}_{X \sim \mu^\dagger}[q(X, g^\dagger(X))]. \tag{2.13}$$

The probability measure $\mu^\dagger$ can be interpreted in either a frequentist or subjectivist (Bayesian) manner, or even just as an abstract probability measure. A typical example is that the event $[g^\dagger(X) \in E]$, for some measurable set $E \subseteq \mathscr{Y}$, constitutes some undesirable "failure" outcome, and it is desired to know the $\mu^\dagger$ probability of failure, in which case q is the indicator function

$$q(x, y) := \begin{cases} 1, & \text{if } y \in E, \\ 0, & \text{if } y \notin E. \end{cases}$$

In practice, the real response function and input distribution pair $(g^\dagger, \mu^\dagger)$ are not known precisely. In such a situation, it is not possible to calculate (2.13) even by approximate methods such as Monte Carlo or other sampling techniques for the simple reason that one does not know *which* probability distribution to sample, and it may be inappropriate to simply assume that a chosen model pair $(g^{\mathrm{m}}, \mu^{\mathrm{m}})$ is $(g^\dagger, \mu^\dagger)$. However, it may be known (perhaps with some degree of statistical confidence) that $(g^\dagger, \mu^\dagger) \in \mathscr{A}$ for some collection $\mathscr{A}$ of pairs of functions $g : \mathscr{X} \to \mathscr{Y}$ and probability measures $\mu \in \mathscr{P}(\mathscr{X})$. If knowledge about which pairs $(g, \mu) \in \mathscr{A}$ are more likely than others to be $(g^\dagger, \mu^\dagger)$ can be encapsulated in a probability measure $\pi \in \mathscr{P}(\mathscr{A})$—what a Bayesian probabilist would call a *prior*—then, instead of (2.13), it makes sense to calculate or estimate

$$\mathrm{e}_{(g,\mu)\sim\pi}\Big[\mathrm{e}_{X\sim\mu}[q(X, g(X))]\Big]. \tag{2.14}$$

(A Bayesian probabilist would also incorporate additional data by conditioning to obtain the *posterior* expected value of q.)

However, in many situations, either due to lack of knowledge or being in a high-consequence regime, it may be either impossible or undesirable to specify such a π. In such situations, it makes sense to adopt a posture of *healthy conservatism*, i.e. to determine the best and worst outcomes consistent with the available knowledge. Hence, instead of (2.13) or (2.14), it makes sense to calculate or estimate

$$\underline{Q}(\mathscr{A}) := \inf_{(g,\mu)\in\mathscr{A}} \mathrm{e}_{X\sim\mu}[q(X, g(X))] \quad \text{and} \tag{2.15a}$$

$$\overline{Q}(\mathscr{A}) := \sup_{(g,\mu)\in\mathscr{A}} \mathrm{e}_{X\sim\mu}[q(X, g(X))]. \tag{2.15b}$$

If the probability distributions μ are interpreted in a Bayesian sense, then this point of view is essentially that of the robust Bayesian paradigm [9] with the addition of uncertainty about the forward model(s) g. Within the operations research and decision theory communities, similar questions have been considered under the name of distributionally robust optimization [17, 32, 76]. Distributional robustness for polynomial chaos methods has been considered in [55]. Our interest lies in providing a UQ analysis for (2.13) by the efficient calculation of the extreme values (2.15).

An important first question is whether the extreme values of the optimization problems (2.15) can be computed at all; since the set $\mathscr{A}$ is generally infinite-dimensional, an essential step is finding finite-dimensional problems that are equivalent to (i.e. have the same extreme values as) the problems (2.15). A strong analogy can be made here with finite-dimensional linear programming: to find the extreme value of a linear functional on a polytope, it is sufficient to search over the extreme points of the polytope; the extremal scenarios of $\mathscr{A}$ turn out to consist of discrete functions and probability measures that are themselves far more singular than would "typically" be encountered "in reality" but nonetheless encode the full range of possible outcomes in much the same way as a polytope is the convex hull of its "atypical" extreme points.

One general setting in which a finite-dimensional reduction can be effected is that in which, for each candidate response function $g\colon \mathscr{X} \to \mathscr{Y}$, the set of input probability distributions $\mu \in \mathscr{P}(\mathscr{X})$ that are admissible in the sense that $(g, \mu) \in \mathscr{A}$ is a (possibly empty) generalized moment class. More precisely, assume that it is known that the $\mu^\dagger$-distributed input random variable X has K independent components $(X_0, \ldots, X_{K-1})$, with each X_k taking values in a Radon space[2] $\mathscr{X}_k$; this is the

[2]This technical requirement is not a serious restriction in practice, since it is satisfied by most common parameter and function spaces. A *Radon space* is a topological space on which every

same as saying that $\mu^\dagger$ is a product of marginal probability measures $\mu_k^\dagger$ on each $\mathscr{X}_k$. By a "generalized moment class", we mean that interval bounds are given for the expected values of finitely many[3] test functions φ against either the joint distribution μ or the marginal distributions μ_k. This setting encompasses a wide spectrum of possible dependence structures for the components of X, all the way from independence, through partial correlation (an inequality constraint on $\mathrm{e}_\mu[X_i X_j]$), to complete dependence (X_i and X_j are treated as a single random variable (X_i, X_j) with arbitrary joint distribution). This setting also allows for coupling of the constraints on g and those on μ (e.g. by a constraint on $\mathrm{e}_\mu[g]$).

To express the previous paragraph more mathematically, we assume that our information about reality $(g^\dagger, \mu^\dagger)$ is that it lies in the set $\mathscr{A}$ defined by

$$\mathscr{A} := \left\{ (g, \mu) \,\middle|\, \begin{array}{c} g\colon \mathscr{X} = \mathscr{X}_0 \times \cdots \times \mathscr{X}_{K-1} \to \mathscr{Y} \text{ is measurable,} \\ \mu = \mu_0 \otimes \cdots \otimes \mu_{K-1} \text{is a product measure on } \mathscr{X}, \\ \langle \text{conditions that constrain } g \text{ pointwise} \rangle \\ \mathrm{e}_\mu[\varphi_j] \leq 0 \text{ for } j = 1, \dots, N, \\ \mathrm{e}_{\mu_k}[\varphi_{k,j_k}] \leq 0 \text{ for } k = 0, \dots, K-1,\, j_k = 1, \dots, N_k \end{array} \right\} \tag{2.16}$$

for some known measurable functions $\varphi_j\colon \mathscr{X} \to \mathbb{R}$ and $\varphi_{k,j_k}\colon \mathscr{X}_k \to \mathbb{R}$. In this case, the following reduction theorem holds:

Theorem 2.6.1 ([66, §4]) *Suppose that $\mathscr{A}$ is of the form* (2.16). *Then*

$$\underline{Q}(\mathscr{A}) = \underline{Q}(\mathscr{A}_\Delta) \quad \textit{and} \quad \overline{Q}(\mathscr{A}) = \overline{Q}(\mathscr{A}_\Delta), \tag{2.17}$$

where

$$\mathscr{A}_\Delta := \left\{ (g, \mu) \in \mathscr{A} \,\middle|\, \begin{array}{c} \textit{for } k = 0, \dots, K-1, \\ \mu_k = \sum_{i_k=0}^{N+N_k} w_{k,i_k} \delta_{x_{k,i_k}} \\ \textit{for some } x_{k,1}, x_{k,2}, \dots, x_{k_{N+N_k}} \in \mathscr{X}_k \\ \textit{and } w_{k,1}, w_{k,2}, \dots, w_{k_{N+N_k}} \geq 0 \\ \textit{with } w_{k,1} + w_{k,2} + \cdots + w_{k_{N+N_k}} = 1 \end{array} \right\}. \tag{2.18}$$

Informally, Theorem 2.6.1 says that if all one knows about the random variable $X = (X_0, \dots, X_{K-1})$ is that its components are independent, together with inequalities on N generalized moments of X and N_k generalized moments of each X_k, then for the purposes of solving (2.15) it is legitimate to consider each X_k to be a *discrete* random variable that takes at most $N + N_k + 1$ distinct values $x_{k,0}, x_{k,1}, \dots, x_{k,N+N_k}$;

Borel probability measure μ is *inner regular* in the sense that, for every measurable set E, $\mu(E) = \sup\{\mu(K) \mid K \subseteq E \text{ is compact}\}$. A simple example of a non-Radon space is the unit interval $[0, 1]$ with the lower limit topology [78, Example 51]: this topology generates the same σ-algebra as does the usual Euclidean topology, and admits the uniform (Lebesgue) probability measure, yet the only compact subsets are countable sets, which necessarily have measure zero.

[3]This is a "philosophically reasonable" position to take, since one can verify finitely many such inequalities in finite time.

those values $x_{k,i_k} \in \mathscr{X}_k$ and their corresponding probabilities $w_{k,i_k} \geq 0$ are the optimization variables.

For the sake of concision and to reduce the number of subscripts required, multi-index notation will be used in what follows to express the product probability measures μ of the form

$$\mu = \bigotimes_{k=0}^{K-1} \sum_{i_k=0}^{N+N_k} w_{k,i_k} \delta_{x_{k,i_k}}$$

that arise in the finite-dimensional reduced feasible set $\mathscr{A}_\Delta$ of (2.18). Write $i := (i_0, \dots, i_{K-1}) \in \mathbb{N}_0^K$ for a multi-index, let $0 := (0, \dots, 0)$, and let

$$M := (M_0, \dots, M_{K-1}) := (N + N_0, \dots, N + N_{K-1}).$$

Let $\#M := \prod_{k=1}^{K} (M_k + 1)$. With this notation, the $\#M$ support points of the measure μ, indexed by $i = 0, \dots, M$, will be written as

$$x_i := (x_{1,i_1}, x_{2,i_2}, \dots, x_{K,i_K}) \in \mathscr{X}$$

and the corresponding weights as

$$w_i := w_{1,i_1} w_{2,i_2} \dots w_{K,i_K} \geq 0,$$

so that

$$\mu = \bigotimes_{k=0}^{K-1} \sum_{j_k=0}^{N+N_k} w_{k,j_k} \delta_{x_{k,j_k}} = \sum_{i=0}^{M} w_i \delta_{x_i}. \tag{2.19}$$

It follows from (2.19) that, for any integrand $f : \mathscr{X} \to \mathbb{R}$, the expected value of f under such a discrete measure μ is the finite sum

$$\mathrm{e}_\mu[f] = \sum_{i=0}^{M} w_i f(x_i) \tag{2.20}$$

(It is worth noting in passing that conversion from product to sum representation and back as in (2.19) is an essential task in the numerical implementation of these UQ problems, because the product representation captures the independence structure of the problem, whereas the sum representation is best suited to integration (expectation) as in (2.20).)

Furthermore, not only is the search over μ effectively finite-dimensional, as guaranteed by Theorem 2.6.1, but so too is the search over g: since integration against a measure requires knowledge of the integrand *only* at the support points of the measure, only the $\#M$ values $y_i := g(x_i)$ of g at the support points $\{x_i \mid i = 0, \dots, M\}$ of μ need to be known. So, for example, if $g^\dagger$ is known, then it is only necessary to evaluate it on the finite support of μ. Another interesting situation of this type is considered in [82], in which $g^\dagger$ is not known exactly, but is known via legacy data at some points of $\mathscr{X}$ and is also known to satisfy a Lipschitz condition—in which case the space of admissible g is infinite-dimensional before reduction to the support of μ, but the finite-dimensional collection of admissible values $(y_0, \dots, y_M)$ has a polytope-like structure.

Theorem 2.6.1, formulae (2.19)–(2.20), and the remarks of the previous paragraph imply that $\overline{Q}(\mathscr{A})$ is found by solving the following finite-dimensional maximization problem (and $\underline{Q}(\mathscr{A})$ by the corresponding minimization problem):

$$
\begin{aligned}
\text{maximize:}\quad & \sum_{i=0}^{M} w_i q(x_i, y_i); \\
\text{among:}\quad & y_i \in \mathscr{Y} \text{ for } i = 0, \dots, M, \\
& w_{k,i_k} \in [0,1] \text{ for } k = 0, \dots, K-1 \text{ and } i_k = 0, \dots, M_k, \\
& x_{k,i_k} \in \mathscr{X}_k \text{ for } k = 0, \dots, K-1 \text{ and } i_k = 0, \dots, M_k; \\
\text{subject to:}\quad & y_i = g(x_i) \text{ for some } \mathscr{A}\text{-admissible } g\colon \mathscr{X} \to \mathscr{Y}, \\
& \sum_{i=0}^{M} w_i \varphi_j(x_i) \le 0 \text{ for } j = 1, \dots, N, \\
& \sum_{i_k=0}^{M_k} w_{k,i_k} \varphi_{k,j_k}(x_{k,i_k}) \le 0 \text{ for } k = 0, \dots, K-1 \text{ and } j_k = 1, \dots, N_k, \\
& \sum_{i_k=0}^{M_k} w_{k,i_k} = 1 \text{ for } k = 0, \dots, K-1.
\end{aligned}
\tag{2.21}
$$

Generically, the reduced OUQ problem (2.21) is non-convex, although there are special cases that can be treated using the tools of convex optimization and duality [10, 17, 76, 86]. Therefore, numerical methods for global optimization must be employed to solve (2.21). Unsurprisingly, the numerical solution of (2.21) is much more computationally intensive when $\#M$ is large—the so-called *curse of dimension*.

2.7 Optimal Design

2.7.1 The Optimal UQ Loop

Earlier, we discussed how the basic inequality

$$\mathscr{L}(\mathscr{A}) \le \mathbb{P}[G \ge a] \le \mathscr{U}(\mathscr{A})$$

provides rigorous optimal certification criteria. The certification process should not be confused with its three possible outcomes (see Table 2.1) which we call "certify" (we assert that the system is safe), "de-certify" (we assert that the system is unsafe) and "cannot decide" (the safety or un-safety of the system is undecidable given the information/assumption set $\mathscr{A}$). Indeed, in the case

$$\mathscr{L}(\mathscr{A}) \leq \varepsilon < \mathscr{U}(\mathscr{A})$$

there exist admissible scenarios under which the system is safe, and other admissible scenarios under which it is unsafe. Consequently, it follows that we can make no definite certification statement for $(G, \mathbb{P})$ without introducing further information/assumptions. If no further information can be obtained, we conclude that we "cannot decide" (this state could also be called "do not decide", because we could (arbitrarily) decide that the system is unsafe due to lack of information, for instance, but do not). However, if sufficient resources exist to gather additional information, then we enter what may be called the *optimal uncertainty quantification loop.*

Experimental Design and Selection of the Most Decisive Experiment

An important aspect of the OUQ loop is the selection of new experiments. Suppose that a number of possible experiments E_i are proposed, each of which will determine some functional $\Phi_i(G, \mathbb{P})$ of G and $\mathbb{P}$. For example, $\Phi_1(G, \mathbb{P})$ could be $\mathbb{e}_{\mathbb{P}}[G]$, $\Phi_2(G, \mathbb{P})$ could be $\mathbb{P}[X \in A]$ for some subset $A \subseteq \mathscr{X}$ of the input parameter space, and so on. Suppose that there are insufficient experimental resources to run all of these proposed experiments. Let us now consider which experiment should be run for the certification problem. Recall that the admissible set $\mathscr{A}$ is partitioned into safe and unsafe subsets as in (2.10). Define $J_{\text{safe},\varepsilon}(\Phi_i)$ to be the closed interval spanned by the possible values for the functional Φ_i over the safe admissible scenarios (i.e. the closed convex hull of the range of Φ_i on $\mathscr{A}_{\text{safe},\varepsilon}$): that is, let

$$J_{\text{safe},\varepsilon}(\Phi_i) := \left[\inf_{(f,\mu)\in\mathscr{A}_{\text{safe},\varepsilon}} \Phi_i(f,\mu), \sup_{(f,\mu)\in\mathscr{A}_{\text{safe},\varepsilon}} \Phi_i(f,\mu) \right] \tag{2.22a}$$

$$J_{\text{unsafe},\varepsilon}(\Phi_i) := \left[\inf_{(f,\mu)\in\mathscr{A}_{\text{unsafe},\varepsilon}} \Phi_i(f,\mu), \sup_{(f,\mu)\in\mathscr{A}_{\text{unsafe},\varepsilon}} \Phi_i(f,\mu) \right]. \tag{2.22b}$$

Note that, in general, these two intervals may be disjoint or may have non-empty intersection; the size of their intersection provides a measure of usefulness of the proposed experiment E_i. Observe that if experiment E_i were run, yielding the value $\Phi_i(G, \mathbb{P})$, then the following conclusions could be drawn:

$$\begin{aligned}
\Phi_i(G,\mathbb{P}) \in J_{\mathrm{safe},\varepsilon}(\Phi_i) \cap J_{\mathrm{unsafe},\varepsilon}(\Phi_i) &\implies \text{no conclusion},\\
\Phi_i(G,\mathbb{P}) \in J_{\mathrm{safe},\varepsilon}(\Phi_i) \setminus J_{\mathrm{unsafe},\varepsilon}(\Phi_i) &\implies \text{the system is safe},\\
\Phi_i(G,\mathbb{P}) \in J_{\mathrm{unsafe},\varepsilon}(\Phi_i) \setminus J_{\mathrm{safe},\varepsilon}(\Phi_i) &\implies \text{the system is unsafe},\\
\Phi_i(G,\mathbb{P}) \notin J_{\mathrm{safe},\varepsilon}(\Phi_i) \cup J_{\mathrm{unsafe},\varepsilon}(\Phi_i) &\implies \text{faulty assumptions},
\end{aligned}$$

where the last assertion (*faulty assumptions*) means that $(G,\mathbb{P}) \notin \mathscr{A}$ and follows from the fact that $\Phi_i(G,\mathbb{P}) \notin J_{\mathrm{safe},\varepsilon}(\Phi_i) \cup J_{\mathrm{unsafe},\varepsilon}(\Phi_i)$ is a contradiction. The validity of the first three assertions is based on the supposition that $(G,\mathbb{P}) \in \mathscr{A}$.

In this way, the computational optimization exercise of finding $J_{\mathrm{safe},\varepsilon}(\Phi_i)$ and $J_{\mathrm{unsafe},\varepsilon}(\Phi_i)$ for each proposed experiment E_i provides an objective assessment of which experiments are worth performing: those for which $J_{\mathrm{safe},\varepsilon}(\Phi_i)$ and $J_{\mathrm{unsafe},\varepsilon}(\Phi_i)$ are nearly disjoint intervals are worth performing since they are likely to yield conclusive results vis-à-vis (de-)certification and conversely, if the intervals $J_{\mathrm{safe},\varepsilon}(\Phi_i)$ and $J_{\mathrm{unsafe},\varepsilon}(\Phi_i)$ have a large overlap, then experiment E_i is not worth performing since it is unlikely to yield conclusive results. Furthermore, the fourth possibility above shows how experiments can rigorously establish that one's assumptions $\mathscr{A}$ are incorrect. See Fig. 2.2 for an illustration.

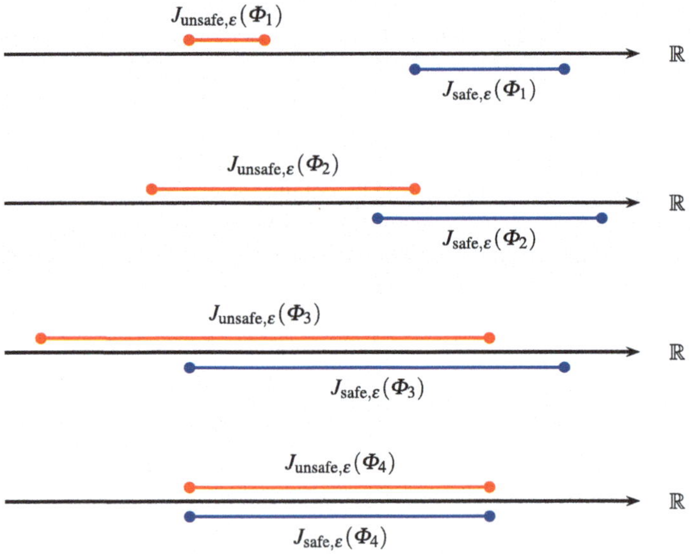

Fig. 2.2 A schematic representation of the intervals $J_{\mathrm{unsafe},\varepsilon}(\Phi_i)$ (in red) and $J_{\mathrm{safe},\varepsilon}(\Phi_i)$ (in blue) as defined by (2.22) for four functionals Φ_i that might be the subject of an experiment. Φ_1 is a good candidate for experiment effort, since the intervals do not overlap and hence experimental determination of $\Phi_1(G,\mathbb{P})$ will certify or de-certify the system; Φ_4 is not worth investigating, since it cannot distinguish safe scenarios from unsafe ones; Φ_2 and Φ_3 are intermediate cases, and Φ_2 is a better prospect than Φ_3

Remark 2.7.1 For the sake of clarity, we have started this description by defining *experiments* as *functionals* Φ_i of $\mathbb{P}$ and G. In practice, some experiments may not be functionals of $\mathbb{P}$ and G but of related objects. Consider, for instance, the situation where (X_1, X_2) is a two-dimensional Gaussian vector with zero mean and covariance matrix C, $\mathbb{P}$ is the probability distribution of X_1, the experiment E_2 determines the variance of X_2 and the information set $\mathscr{A}$ is $C \in B$, where B is a subset of symmetric positive definite 2×2 matrices. The outcome of the experiment E_2 is not a function of the probability distribution $\mathbb{P}$; however, the knowledge of $\mathbb{P}$ restricts the range of possible outcomes of E_2. Hence, for some experiments E_i, the knowledge of $(G, \mathbb{P})$ does not determine the outcome of the experiment, but only the set of possible outcomes. For those experiments, the description given above can be generalized to situations where Φ_i is a *multivalued* functional of $(G, \mathbb{P})$ determining the set of possible outcomes of the experiment E_i. This picture can be generalized further by introducing measurement noise, in which case $(G, \mathbb{P})$ may not determine a deterministic set of possible outcomes, but instead a measure of probability on a set of possible outcomes.

Selection of the Most Predictive Experiment

The computation of safe and unsafe intervals described in the previous paragraph allows of the selection of the most selective experiment. If our objective is to have an "accurate" prediction of $\mathbb{P}[G(X) \geq a]$, in the sense that $\mathscr{U}(\mathscr{A}) - \mathscr{L}(\mathscr{A})$ is small, then one can proceed as follows. Let $\mathscr{A}_{E,c}$ denote those scenarios in $\mathscr{A}$ that are compatible with obtaining outcome c from experiment E. An experiment E^* that is most predictive, even in the worst case, is defined by a minmax criterion: we seek (see Fig. 2.3)

$$E^* \in \underset{\text{experiments E}}{\arg\min} \left(\sup_{\text{outcomes } c} \left(\mathscr{U}(\mathscr{A}_{E,c}) - \mathscr{L}(\mathscr{A}_{E,c}) \right) \right) \tag{2.23}$$

The idea is that, although we can not predict the precise outcome c of an experiment E, we can compute a worst-case scenario with respect to c, and obtain an optimal bound for the minimum decrease in our prediction interval for $\mathbb{P}[G(X) \geq a]$ based

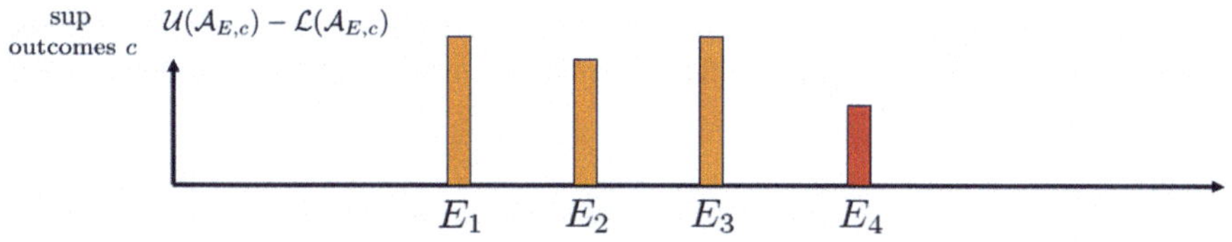

Fig. 2.3 A schematic representation of the size of the prediction intervals $\sup_{\text{outcomes } c} \left(\mathscr{U}(\mathscr{A}_{E,c}) - \mathscr{L}(\mathscr{A}_{E,c}) \right)$ in the worst case with respect to outcome c. E_4 is the most predictive experiment

on the (yet unknown) information gained from experiment E. Again, the theorems given in this paper can be applied to reduce this kind of problem. Finding E^* is a bigger problem than just calculating $\mathscr{L}(\mathscr{A})$ and $\mathscr{U}(\mathscr{A})$, but the presumption is that computer time is cheaper than experimental effort.

2.8 Model-Form Uncertainty

2.8.1 Optimal UQ and Model Error

A traditional way to deal with the missing information to model error has been to generate (possibly probabilistic) models that are compatible with the aspects that are known about the system. A key problem with this approach is that the space of such models typically has infinite dimensions while individual predictions are limited to a single element in that space. Our approach will be based on the *Optimal Uncertainty Quantification* (OUQ) framework [3, 33, 41, 44, 54, 66, 82] detailed in previous sections. In the context of OUQ, model errors can be computing by solving optimization problems (worst case scenarios) with respect to what the true response function and probability distributions could be. Note that models by themselves to not provide information (or hard constraints) on the set of admissible response functions (they are only elements of that set). However the computation of (possibly optimal) bounds on model errors enables the integration of such models with data by constraining the admissible space of underlying response functions and measures as illustrated in [82].

A Reminder of OUQ

Let $\mathscr{X}$ be a measurable space. Let $\mathscr{M}(\mathscr{X})$ be the set of Borel probability measures on $\mathscr{X}$. Let $\mathscr{B}(\mathscr{X};\mathbb{R})$ be the space of real-valued Borel-measurable functions on $\mathscr{X}$, and let $\mathscr{G} \subseteq \mathscr{B}(\mathscr{X};\mathbb{R})$. Let $\mathscr{A}$ be an arbitrary subset of $\mathscr{G} \times \mathscr{M}(\mathscr{X})$, and let $\Phi\colon \mathscr{G} \times \mathscr{M}(\mathscr{X}) \to \mathbb{R}$. In the context of the OUQ framework as described in [66] one is interested in estimating $\Phi(G,\mathbb{P})$, where $(G,\mathbb{P}) \in \mathscr{G} \times \mathscr{M}(\mathscr{X})$ corresponds to an *unknown reality*. If $\mathscr{A}$ represents all that is known on $(G,\mathbb{P})$ (in the sense that $(G,\mathbb{P}) \in \mathscr{A}$ and that any $(f,\mu) \in \mathscr{A}$ could, a priori, be $(G,\mathbb{P})$ given the available information) then [66] shows that the following quantities (2.24) and (2.25) are the optimal (with respect to the available information) upper and lower bounds on the quantity of interest $\Phi(G,\mathbb{P})$:

$$\mathscr{U}(\mathscr{A}) := \sup_{(f,\mu)\in\mathscr{A}} \Phi(f,\mu), \tag{2.24}$$

$$\mathscr{L}(\mathscr{A}) := \inf_{(f,\mu)\in\mathscr{A}} \Phi(f,\mu). \tag{2.25}$$

2.8.2 Game-Theoretic Formulation and Model Error

Since the pioneering work of Von Neumann and Goldstine [88], the prime objective of Scientific Computing has been focused on the efficient numerical evaluation of scientific models and underlying challenges have been defined in terms of the size and complexity of such models. The purpose of such work is to enable computers to develop models of reality based on imperfect and limited information (rather than just run numbers through models developed by humans after a laborious process of scientific investigation). Although the importance of the algorithmic aspects of decision making has been recognized in the emerging field of Algorithmic Decision Theory [68], part of this work amounts to its incorporation in a generalization of Wald's Decision Theory framework [90]. Owhadi et al. has recently laid down the foundations for the scientific computation of optimal statistical estimators (SCOSE) [25, 33, 60, 62–65, 67]. SCOSE constitutes a generalization of the Optimal Uncertainty Quantification (OUQ) framework [3, 41, 44, 54, 66, 82] (to information coming in the form of sample data). This generalization is built upon Von Neumann's Game Theory [89], Nash's non-cooperative games [56, 57], and Wald's Decision Theory [90].

In the presence of data, the notion of optimality is (in this framework) that of the optimal strategy for a non-cooperative game where (1) Player A chooses a (probability) measure $\mu^\dagger$ and a (response) function $f^\dagger$ in an admissible set $\mathscr{A}$ (that is typically infinite dimensional and finite co-dimensional) (2) Player B chooses a function θ of the data d (sampled according to the data generating distribution $\mathbb{D}(f, \mu)$, which depends on (f, μ)) (3) Player A tries to maximize the statistical error $\mathscr{E}$ of the quantity of interest while Player B tries to minimize it. Therefore optimal estimators are obtained as solutions of

$$\min_{\theta} \max_{(f,\mu)\in\mathscr{A}} \mathrm{e}_{d\sim\mathbb{D}(f,\mu)}\big[\mathscr{E}(\theta(d), \Phi(f, u))\big] \tag{2.26}$$

The particular choice of the cost function $\mathscr{E}$ determines the specific quantification of uncertainties (e.g., the derivation of optimal intervals of confidence, bounds on the probability or detection of rare events, etc.). If θ^* is an arbitrary model (not necessarily optimal) then

$$\max_{(f,\mu)\in\mathscr{A}} \mathrm{e}_{d\sim\mathbb{D}(f,\mu)}\big[\mathscr{E}(\theta^*(d), \Phi(f, u))\big] \tag{2.27}$$

provides a rigorous and optimal bound on its statistical error.

By minimizing this optimal bound over θ (as formulated in (2.26)) one obtains an optimal (statistical) model, which could be used to facilitate the extraction of as much information as possible from the available data. This is specially important when the amount of sample data is limited and each new set of data requires an expensive experiment. Once an optimal estimator has been computed, it can be turned into a digital statistical table accessible to the larger scientific, medical, financial and engineering communities.

Although the min max optimization problem (2.26) requires searching the space of all functions of the data, since it is a zero sum game [89], under mild conditions (compactness of the decision space [90]) it can be approximated by a finite game where optimal solutions are mixed strategies [56, 57] and live in the Bayesian class of estimators, i.e. the optimal strategy for player A (the adversary) is to place a prior distribution π' over $\mathscr{A}$ and select (f, μ) at random, while the optimal strategy for player B (the model/estimator builder) is to assume that player A has selected such a strategy, and place a prior distribution π over $\mathscr{A}$ and derive θ as the Bayesian estimator $\theta_\pi(d) = \mathrm{e}_{(f,\mu)\sim\pi,d'\sim\mathbb{D}(f,\mu)}[\Phi(f,\mu)|d' = d]$. Therefore optimal strategies can be obtained by reducing (2.26) to a min max optimization over prior distributions on $\mathscr{A}$. Furthermore, under the same mild conditions [56, 57, 90], duality holds, and allows us to show that the optimal strategy for player B corresponds to the worst Bayesian prior, i.e. the solution of the max problem: $\max_{\pi\in\mathscr{A}} \mathrm{e}_{(f,\mu)\sim\pi,d\sim\mathbb{D}(f,\mu)}\big[\mathscr{E}(\theta_\pi(d), \Phi(f,u))\big]$. Although this is an optimization problem over measures and functions, it has been shown in [64] that analogous problems can be reduced to a nesting of optimization problems of measures (and functions) amenable to finite-dimensional reduction by the techniques developed by Owhadi et al. in the context of stochastic optimization [33, 65, 66, 82]. Therefore, although the computation of optimal (statistical) models (estimators) requires, at an abstract level, the manipulation of measures on infinite dimensional spaces of measures and functions, they can be reduced to the manipulation of discrete finite-dimensional objects through a form of calculus manipulating the codimension of the information set (what is known). Observe also that an essential difference with Bayesian Inference is that the game (2.26) is non cooperative (players A and B may have different prior distributions) and an optimization problem has to be solved to find the prior leading to the optimal estimator/model.

2.9 Design and Decision-Making Under Uncertainty

2.9.1 Optimal UQ for Vulnerability Identification

The extremizers of OUQ optimization problems are singular probability distributions with support points on the key players, i.e. weak points of the system. Therefore by solving optimization problems corresponding to worst case scenarios these *key characteristic descriptors* will naturally be identified. This analysis can sometimes give surprising and non-intuitive insights into what are the critical variables and critical missing information to regularize and improve a model, guiding us towards determining which missing experiments or information are *on the critical path* for reducing uncertainties in the model and risk in the system. Note that the same analysis can also be performed in the game theoretic formulation (2.26) where the extremizers of (2.27) identify (in presence of sample data) admissible response functions and probability distributions maximizing risk.

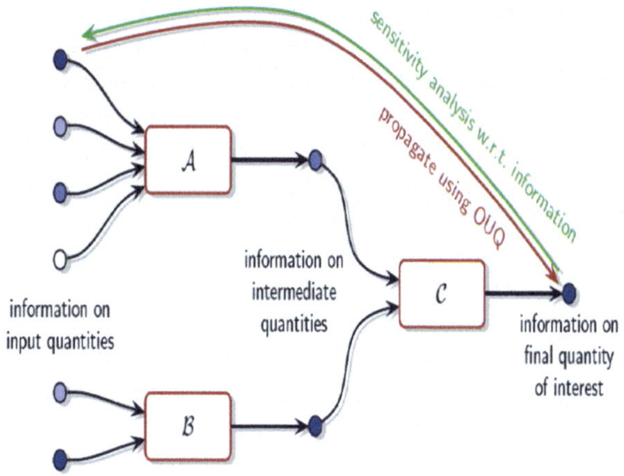

Fig. 2.4 Because OUQ is a sharp information propagation scheme, the results of sensitivity analysis (inverse OUQ) give non-trivial insights into the roles of the various pieces of input information. Some inputs may even be irrelevant!

In presence of sample data, the safety or vulnerability of a system can be assessed as an (optimal) classification problem (safe vs. unsafe). The Standard Classification Problem of Machine Learning was heavily researched for several decades, but it was Vapnik's introduction of the Support Vector Machine, see e.g. Cortes and Vapnik [15], that promised the simultaneous achievement of good performance and efficient computation. See Christmann and Steinwart [79] for both the history and a comprehensive treatment. However, to our knowledge, a complete formulation of the Classification Problem in Wald's theory of Optimal Statistical Decisions has yet to be accomplished.

2.9.2 Data Collection for Design Optimization

The OUQ framework allows the development of an OUQ loop that can be used for experimental design and design optimization [66]. The problem of predicting optimal bounds on the results of experiments under the assumption that the system is safe (or unsafe) is well-posed and benefits from similar reduction properties. Best experiments are then naturally identified as those whose predicted ranges have minimal overlap between safe and unsafe systems.

Another component of SCOSE is the application of the game theoretic framework to data collection and design optimization. Note that if the model is exact (and if the underlying probability distributions are known) then the design problem is, given a loss function (such as probability of failure), a straightforward optimization problem that can potentially be handled via `mystic`. The difficulty of this design problem lies in the facts that the model is not perfect and the true response function and data generating distribution are imperfectly known. If safety is to be privileged, then this design under incomplete formulation can be formulated as a non-cooperative game when player A chooses the true response function and data generating distribution and player B chooses the model and a resulting design (derived from the combination of the model of the data). In this adversarial game player A tries to maximize the loss function (e.g. probability of failure) while player B tries to minimize, and the resulting design is optimal given available information. Since the resulting optimization can (even after reduction) be highly non-linear and highly constrained our approach is hierarchical and based on non-cooperative information games played at different levels of complexity (i.e. the idea is to solve the design problem at different levels of complexity (Fig. 2.4). Recent work has shown this facilitation of the design process is not only possible but could also automate the process of scientific discovery [60, 63]. In particular, we refer to [61] for an illustration of an application of this framework to the automation of the design and discovery of interpolation operators for multigrid methods (for PDEs with rough coefficients, a notoriously difficult open problem in the CSE community) and to the automation of orthogonal multi-resolution operator decomposition.

2.10 A Software Framework for Optimization and UQ in Reduced Search Space

2.10.1 Optimization and UQ

A rigorous quantification of uncertainty can easily require several thousands of model evaluations $f(x)$. For all but the smallest of models, this requires significant clock time—a model requiring 1 min of clock time evaluated 10,000 times in a global optimization will take 10,000 min ($\sim$7 days) with a standard optimizer. Furthermore, realistic models are often high-dimensional, highly-constrained, and may require several hours to days even when run on a parallel computer cluster. For studies of this size or larger to be feasible, a fundamental shift in how we build optimization algorithms is required. The need to provide support for parallel and distributed computing at the lowest level—within the optimization algorithm—is clear. Standard optimization algorithms must be extended to parallel. The need for new massively-parallel optimization algorithms is also clear. If these parallel optimizers are not also seamlessly extensible to distributed and heterogeneous computing, then the scope of problems that can be addressed will be severely limited.

While several robust optimization packages exist [27, 39], there are very few that provide massively-parallel optimization [8, 23, 37]—the most notable effort being DAKOTA [2], which also includes methods for uncertainty quantification. A rethinking of optimization algorithms, from the ground up, is required to dramatically lower the barrier to massively-parallel optimization and rigorous uncertainty quantification. The construction and tight integration of a framework for heterogeneous parallel computing is required to support such optimizations on realistic models. The goal should be to enable widespread availability of these tools to scientists and engineers in all fields.

2.10.2 A Highly-Configurable Optimization Framework

We have built a robust optimization framework (`mystic`) [52] that incorporates the mathematical framework described in [66], and have provided an interface to prediction, certification, and validation as a framework service. The `mystic` framework provides a collection of optimization algorithms and tools that lowers the barrier to solving complex optimization problems. `mystic` provides a selection of optimizers, both global and local, including several gradient solvers. A unique and powerful feature of the framework is the ability to apply and configure solver-independent termination conditions—a capability that greatly increases the flexibility for numerically solving problems with non-standard convergence profiles. All of `mystic`'s solvers conform to a solver API, thus also have common method calls to configure and launch an optimization job. This allows any of `mystic`'s solvers to be easily swapped without the user having to write any new code.

The minimal solver interface:

```
# the function to be minimized and the initial values
from mystic.models import rosen as my_model
x0 = [0.8, 1.2, 0.7]

# configure the solver and obtain the solution
from mystic.solvers import fmin
solution = fmin(my_model, x0)
```

The criteria for when and how an optimization terminates are of paramount importance in traversing a function's potential well. Standard optimization packages provide a single convergence condition for each optimizer. `mystic` provides a set of fully customizable termination conditions, allowing the user to discover how to better navigate the optimizer through difficult terrain.

The expanded solver interface:

```
# the function to be minimized and initial values
from mystic.models import rosen as my_model
x0 = [0.8, 1.2, 0.7]

# get monitor and termination condition objects
from mystic.monitors import Monitor, VerboseMonitor
stepmon = VerboseMonitor(5)
evalmon = Monitor()
from mystic.termination import ChangeOverGeneration
terminate = ChangeOverGeneration()

# instantiate and configure the solver
from mystic.solvers import NelderMeadSimplexSolver
```

```
solver = NelderMeadSimplexSolver(len(x0))
solver.SetInitialPoints(x0)
solver.SetGenerationMonitor(stepmon)
solver.SetEvaluationMonitor(evalmon)
solver.Solve(my_model, terminate)

# obtain the solution
solution = solver.bestSolution

# obtain diagnostic information
function_evals = solver.evaluations
iterations = solver.generations
cost = solver.bestEnergy

# modify the solver configuration, then restart
from mystic.termination import VTR, Or
terminate = ChangeOverGeneration(tolerance=1e-8)
solver.Solve(my_model, Or(VTR(), terminate))

# obtain the new solution
solution = solver.bestSolution
```

2.10.3 Reduction of Search Space

`mystic` provides a method to constrain optimization to be within an N-dimensional box on input space, and also a method to impose user-defined parameter constraint functions on any cost function. Thus, both *bounds constraints* and *parameter constraints* can be generically applied to any of `mystic`'s unconstrained optimization algorithms. Traditionally, constrained optimization problems tend to be solved iteratively, where a penalty is applied to candidate solutions that violate the constraints. Decoupling the solving of constraints from the optimization problem can greatly increase the efficiency in solving highly-constrained nonlinear problems—effectively, the optimization algorithm only selects points that satisfy the constraints. Constraints can be solved numerically or algebraically, where the solving of constraints can itself be cast as an optimization problem. Constraints can also be dynamically applied, thus altering an optimization in progress.

Penalty methods apply an energy barrier $\Delta E = k \cdot p(\mathbf{x})$ to the unconstrained cost function $f(\mathbf{x})$ when the constraints are violated. The modified cost function ϕ is thus written as:

$$\phi(\mathbf{x}) = f(\mathbf{x}) + k \cdot p(\mathbf{x}) \tag{2.28}$$

Alternately, kernel methods apply a transform c that maps or reduces the search space so that the optimizer will only search over the set of candidates that satisfy the constraints. The transform has an interface $\mathbf{x}' = c(\mathbf{x})$, and the cost function becomes:

$$\phi(\mathbf{x}) = f(c(\mathbf{x})) \tag{2.29}$$

Adding penalties or constraints to a solver is done with the `penalty` or `constraint` keyword (or with the `SetConstraints` and `SetPenalty` methods in the expanded interface).

```
from mystic.math.measures import mean, spread
from mystic.constraints import with_penalty, with_mean
from mystic.constraints import quadratic_equality

# build a penalty function
@with_penalty(quadratic_equality, kwds={'target':5.0})
def penalty(x, target):
  return mean(x) - target

# define an objective
def cost(x):
  return abs(sum(x) - 5.0)

# solve using a penalty
from mystic.solvers import fmin
x = array([1,2,3,4,5])
y = fmin(cost, x, penalty=penalty)

# build a kernel transform
@with_mean(5.0)
def constraint(x):
return x

# solve using constraints
y = fmin(cost, x, constraint=constraint)
```

`mystic` provides a simple interface to a lot of underlying complexity—enabling a non-specialist user to easily access optimizer configurability and high-performance computing without a steep learning curve. `mystic` also provides a simple interface to the application of constraints on a function or measure. The natural syntax for a constraint is one of symbolic math, hence `mystic` leverages SymPy [13] to construct a symbolic math parser for the translation of symbolic text input into functioning constraint code objects:

```
"""
    Minimize: f = 2*x[0] + 1*x[1]

    Subject to:  -1*x[0] + 1*x[1] <= 1
                  1*x[0] + 1*x[1] >= 2
                           1*x[1] >= 0
                  1*x[0] - 2*x[1] <= 4

    where:  -inf <= x[0] <= inf
"""

def objective(x):
    x0,x1 = x
    return 2*x0 + x1

equations = """
-x0 + x1 - 1.0 <= 0.0
-x0 - x1 + 2.0 <= 0.0
x0 - 2*x1 - 4.0 <= 0.0
"""
bounds = [(None, None),(0.0, None)]

# parse the equations into penalties and/or constraints
from mystic.symbolic import generate_conditions, simplify,
     generate_penalty, generate_constraint, generate_solvers
pf = generate_penalty(generate_conditions(equations), k=1e3)
cf = generate_constraint(generate_solvers(simplify(equations)))

from mystic.solvers import fmin_powell
result = fmin_powell(objective, x0=[0.0,0.0], bounds=bounds,
     constraint=cf, penalty=pf, disp=True, gtol=3)
```

The constraints parser can parse multiple and nonlinear constraints, and equality or inequality constraints. Similarly for the penalty parser. Available penalty methods include the exterior penalty function method [87], the augmented Lagrange multiplier method [42], and the logarithmic barrier method [38]. Available transforms include range constraints, uniqueness and set-membership constraints, probabilistic and statistical constraints, constraints imposing sampling statistics, inputs from sampling distributions, constraints from legacy data, constraints from models and distance metrics, constraints on measures, constraints on support vectors, and so on.

It is worth noting that the use of a transform c does not require the constraints be bound to the cost function. The evaluation of the constraints are decoupled from the evaluation of the cost function—hence, with `mystic`, highly-constrained optimization decomposes to the solving of K independent constraints, followed by an unconstrained optimization over only the set of valid points. This method has been shown effective for solving optimization problems where $K \approx 200$ [66].

2.10.4 New Massively-Parallel Optimization Algorithms

In `mystic`, optimizers have been extended to parallel whenever possible. To have an optimizer execute in parallel, the user only needs to provide the solver with a parallel map. For example, extending the Differential Evolution [80] solver to parallel is involves passing a `Map` to the `SetEvaluationMap` method. In the example below, each generation has 20 candidates, and will execute in parallel with 4 workers:

```
# the function to be minimized and the bounds
from mystic.models import rosen as my_model
lb = [0.0, 0.0, 0.0]; ub = [2.0, 2.0, 2.0]

# get termination condition object
from mystic.termination import ChangeOverGeneration
terminate = ChangeOverGeneration()

# select the parallel launch configuration
from pathos.pools import ProcessPool
my_map = ProcessPool(4).map

# instantiate and configure the solver
from mystic.solvers import DifferentialEvolutionSolver
solver = DifferentialEvolutionSolver(len(lb), 20)
solver.SetRandomInitialPoints(lb, ub)
solver.SetStrictRanges(lb, ub)
solver.SetEvaluationMap(my_map)
solver.Solve(my_model, terminate)

# obtain the solution
solution = solver.bestSolution
```

Another type of new parallel solver utilizes the `SetNestedSolver` method to stage a parallel launch of N optimizers, each with different initial conditions (Fig. 2.5). The following code shows the `BuckshotSolver` scheduling a launch of $N = 20$ optimizers in parallel to the default queue, where 5 nodes each with 4 processors have been requested:

```
# the function to be minimized and the bounds
from mystic.models import rosen as my_model
lb = [0.0, 0.0, 0.0]; ub = [2.0, 2.0, 2.0]

# get monitor and termination condition objects
from mystic.monitors import LoggingMonitor
stepmon = LoggingMonitor(1, 'log.txt')
```

```
from mystic.termination import ChangeOverGeneration
terminate = ChangeOverGeneration()

# select the parallel launch configuration
from pyina.launchers import TorqueMpi
my_map = TorqueMpi('5:ppn=4').map

# instantiate and configure the nested solver
from mystic.solvers import PowellDirectionalSolver
my_solver = PowellDirectionalSolver(len(lb))
my_solver.SetStrictRanges(lb, ub)
my_solver.SetEvaluationLimits(1000)

# instantiate and configure the outer solver
from mystic.solvers import BuckshotSolver
solver = BuckshotSolver(len(lb), 20)
solver.SetRandomInitialPoints(lb, ub)
solver.SetGenerationMonitor(stepmon)
solver.SetNestedSolver(my_solver)
solver.SetSolverMap(my_map)
solver.Solve(my_model, terminate)

# obtain the solution
solution = solver.bestSolution
```

Instead of using ensemble optimizers to search for the global minimum, an ensemble of optimizers can just as easily be configured to search for all critical points of an unknown surface. In this mode, batches of ensemble solvers are launched until no more critical points are found—afterward, an accurate surrogate for the unknown surface can be interpolated from the critical points and other points the optimizers have visited. In materials science, the typical approach for calculating a unknown potential energy surface is to find the global minimum and then perform random walks in hope to discover the unknown energy surface. The ensemble optimizer approach discussed here provides several advantages, as it is embarrassingly parallel, and also does not have the single point of failure (solving for the global minimum) that traditional methods have.

2.10.5 Probability and Uncertainty Tooklit

OUQ problems can be thought of optimization problems where the goal is to find the global maximum of a probability function $\mu[H \leq 0]$, where $H \leq 0$ is a failure criterion for the model response function H. Additional conditions in an OUQ problem

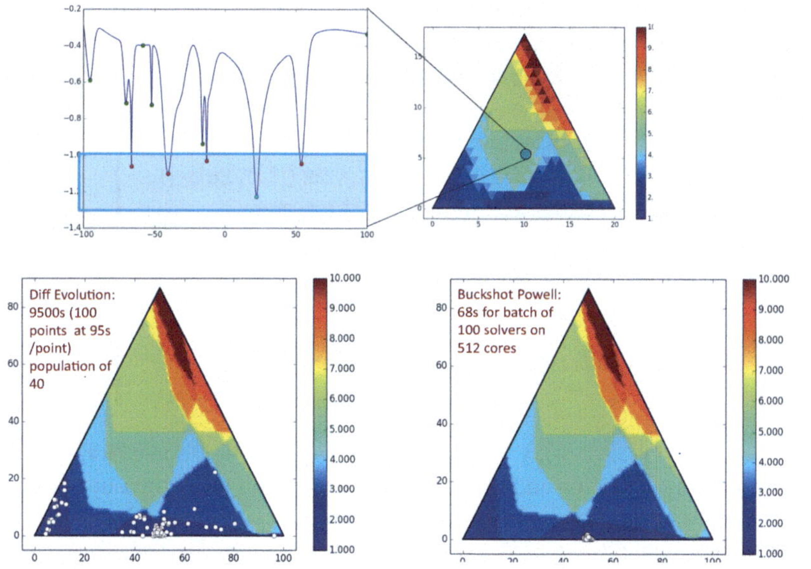

Fig. 2.5 Solutions (white dots) of a model nanostructure problem (top), using a Differential Evolution solver (left) and a Buckshot-Powell ensemble solver (right). The color scale indicates the number of degenerate local minima near in the neighborhood of the global minimum. Note that the ensemble optimizer solutions are more accurate and converge much more quickly than the traditional global optimizer

are provided as constraints on the information set. For example, a condition such as a mean constraint on H, $m_1 \leq \mathbb{E}_\mu[H] \leq m_2$, will be imposed on the maximization. After casting the OUQ problem in terms of optimization and constraints, we can plug these terms into the infrastructure provided by `mystic`.

Optimal uncertainty quantification (OUQ) typically involves a maximization over a probability distribution, thus the objective is not a simple metric on the user-provided model function, but is instead a statistical quantity operating on a constrained probability measure. For example, a discrete measure is represented by a collection of support points, each with an accompanying weight. Measures have methods for calculating the mass, range, mean, and other moments of the measure, and also for imposing a mass, range, mean, and other moments on the measure. Discrete measures also provide basic operations, including point addition and subtraction, and the formation of product measures and data sets.

Global optimizations used in solving OUQ problems are composed in the same manner as shown above for the `DifferentialEvolutionSolver`. The cost function, however, is not formulated as in the examples above—OUQ is an optimization over product measures, and thus uses `mystic`'s `product_measure` class as the target of the optimization. Also as shown above, the bounds constraints are imposed with the `SetStrictRanges` method, while parameter constraints (composed as below) are imposed with the SetConstraints method. The union set of these constraints defines the set $\mathscr{A}$.

So for example, let us define the feasable set

$$\mathscr{A} = \left\{ (f, \mu) \,\middle|\, \begin{array}{c} f = \mathtt{my\_model} : \prod_{i=1}^{3}[\mathtt{lb_i}, \mathtt{ub_i}] \to \mathbb{R}, \\ \mu = \bigotimes_{i=1}^{3} \mu_i \in \bigotimes_{i=1}^{3} \mathscr{M}([\mathtt{lb_i}, \mathtt{ub_i}]), \\ \mathtt{m_{lb}} \leq \mathbb{E}_{\mu}[\mathtt{f}] \leq \mathtt{m_{ub}} \end{array} \right\} \tag{2.30}$$

which reduces to the finite-dimensional subset

$$\mathscr{A}_{\Delta} = \left\{ (f, \mu) \in \mathscr{A} \,\middle|\, \begin{array}{c} \text{for } \mathbf{x} \text{ and } \mathbf{y} \in \prod_{i=1}^{3}[\mathtt{lb_i}, \mathtt{ub_i}], \\ \text{and } \mathbf{w} \in [0, 1], \\ \mu_i = w_i \delta_{x_i} + (1 - w_i)\delta_{y_i} \end{array} \right\} \tag{2.31}$$

where $\mathbf{x} = \text{some } (x_1, x_2, x_3)$, $\mathbf{y} = \text{some } (y_1, y_2, y_3)$, and $\mathbf{w} = \text{some } (w_1, w_2, w_3)$.

The constraints function and the cost function are built using measure mathematics, and can be passed to any of mystic's optimizers. For, say, three Dirac delta masses in each direction, we would define npts = (3, 3, 3). The lb would then be defined as a sequence (of length 9) of lower bounds for each Dirac delta mass, and similarly for upper bounds in ub. The optimization parameters param would correspondingly be a sequence of length 9.

The first block of constraints below check if measure.mass $\approx$ 1.0; and if not, the measure's mass is normalized to 1.0. The second block of constraints below check if $m_{lb} \leq \mathbb{E}_{\mu}[H] \leq m_{ub}$, where m_{lb} = target_mean − error and m_{ub} = target_mean + error; and if not, an optimization is performed to satisfy this mean constraint. The product_measure is built (with load) from the optimization parameters param, and after all the constraints are applied, flatten is used to extract the updated param:

```
from mystic.math.measures import split_param
from mystic.math.discrete import product_measure
from mystic.math import almostEqual

# split bounds into weight-only & sample-only
w_lb, m_lb = split_param(lb, npts)
w_ub, m_ub = split_param(ub, npts)

# generate constraints function
def constraints(param):
  c = product_measure().load(param, npts)

  # impose norm on measures
  for measure in c:
    if not almostEqual(float(measure.mass), 1.0):
      measure.normalize()

  # impose expectation on product measure
  E = float(c.expect(my_model))
```

```
    if not (E <= float(target_mean + error)) \
    or not (float(target_mean - error) <= E):
      c.set_expect((target_mean, error), my_model, (m_lb,m_ub))

  # extract weights and positions
    return c.flatten()
```

The cost function calculates probability of failure, using the `pof` method:

```
  # generate maximizing function
  def cost(param):
    return MINMAX * product_measure().load(param, npts).pof(my_model)
```

When `MINMAX=-1`, we are seeking the *supremum*, and upon solution, the function maximum is `-solver.bestEnergy`. Alternatively, with `MINMAX=1`, we are seeking the *infimum*, and upon solution, the function minimum is `solver.bestEnergy`.

To solve this OUQ problem, we first write the code for the bounds, cost function, and constraints—then we plug this code into a global optimization script, as noted above.

2.11 Scalability

2.11.1 Scalability Through Asynchronous Parallel Computing

Global optimizations with dynamically spawned constraints solvers are possible because `mystic` provides a functional abstract programming interface (API) for constraints solvers and optimization algorithms. The optimizer working against the objective function drives the execution logic, and through the functional API, can dynamically stage and launch several thousand nested optimizer and/or constraints solver instances. A full suite of blocking, iterative non-blocking, and asynchronous maps and pipes have been developed for multiprocessing, MPI, and IPC-based distributed computing. The `mystic` framework provides optimization algorithms that can leverage parallel computing at several levels–for example, population-based solvers [80] that can launch function evaluations in parallel [54, 66], and ensemble solvers that can launch multiple optimizers in parallel [3, 44, 54]. All of `mystic`'s optimizers have been adapted for asynchronous computing, where each optimizer is capable of saving its full state at each iteration in the optimization. While most optimization frameworks implement optimization algorithms that block until complete, `mystic`'s asynchronous solvers enable optimizations to proceed step-by-step, with full restart capabilities at each iteration. Because `mystic`'s solvers are also serializable, an optimizer state can be shipped off to a database by logging the optimizer itself into the database. Additionally, this serialization enables an optimizer to be stopped, saved, shipped to another resource, and then restarted without any loss of information, accuracy, or progress.

The complexity required by hierarchical optimizations is managed by the underlying graph execution and management framework, *pathos* [51]. The *pathos* framework provides an abstraction layer on programming models for heterogeneous parallel and distributed computing. Currently, *pathos* has multi-core, multi-node, multi-thread, and multi-cluster launching capabilities.

`mystic` and *pathos* originate in the DANSE project ($15M, NSF), and were first integrated in the PSAAP project ($17M, NNSA) in support of PSAAP's uncertainty quantification objective. Development of the `mystic` and *pathos* frameworks continued under the ExMatEx project (0.6M$ subk, DOE/BES), the Complex Modelling Initiative (0.1M$ subk, DOE-BES), and with funding from the AFOSR (0.9M$). With an eye on the exascale, *pathos* was redesigned to provide heterogeneous asynchronous parallel and distributed computing that is robust against failure. *pathos* was extended to seamlessly utilize local memory cache and common archival storage (e.g. on-disk or in-database), thus providing efficient storage and retrieval of results while minimizing recomputation across a computational campaign (i.e. related runs potentially separated by several hours to years). `mystic`'s optimizers were partly extended to asynchronous parallel computing, where optimizers can be controlled and configured at each solver iteration. With this, optimizers are fully state-preserving and restartable, and thus optimizations can run as daemons (long lived processes) reacting dynamically to evolving constraints. Essentially, whenever new information is provided, the optimizer can halt, add the relevant constraints, and then proceed from exactly where it was halted. If there is no new information, and the optimizer reaches a termination condition, the optimization daemon can go to sleep until new information is provided.

Both `mystic` and *pathos* have demonstrated capabilities for hierarchical and heterogeneous asynchronous parallel and distributed computing; however, many of these capabilities have not been battle-tested in large-scale hierarchical UQ optimizations, and definitely not in the context of materials discovery and design.

`mystic` and *pathos* are distributed as several standalone software packages, each representing a portion of the full framework's capacity. The latest stable releases and development branches of all packages in the `mystic` and *pathos* framework are available on github [1], and are BSD licensed.

References

1. http://github.com/uqfoundation
2. B. Adams, W. Bohnhoff, K. Dalbey, J. Eddy, M. Eldred, D. Gay, K. Haskell, P. Hough, L. Swiler, DAKOTA, a multilevel parallel object-oriented framework for design optimization, parameter estimation, uncertainty quantification, and sensitivity analysis: Version 5.0 user's manual. Technical report, Dec 2009. Sandia Technical Report SAND2010-2183

3. M. Adams, A. Lashgari, B. Li, M. McKerns, J. Mihaly, M. Ortiz, H. Owhadi, A.J. Rosakis, M. Stalzer, T.J. Sullivan, Rigorous model-based uncertainty quantification with application to terminal ballistics. Part II. Systems with uncontrollable inputs and large scatter. J. Mech. Phys. Solids **60**(5), 1002–1019 (2012)
4. I. Babuška, F. Nobile, R. Tempone, A stochastic collocation method for elliptic partial differential equations with random input data. SIAM Rev. **52**(2), 317–355 (2010)
5. R.E. Barlow, F. Proschan, Mathematical theory of reliability, in *Classics in Applied Mathematics*, vol. 17 (Society for Industrial and Applied Mathematics (SIAM), Philadelphia, PA, 1996). With contributions by L. C. Hunter, Reprint of the 1965 original [MR 0195566]
6. J.L. Beck, Bayesian system identification based on probability logic. Struct. Control Health Monit. **17**, 825–847 (2010)
7. J.L. Beck, L.S. Katafygiotis, Updating models and their uncertainties: Bayesian statistical framework. J. Eng. Mech. **124**(4), 455–461 (1998)
8. S. Benson, L. Curfman McInnes, J. More, T. Munson, J. Sarich, TAO user manual (revision 1.10.1) (2010)
9. J.O. Berger, An overview of robust Bayesian analysis. Test **3**(1), 5–124 (1994). With comments and a rejoinder by the author
10. D. Bertsimas, I. Popescu, Optimal inequalities in probability theory: a convex optimization approach. SIAM J. Optim. **15**(3), 780–804 (electronic) (2005). https://doi.org/10.1137/S1052623401399903
11. M. Bieri, C. Schwab, Sparse high order FEM for elliptic sPDEs. Comput. Methods Appl. Mech. Eng. **198**(13–14), 1149–1170 (2009)
12. Boeing, Statistical summary of commercial jet airplane accidents worldwide operations 1959–2009. Technical report, Aviation Safety Boeing Commercial Airplanes, Seattle, Washington 98124-2207, U.S.A., July 2010
13. A. Certik et al., SymPy: Python Library for Symbolic Mathematics (2011)
14. M.J. Cliffe, M.T. Dove, D.A. Drabold, A.L. Goodwin, Phys. Rev. Lett. **104**, 125501 (2010)
15. C. Cortes, V. Vapnik, Support-vector networks. Mach. Learn. **20**(3), 273–297 (1995)
16. W.I.F. David, K. Shankland, L.B. McCusker, C. Baerlocher (eds.), *Structure Determination from Powder Diffraction Data* (Oxford University Press, Oxford, 2002)
17. E. Delage, Y. Ye, Distributionally robust optimization under moment uncertainty with application to data-driven problems. Oper. Res. **58**(3), 595–612 (2010). https://doi.org/10.1287/opre.1090.0741
18. P. Diaconis, D. Freedman, On the consistency of Bayes estimates. Ann. Stat. **14**(1), 1–67 (1986). With a discussion and a rejoinder by the authors
19. P.W. Diaconis, D. Freedman, Consistency of Bayes estimates for nonparametric regression: normal theory. Bernoulli **4**(4), 411–444 (1998)
20. A. Doostan, H. Owhadi, A non-adapted sparse approximation of PDEs with stochastic inputs (2010)
21. R.F. Drenick, P.C. Wang, C.B. Yun, A.J. Philippacopoulos, Critical seismic response of nuclear reactors. J. Nucl. Eng. Des. **58**(3), 425–435 (1980)
22. T. Egami, S.J.L. Billinge, *Underneath the Bragg Peaks: Structural Analysis of Complex Materials* (Pergamon Press, Oxford, 2003)
23. I. Egorov, G. Kretinin, I. Leshchenko, S. Kuptzov, IOSO optimization toolkit—novel software to create better design (2002)
24. I.H. Eldred, C.G. Webster, P.G. Constantine, Design under uncertainty employing stochastic expansion methods. American Institute of Aeronautics and Astronautics Paper 2008–6001 (2008)
25. H. Owhadi, C. Scovel, Toward machine wald. in *Handbook of Uncertainty Quantification* (2016)
26. L. Esteva, Seismic risk and seismic design, in *Seismic Design for Nuclear Power Plants* (The M.I.T. Press, 1970), pp. 142–182
27. D.L. Kroshko et al., OpenOpt
28. O. Gereben, L. Pusztai, Phys. Rev. B **50**, 14136 (1994)

29. R. Ghanem, Ingredients for a general purpose stochastic finite elements implementation. Comput. Methods Appl. Mech. Eng. **168**(1–4), 19–34 (1999)
30. R. Ghanem, S. Dham, Stochastic finite element analysis for multiphase flow in heterogeneous porous media. Transp. Porous Media **32**(3), 239–262 (1998)
31. W.D. Gillford, Risk analysis and the acceptable probability of failure. Struct. Eng. **83**(15), 25–26 (2005)
32. J. Goh, M. Sim, Distributionally robust optimization and its tractable approximations. Oper. Res. **58**(4, part 1), 902–917 (2010). https://doi.org/10.1287/opre.1090.0795
33. S. Han, M. Tao, U. Topcu, H. Owhadi, R.M. Murray, Convex optimal uncertainty quantification (2014). arXiv:1311.7130
34. W. Hoeffding, On the distribution of the number of successes in independent trials. Ann. Math. Stat. **27**(3), 713–721 (1956)
35. W. Hoeffding, The role of assumptions in statistical decisions, in *Proceedings of the Third Berkeley Symposium on Mathematical Statistics and Probability, 1954–1955*, vol. I (University of California Press, Berkeley and Los Angeles, 1956), pp. 105–114
36. W.A. Hustrulid, M. McCarter, D.J.A. Van Zyl (eds.), *Slope Stability in Surface Mining* (Society for Mining Metallurgy & Exploration, 2001)
37. The MathWorks Inc., Technical report, Mar 2009. Technical Report 91710v00
38. P. Jensen, J. Bard, Algorithms for constrained optimization (supplement to: Operations research models and methods) (2003)
39. E. Jones et al., *SciPy: Open Source Scientific Tools for Python* (2001)
40. P. Juhas, L. Granlund, S.R. Gujarathi, P.M. Duxbury, S.J.L. Billinge, Crystal structure solution from experimentally determined atomic pair distribution functions. J. Appl. Cryst. **43**, 623–629 (2010)
41. P.-H.T. Kamga, B. Li, M. McKerns, L.H. Nguyen, M. Ortiz, H. Owhadi, T.J. Sullivan, Optimal uncertainty quantification with model uncertainty and legacy data. J. Mech. Phys. Solids **72**, 1–19 (2014)
42. B.K. Kanna, S. Kramer, An augmented Lagrange multiplier based method for mixed integer discrete continuous optimization and its applications to mechanical design. J. Mech. Des. **116**, 405 (1994)
43. K.E. Kelly, The myth of 10^{-6} as a definition of acceptable risk, in *Proceedings of the International Congress on the Health Effects of Hazardous Waste*, Atlanta (Agency for Toxic Substances and Disease Registry, 1993)
44. A. Kidane, A. Lashgari, B. Li, M. McKerns, M. Ortiz, H. Owhadi, G. Ravichandran, M. Stalzer, T.J. Sullivan, Rigorous model-based uncertainty quantification with application to terminal ballistics. Part I: Systems with controllable inputs and small scatter. J. Mech. Phys. Solids **60**(5), 983–1001 (2012)
45. T. Leonard, J.S.J. Hsu, *Bayesian Methods: An Analysis for Statisticians and Interdisciplinary Researchers*. Cambridge Series in Statistical and Probabilistic Mathematics, vol. 5 (Cambridge University Press, Cambridge, 1999)
46. J.S. Liu, *Monte Carlo Strategies in Scientific Computing*, Springer Series in Statistics (Springer, New York, 2008)
47. L.J. Lucas, H. Owhadi, M. Ortiz, Rigorous verification, validation, uncertainty quantification and certification through concentration-of-measure inequalities. Comput. Methods Appl. Mech. Eng. **197**(51–52), 4591–4609 (2008)
48. N. Mantel, W.R. Bryan, "Safety" testing of carcinogenic agents. J. Natl. Cancer Inst. **27**, 455–470 (1961)
49. C. McDiarmid, On the method of bounded differences, in *Surveys in Combinatorics, 1989 (Norwich, 1989)*, London Mathematical Society Lecture Note Series, vol. 141 (Cambridge University Press, Cambridge, 1989), pp. 148–188
50. C. McDiarmid, Concentration, in *Probabilistic Methods for Algorithmic Discrete Mathematics*, Algorithms and Combinatorics, vol. 16 (Springer, Berlin, 1998), pp. 195–248
51. M. McKerns, M. Aivazis, Pathos: A framework for heterogeneous computing (2010)

52. M. McKerns, P. Hung, M. Aivazis, Mystic: a simple model-independent inversion framework (2009)
53. M. McKerns, H. Owhadi, C. Scovel, T.J. Sullivan, M. Ortiz, The optimal uncertainty algorithm in the mystic framework. Caltech CACR Technical Report, Aug 2010. http://arxiv.org/pdf/1202.1055v1
54. M. McKerns, L. Strand, T.J. Sullivan, A. Fang, M. Aivazis, Building a framework for predictive science, in *Proceedings of the 10th Python in Science Conference (SciPy 2011)*, June 2011 (2011), pp. 67–78. http://arxiv.org/pdf/1202.1056
55. A.C. Narayan, D. Xiu, Distributional sensitivity for uncertainty quantification. Commun. Comput. Phys. **10**(1), 140–160 (2011). http://dx.doi.org/10.4208/cicp.160210.300710a
56. J.F. Nash Jr., Equilibrium points in n-person games. Proc. Natl. Acad. Sci. USA **36**, 48–49 (1950)
57. J. Nash, Non-cooperative games. Ann. Math. **2**(54), 286–295 (1951)
58. H. Niederreiter, *Random Number Generation and Quasi-Monte Carlo Methods*. CBMS-NSF Regional Conference Series in Applied Mathematics, vol. 63 (Society for Industrial and Applied Mathematics (SIAM), Philadelphia, PA, 1992)
59. H. Nyquist, IEEE Trans. **47**, 617 (1928)
60. H. Owhadi, Bayesian numerical homogenization (2014). arXiv:1406.6668
61. H. Owhadi, Multi-grid with rough coefficients and multiresolution operator decomposition from hierarchical information games (2015) (To appear)
62. H. Owhadi, C. Scovel, *Qualitative Robustness in Bayesian Inference* (2014). arXiv:1411.3984
63. H. Owhadi, C. Scovel, Brittleness of Bayesian inference and new Selberg formulas. Commun. Math. Sci. (2015). arXiv:1304.7046
64. H. Owhadi, C. Scovel, T.J. Sullivan, *On the Brittleness of Bayesian Inference* (2013). arXiv:1308.6306
65. H. Owhadi, C. Scovel, T.J. Sullivan, Brittleness of Bayesian inference under finite information in a continuous world. Electron. J. Stat. **9**, 1–79 (2015). arXiv:1304.6772
66. H. Owhadi, C. Scovel, T.J. Sullivan, M. McKerns, M. Ortiz, Optimal uncertainty quantification. SIAM Rev. **55**(2), 271–345 (2013)
67. H. Owhadi, L. Zhang, L. Berlyand, Polyharmonic homogenization, rough polyharmonic splines and sparse super-localization. ESAIM Math. Model. Numer. Anal. **48**(2), 517–552 (2014)
68. F. Rossi, A. Tsoukias (eds.), *Algorithmic Decision Theory, First International Conference*. Lecture Notes in Computer Science, vol. 5783 (Springer, 2009)
69. A. Saltelli, K. Chan, E.M. Scott (eds.), *Sensitivity Analysis*, Wiley Series in Probability and Statistics (Wiley, Chichester, 2000)
70. A. Saltelli, M. Ratto, T. Andres, F. Campolongo, J. Cariboni, D. Gatelli, M. Saisana, S. Tarantola, *Global Sensitivity Snalysis. The Primer* (Wiley, Chichester, 2008)
71. C. Scovel, I. Steinwart, Hypothesis testing for validation and certification. J. Complex. (2010) (Submitted)
72. C.E. Shannon, Proc. IRE **37**, 10 (1949)
73. X. Shen, L. Wasserman, Rates of convergence of posterior distributions. Ann. Stat. **29**(3), 687–714 (2001)
74. I.H. Sloan, Sparse sampling techniques, *Presented at 2010 ICMS Uncertainty Quantification workshop* (2010)
75. I.H. Sloan, S. Joe, *Lattice Methods for Multiple Integration* (Oxford Science Publications, The Clarendon Press Oxford University Press, New York, 1994)
76. J.E. Smith, Generalized Chebychev inequalities: theory and applications in decision analysis. Oper. Res. **43**(5), 807–825 (1995). https://doi.org/10.1287/opre.43.5.807
77. H.M. Soekkha (ed.), *Aviation Safety: Human Factors, System Engineering, Flight Operations, Economics, Strategies, Management* (VSP, 1997)
78. L.A. Steen, J.A. Seebach Jr., *Counterexamples in Topology* (Dover Publications Inc., Mineola, NY, 1995). Reprint of the second (1978) edition
79. I. Steinwart, A. Christmann, *Support Vector Machines. Information Science and Statistics* (Springer, New York, 2008)

80. R.M. Storn, K.V. Price, Differential evolution—a simple and efficient heuristic for global optimization over continuous spaces. J. Global Optim. **11**(4), 341–359 (1997)
81. A.M. Stuart, Inverse problems: a Bayesian perspective. Acta Numer. **19**, 451–559 (2010)
82. T.J. Sullivan, M. McKerns, D. Meyer, F. Theil, H. Owhadi, M. Ortiz, Optimal uncertainty quantification for legacy data observations of Lipschitz functions. ESAIM Math. Model. Numer. Anal. **47**(6), 1657–1689 (2013)
83. T.J. Sullivan, U. Topcu, M. McKerns, H. Owhadi, Uncertainty quantification via codimension-one partitioning. Int. J. Numer. Meth. Eng. **85**(12), 1499–1521 (2011)
84. B.H. Toby, N. Khosrovani, C.B. Dartt, M.E. Davis, J.B. Parise, Structure-directing agents and stacking faults in the con system: a combined crystallographic and computer simulation study. Microporous Mesoporous Mater. **39**(1–2), 77–89 (2000)
85. R.A. Todor, C. Schwab, Convergence rates for sparse chaos approximations of elliptic problems with stochastic coefficients. IMA J. Numer. Anal. **27**(2), 232–261 (2007)
86. L. Vandenberghe, S. Boyd, K. Comanor, Generalized Chebyshev bounds via semidefinite programming. SIAM Rev. **49**(1), 52–64 (2007)
87. P. Venkataraman, *Applied Optimization with MATLAB Programming* (Wiley, Hoboken, NJ, 2009)
88. J. Von Neumann, H.H. Goldstine, Numerical inverting of matrices of high order. Bull. Am. Math. Soc. **53**, 1021–1099 (1947)
89. J. von Neumann, O. Morgenstern, *Theory of Games and Economic Behavior* (Princeton University Press, Princeton, New Jersey, 1944)
90. A. Wald, Statistical decision functions which minimize the maximum risk. Ann. Math. **2**(46), 265–280 (1945)
91. D. Xiu, Fast numerical methods for stochastic computations: a review. Commun. Comput. Phys. **5**(2–4), 242–272 (2009)

Chapter 3
Importance of Feature Selection in Machine Learning and Adaptive Design for Materials

Prasanna V. Balachandran, Dezhen Xue, James Theiler, John Hogden, James E. Gubernatis and Turab Lookman

Abstract In materials informatics, features (or descriptors) that capture trends in the structure, chemistry and/or bonding for a given chemical composition are crucial. Here, we explore their role in the accelerated search for new materials using machine learning adaptive design. We focus on a specific class of materials referred to as apatites [$A_{10}(BO_4)_6X_2$] and our objective is to identify an apatite compound with the largest band gap (E_g) without performing density functional theory calculations over the entire composition space. We construct three datasets that use three sets of features of the A, B, and X-ions (ionic radii, electronegativities, and the combination of both) and independently track which of these sets finds most rapidly the composition with the largest E_g. We find that the combined feature set performs best, followed by the ionic radii feature set. The reason for this ranking is the B-site ionic radius, which is the key E_g-governing feature and appears in both the ionic radii and combined feature sets. Our results show that a relatively poor ML model with large error but one that contains key features can be more efficient in accelerating the search than a low-error model that lack such features.

P. V. Balachandran (✉) · J. Theiler · J. Hogden · J. E. Gubernatis · T. Lookman
Los Alamos National Laboratory, Los Alamos, NM 87545, USA
e-mail: pvb5e@virginia.edu

T. Lookman
e-mail: txl@lanl.gov

D. Xue
State Key Laboratory for Mechanical Behavior of Materials,
Xi'an Jiaotong University, X'ian 710049, China

P. V. Balachandran
Department of Materials Science and Engineering, Department of Mechanical and Aerospace Engineering, University of Virginia, Charlottesville, VA 22904, USA

T. Lookman et al. (eds.), *Materials Discovery and Design*, Springer Series in Materials Science 280, https://doi.org/10.1007/978-3-319-99465-9_3

3.1 Introduction

Many critical technologies, including energy, electronics, security, and environment, rely on the design and discovery of advanced materials. Typically, these materials are multicomponent by design and have enormous complexities at the atomic and mesoscale levels. It is not practical for intuition or trial-and-error methods to navigate the vast search space in an effort to find new materials with targeted properties. Thus, predictive computational strategies that identify promising candidates with desired response for experimental synthesis and characterization are pivotal to accelerate the discovery and realization of novel advanced materials. Traditionally, computational methods based on density functional theory [1], molecular dynamics [2] and phase field modeling [3] (to name a few) have enabled rational design. More recently, machine learning methods have also gained importance in guiding computations and experiments [4].

In the application of machine learning methods for accelerated materials design and discovery, an emerging theme is adaptive design [4–10]. In this approach, we start by constructing a *database* of known materials with features that are expected to affect the desired property. The source for data can be experiments or computer simulations. Once the dataset is ready, we build *machine learning* (ML) models that capture the relationship between the features and desired property. In adaptive design, the ML models must also be able to quantify uncertainties associated with each data sample. The performance of the trained models are evaluated using standard practices such as k-fold cross validation, independent test sets or re-substitution error. After evaluation, these ML models are applied to predict the response, with associated uncertainties, of a vast library of unknown, missing, or yet-to-be synthesized materials. The next step is *design*, which allows us to choose the "promising" material(s) from this vast space and recommend it (them) for experimental or computational *validation*. The final step is *feedback*, where the new data is then augmented to the database and the process repeats until the optimal material with the desired response is found. We recently demonstrated the efficacy of this strategy by rationally guiding experiments that have led to the discovery of novel shape memory alloys [5, 11] and lead-free piezoelectric perovskites [12].

In contrast to previous studies [9] that have focused on the choice of ML and experimental design methods, we explore here the influence of features on the accelerated search in the course of adaptive design. It is through the features that we can build the materials understanding into the database to inform the ML to establish the chemistry-structure-property relationships. As a result, there is a concerted effort within the community towards representation learning [13–18], where the goal is to engineer representations of features that makes it easier to extract useful, yet physically meaningful, information for building robust ML models [19]. Our objective here is to evaluate adaptive design from the viewpoint of three different datasets. The datasets we consider contain the same set of chemical compositions and response (or target) variable, but they differ in the choice of features that represent these compositions. Since the performance of ML methods hinges critically on the features

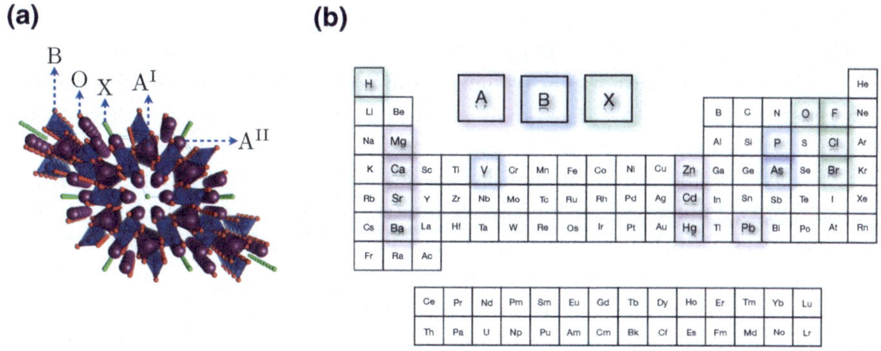

Fig. 3.1 **a** Crystal structure of apatites with chemical formula $A_{10}(BO_4)_6X_2$ in the aristotype hexagonal $P6_3/m$ (# 176) space group. There are two crystallographically distinct A-sites (labeled A^I and A^{II}) in the aristotype structure. **b** The chemical space of apatite crystal chemistries considered in this work with three degrees of chemical freedom: A, B and X. In this paper, we have constrained our chemical space such that same chemical element occupies both the A^I and A^{II} sites in the lattice. Our overall chemical space span 96 unique stoichiometric compositions

used, we anticipate the three sets of features to have varied impact on the accelerated search process.

We demonstrate our approach on a computational dataset generated from density functional theory (DFT) calculations (cf. Computational Details). We focus on a particular class of compounds referred to as "apatites" with chemical formula $A_{10}(BO_4)_6X_2$, where A and B are divalent and pentavalent cations, respectively, and X is an anion. The aristotype structure of a typical $A_{10}(BO_4)_6X_2$ apatite belongs to the space group $P6_3/m$ (# 176) as shown in Fig. 3.1a, and there are two crystallographically distinct A-sites associated with the structure (shown as A^I and A^{II} in Fig. 3.1a). The complex structure can be further decomposed into three basic building (or structural) units based on the principles of coordination polyhedra: A^IO_6 metaprism, BO_4 tetrahedra and $A^{II}O_6X_{1,2}$ polyhedron [20–22]. The unit cell of a prototypical fluorapatite [e.g. $Ca_{10}(PO_4)_6F_2$], where X = F anion, consists of 42 atoms, whereas in the ground state monoclinic structure of hydroxyapatites (where X = OH anion) there are 88 atoms per unit cell. These materials are typically wide band gap (E_g) insulators and possess properties important for many applications as biomaterials, luminescent materials, and host lattices for immobilizing heavy and toxic elements and radiation tolerant materials [23].

One of the intriguing characteristics of an apatite host lattice is its chemical flexibility and structural diversity. In Fig. 3.1b, we show a partial collection of the chemical elements that can occupy various atomic sites in the apatite lattice as considered in this work. We have A = {Mg, Ca, Sr, Ba, Zn, Cd, Hg or Pb}, B = {P, As or V}, and X = {F, Cl, Br or OH}. Overall, there are 96 unique $A_{10}(BO_4)_6X_2$ chemical compositions that span the chemical space. Our materials design objective is to find an apatite composition with the largest E_g in the above considered composition space.

We calculated the E_g for 13 randomly chosen apatites at the DFT-GGA level (cf. Computational Details) in their lowest energy structures. We treat this as a small data problem to mimic a scenario that is common when performing ML studies with experimental data [5, 24, 25] and use our adaptive ML strategy to discover the composition (among the remaining 83 compositions) with the largest E_g in as few iterations as possible. DFT calculations serve as the *oracle* to validate our predictions and provide feedback. We utilize Shannon's ionic radii (r) and Pauling electronegativity (EN) differences as *features sets* for building the ML models [26, 27]. More specifically, we considered the following seven features: (i) r_A, (ii) r_B, (iii) r_X, (iv) A_{EN}-O_{EN}, (v) B_{EN}-O_{EN}, (vi) A_{EN}-X_{EN} and (vii) A_{EN}-B_{EN}. Balachandran and Rajan [22] have shown that these features correlate strongly with various bond geometrical characteristics of the apatite lattice. However, the correlations between these features and E_g are not known. We further divided these seven features into two sets: (i) ionic radii set (r_A, r_B, and r_X) and (ii) electronegativity set (A_{EN}-O_{EN}, B_{EN}-O_{EN}, A_{EN}-X_{EN}, and A_{EN}-B_{EN}). The key question is the following—*Which amongst the two feature sets would efficiently guide the DFT to find the optimal composition with the largest E_g in as few iterations as possible?* Addressing this question then allows us to glean important insights into the adaptive design strategy that are crucial to understand the accelerated search process. We also considered a third feature set, the "combined set", where all seven features are used for adaptive design.

Our findings indicate a subtle interplay between essential features and ML model quality in impacting the accelerated search. Although the quality of the ML models trained on the electronegativity feature set had lower average error compared to the ionic radii feature set, the electronegativity-based models required more iterations to find the optimal composition with the largest E_g. The reason is attributed to the B-site ionic radius (r_B), which is identified from DFT calculations as the key E_g-governing feature (discussed in detail in the Sect. 3.3 Results) in our chemical space and this feature is found in both ionic radii and combined feature sets. Also, a relatively poor ML model (with large error) that contains essential features is more efficient than a good model (with small error) that does not have those features. The optimal scenario is the one where we have both key features and a good ML model (with small error), as exemplified with the combined feature set. The broader validity of these findings, beyond the results reported in this work, is still not fully resolved and we hope our study will motivate further research in this area.

3.2 Computational Details

3.2.1 Density Functional Theory

Density functional theory (DFT) calculations for the apatites were performed within the generalized gradient approximation (GGA) as implemented in QUANTUM ESPRESSO [28]. The PBEsol exchange-correlation functional [29] was used and the core and valence electrons were treated with ultrasoft pseudopotentials [30]. The Brillouin zone integration was performed using a $2 \times 2 \times 4$ Monkhorst-Pack k-point mesh [31] centered at Γ and 60 Ry plane-wave cutoff for wavefunctions (600 Ry kinetic energy cutoff for charge density and potential). Non self-consistent field (NSCF) calculations were performed using a $4 \times 4 \times 6$ Monkhorst-Pack k-point mesh (unshifted). The scalar relativistic pseudopotentials were taken from the PSLIBRARY [32]. The atomic positions and the cell volume were allowed to relax until an energy convergence threshold of 10^{-8} eV and Hellmann-Feynman forces less than 2 meV/Å, respectively, were achieved. We also considered the following crystal symmetries or space groups to determine the ground state: (i) in the case of fluorapatites (X = F), calculations were done for the hexagonal ($P6_3/m$) and triclinic ($P\bar{1}$) structures, (ii) in the case of chlorapatites (X = Cl) and bromapatites (X = Br), calculations were done for the hexagonal ($P6_3/m$) and monoclinic ($P2_1/b$) structures, and (iii) in the case of hydroxyapatites (X = OH), DFT calculations were performed for the monoclinic ($P2_1/b$ and $P2_1$) and hexagonal ($P6_3$) crystal symmetries. The choice of these symmetries was motivated by the earlier work in the literature [23, 33]. Only the E_g associated with the lowest energy structure is considered for ML. The space groups of the optimized structures were determined using FINDSYM [34] and the resulting crystal structures were visualized in VESTA [35].

3.2.2 Machine Learning

We use ε-support vector regression with non-linear Gaussian radial basis function kernel (SVR_{RBF}) as implemented in the `e1071` package [36] within the RSTUDIO environment [37]. The SVR_{RBF} ML method establishes the relationship between the features and E_g. The hyperparameters for the SVR_{RBF} were optimized using leave-one-out cross-validation method. Error bars for each prediction were estimated using the bootstrap resampling method [38]. We then use those SVR_{RBF} models to predict the E_g of compositions in the dataset. From 100 SVR_{RBF} models, we have 100 predicted E_g values for each composition. The mean (μ) and standard deviations (error bar, σ) are estimated from the 100 SVR_{RBF} models.

3.2.3 Design

We utilize the efficient global optimization (EGO) method [39] for design. In this approach, we estimate the "expected improvement, E(I)" for each composition in the dataset (whose E_g is not known) from the trained ML models. We calculate E(I) using the formula, $\sigma[\phi(z) + z\Phi(z)]$, where $z = (\mu - \mu^*)/\sigma$ and μ^* is the maximum value observed so far in the current training set, $\phi(z)$ and $\Phi(z)$ are the standard normal density and cumulative distribution functions, respectively. Here, E(I) balances the tradeoff between "exploitation" and"exploration" of the ML model. At the end of each iteration, our design returns a score for E(I) for each unmeasured composition, whose relative magnitude depends on the ML predicted (μ, σ) pair for those compositions and the value of μ^* in the training set. We then pick the composition with the maximum E(I) [*max* E(I)] and recommend it for validation and feedback. Here, we also track *max* E(I) as a function of number of iterations for both feature sets to further understand the evolution of the adaptive design process.

3.3 Results

We begin our analysis with the ionic radii and electronegativity feature sets. In Fig. 3.2, we show the performance of SVR_{RBF} ML models on the initial training set for the ionic radii and electronegativity feature sets. Note that the initial training set has a total of 13 compositions for which the E_g's are known and we have a list of 83 compositions for which the E_g's are not known a priori. The largest E_g in the training set is 5.35 eV and this belongs to $Sr_{10}(PO_4)_6F_2$ (SrPF) with $P6_3/m$ crystal symmetry. In the case of ionic radii feature set (Fig. 3.2a), we find that the ML models overestimate and underestimate the E_g for the small and large E_g compositions, respectively. As a result, we have relatively large error bars at both extremities of the ML model in the training set. The mean squared error is estimated to be 0.54 eV/composition. In contrast, the ML predicted E_g values were relatively closer to the E_g data when these models were trained on the electronegativity feature set (Fig. 3.2b). However, the error bars were found to be large for compositions whose E_g fall in the range 3–4 eV for the electronegativity feature set. The mean squared error is estimated to be 0.19 eV/composition. Thus, the electronegativity-based ML models achieve

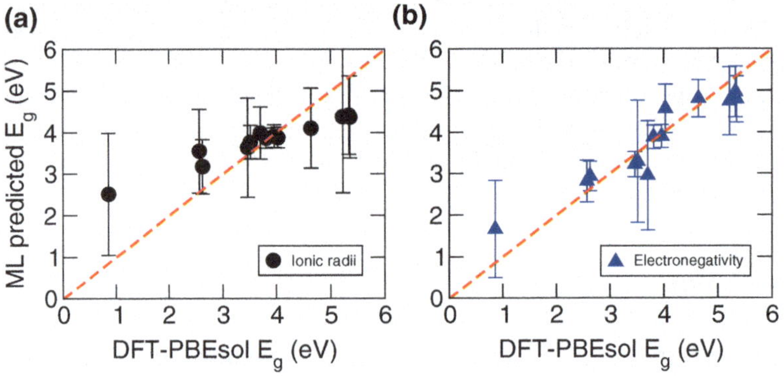

Fig. 3.2 Performance of the SVR_{RBF} ML models on the apatite data trained using **a** Ionic radii feature set (black circles) and **b** Electronegativity feature set (blue triangles). In the x- and y-axis we plot the DFT-PBEsol band gap (in eV) and machine learning (ML) predicted band gap (in eV), respectively. The uncertainties (error bars) correspond to the standard deviation from the 100 bootstrap models. Red dashed line indicate the x = y line, where the predictions from ML and DFT E_g data exactly coincide

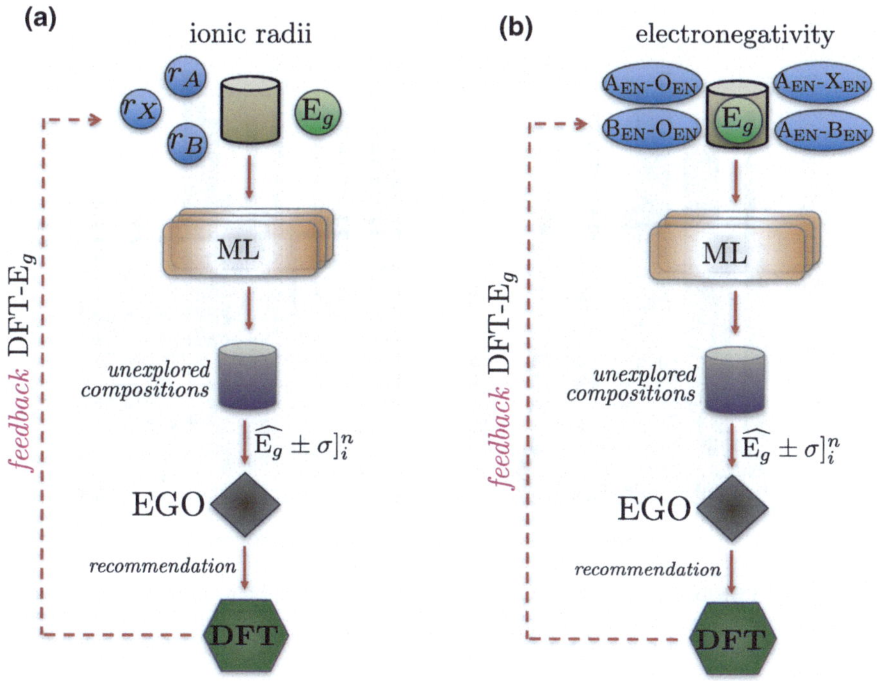

Fig. 3.3 Adaptive design strategy in search of apatite compositions with the largest E_g. **a** Machine learning model trained using ionic radii feature sets (r_A, r_B, and r_X) and **b** Machine learning (ML) model trained using electronegativity feature sets (A_{EN}-O_{EN}, B_{EN}-O_{EN}, A_{EN}-X_{EN}, and A_{EN}-B_{EN}). The only difference between **a** and **b** is in the choice of features. EGO stands for efficient global optimization, which evaluates the tradeoff between "exploration" and "exploitation" to recommend the next composition for DFT validation. We ran this loop for a total of 25 times to gain insights into the strategy

lower error compared to the ionic radii-based ML models. Now, with the help of these two ML models, we independently explore the adaptive design strategy for the two feature sets with the objective of finding an apatite composition with the largest E_g in our chemical space.

Our adaptive design strategy is schematically shown in Fig. 3.3. We independently run our iterative design loop for both feature sets. At the end of each iteration, we evaluate the E_g using DFT-PBEsol calculations for the composition recommended by design [max E(I)] and augment our training set with this new composition. We then retrain our ML model with (now) 14 data points and pick the next composition from a pool of 82 compositions. We continue iterating this loop 25 times. In Fig. 3.4, we show the DFT-PBEsol calculated E_g data for the compositions recommended by our adaptive design at the end of each iteration until we reached our 25 iterations limit. With the ionic radii feature set (Fig. 3.4a), it can be seen that our approach found the optimal composition [$Ca_{10}(PO_4)_6F_2$ (CaPF) in the $P6_3/m$ crystal symmetry] with the largest E_g of 5.67 eV in the 9th iteration. We confirm that CaPF is the optimal

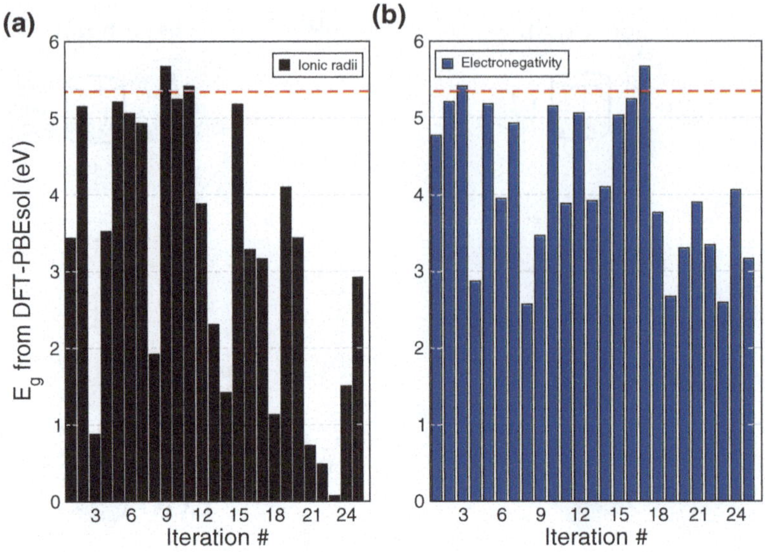

Fig. 3.4 DFT-PBEsol calculated E_g (in eV) as a function of number of iterations for the compositions recommended by adaptive design using **a** ionic radii and **b** electronegativity feature sets. The horizontal red dashed line represent the maximum value of E_g found in the original training set (iteration #0)

composition with the largest E_g (5.67 eV) in our search space, because the remaining 58 compositions had either V- or As-atom occupying the B-site. Both V- and As-containing apatites have smaller E_g compared to their P-containing counterparts [40]. In contrast, with the electronegativity feature set (Fig. 3.4b) we found the optimal composition only in the 17th iteration. Furthermore, we find that the ionic radii feature set has resulted in a far greater exploration of the E_g space relative to that of the electronegativity feature set, where no new compositions (other than those that are already present in the initial training set) were found with DFT-PBEsol $E_g < 2.5$ eV using our design. Although both feature sets identified the optimal composition in relatively few iterations (and not requiring a total of 83 iterations), our results clearly demonstrate that the choice of the feature sets have an important role in determining not only the efficacy but also the trajectory in which the accelerated search process has evolved. We now take a closer look at the evolution of the search process to further understand the adaptive design.

In Table 3.1, we provide the list of chemical compositions as recommended by *max* μ-ML (i.e. composition with the largest predicted E_g by ML) and *max* E(I) at the end of each iteration (until the optimal composition is found) for both feature sets. We observe two scenarios in Table 3.1: (i) Both *max* μ-ML and *max* E(I) recommend the same composition, and (ii) *max* μ-ML and *max* E(I) recommend different compositions. As indicated in the Methods section, E(I) is calculated using the formula, $\sigma[\phi(z) + z\Phi(z)]$, where $z = (\mu - \mu^*)/\sigma$ and μ^* is the maximum value observed so far in the current training set, $\phi(z)$ and $\Phi(z)$ are the standard normal

Table 3.1 List of chemical compositions recommended by *max* μ-ML and *max* E(I) at the end of each iteration in our adaptive feedback loop for the two feature sets (ionic radii and electronegativity). For DFT-PBEsol validation and feedback, we chose the recommended compositions from *max* E(I). For simplicity, we follow the ABX notation to label each composition [e.g. ZnPOH stands for $Zn_{10}(PO_4)_6(OH)_2$ and CaPF stands for $Ca_{10}(PO_4)_6F_2$]. The composition CaPF is highlighted in bold font to indicate that it has the largest DFT-PBEsol E_g in the composition space explored in this work

Iteration number	Ionic radii		Electronegativity	
	max μ-ML	*max* E(I)	*max* μ-ML	*max* E(I)
1	ZnPOH	ZnPOH	SrPOH	CaPBr
2	HgPOH	BaPCl	SrPOH	BaPF
3	MgPCl	HgPOH	SrPOH	CaPCl
4	MgPBr	MgPCl	SrPCl	PbAsBr
5	BaPF	BaPF	SrPCl	SrPCl
6	BaPBr	BaPBr	SrPOH	CaAsOH
7	MgPF	SrPOH	BaPCl	SrPOH
8	MgVBr	MgPBr	BaPCl	PbVBr
9	MgPF	**CaPF**	SrPBr	CaAsCl
10	–	–	BaPCl	BaPCl
11	–	–	BaPBr	BaVBr
12	–	–	BaPBr	BaPBr
13	–	–	SrPBr	BaVCl
14	–	–	SrPBr	BaAsF
15	–	–	SrPBr	SrPF
16	–	–	**CaPF**	MgPF
17	–	–	**CaPF**	**CaPF**

density and cumulative distribution functions, respectively. In scenario (i), μ from the ML models dominate in the calculation of E(I), which leads to both *max* μ-ML and *max* E(I) recommending the same composition. Whereas in scenario (ii), σ from the ML models dominate in the calculation of E(I) and thus, both *max* μ-ML and *max* E(I) recommend different compositions. Note that, in this work, we only select the compositions recommended by *max* E(I) for DFT-PBEsol validation. In the case of ionic radii feature set, it is clear that the *max* E(I) in Table 3.1 recommended only those compositions that contain P-atom (phosphorus atom) in the B-site of the apatite lattice. In contrast, the electronegativity feature set had to explore compositions that involve all three chemical sites (A, B and X in the apatite lattice) before it found the optimal composition in the 17th iteration.

Note that our chemical space allows for three different atoms (namely P, V, or As) to occupy the B-site of the apatite lattice (also see Fig. 3.1b). To further understand the role of these B-atoms on the electronic structure, we calculated the density of states (DOS) and partial DOS for $Sr_{10}(BO_4)_6F_2$, where B = P, V, or As, in the $P6_3/m$ crystal symmetry, which is also the lowest energy crystal structure for these

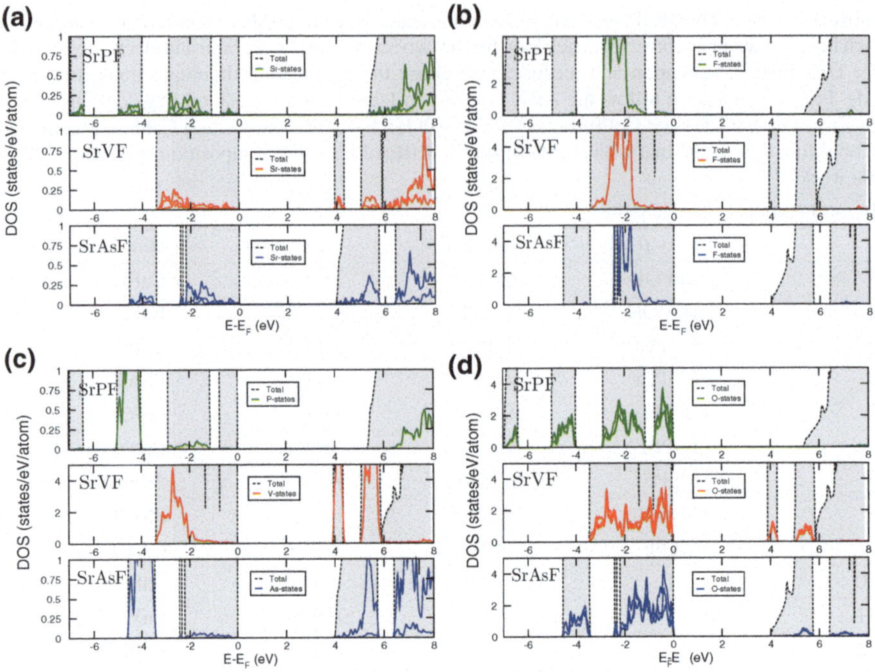

Fig. 3.5 Density of states (DOS) and atom-projected partial DOS data for $Sr_{10}(BO_4)_6F_2$, where B = P (green, top panel), V (red, middle panel), or As (blue, bottom panel), in the $P6_3/m$ space group. **a** Sr-states, **b** F-states, **c** B-states, where B = P, V or As, and **d** O-states. The total DOS is given as dashed black line and the area under the curve is shaded in color grey. E_F is the Fermi level (in eV). SrPF, SrVF, and SrAsF stands for $Sr_{10}(PO_4)_6F_2$, $Sr_{10}(VO_4)_6F_2$, and $Sr_{10}(AsO_4)_6F_2$, respectively

compositions. The electronic configurations for P, V and As atoms can be written as [Ne]$3s^2 3p^3$, [Ar]$3d^3 4s^2$ and [Ar]$3d^{10} 4s^2 4p^3$, respectively. In Fig. 3.5a–d, we show the total DOS and atom projected partial DOS for the three compositions. The overall E_g trend can be described as follows: $E_g^P > E_g^{As} \approx E_g^V$. The Shannon's ionic radius for P^{5+}, V^{5+}, and As^{5+} cations in the four-fold coordination is 0.17, 0.335, and 0.355 Å, respectively. The Pauling electronegativity for P, V, and As atoms is 2.19, 1.63, and 2.18, respectively. Thus, apatites with smaller B-ionic radii (i.e. P-atoms) have larger E_g, when the A- and X-sites are fixed and within the constraints of the composition space explored in this work. In another independent DFT study, Zheng et al. showed that the E_g of $Ca_{10}(PO_4)_6(OH)_2$ (5.25 eV) is greater than that of $Ca_{10}(AsO_4)_6(OH)_2$ (3.95 eV)[40]. In addition, from our own DFT calculations of ≈40 apatites, we find that $Ba_{10}(AsO_4)_6F_2$ has the largest E_g of 4.1 eV among V- or As-containing apatites. Thus, we infer that the ionic radii of the B-site (r_B) is a key feature for distinguishing large E_g (P-apatite) from small E_g (V- or As-apatite) compositions.

The DOS and partial DOS data also provides us with insights for explaining the $E_g^P > E_g^{As} \approx E_g^V$ trend in $Sr_{10}(BO_4)_6F_2$ compounds. In SrPF and $Sr_{10}(AsO_4)_6F_2$

(SrAsF), we find that the bottom of the conduction bands are occupied by Sr-states (Fig. 3.5a). The center of mass for the F-states can be found at about 2 eV below the Fermi level (E_F) in the energy window shown in Fig. 3.5b. In SrAsF, in addition to Sr-states we also find some contributions from the As *s*-states (Fig. 3.5c, bottom panel) on the bottom of the conduction bands. In the case of $Sr_{10}(VO_4)_6F_2$ (SrVF), there is a strong contribution from the V *d*-states (Fig. 3.5c, middle panel) to the bottom of the conduction bands. In all three compositions, the top of the valence band is occupied by the O *p*-states (Fig. 3.5d).

In Fig. 3.6a–c, we compare the performance of our ML models that were trained on the ionic radii feature set with respect to the "ground truth" DFT-PBEsol E_g data for the first 23 iterations of our adaptive design. The filled magenta square data points (that are shown from iteration 1 onwards) represent the compositions recommended by our design [*max* E(I)] for DFT-PBEsol validation based on the ML models from the previous iteration. During the initial few iterations (especially, see iterations 1, 3, and 4 in Fig. 3.6a), we observe that the recommendations from design did not

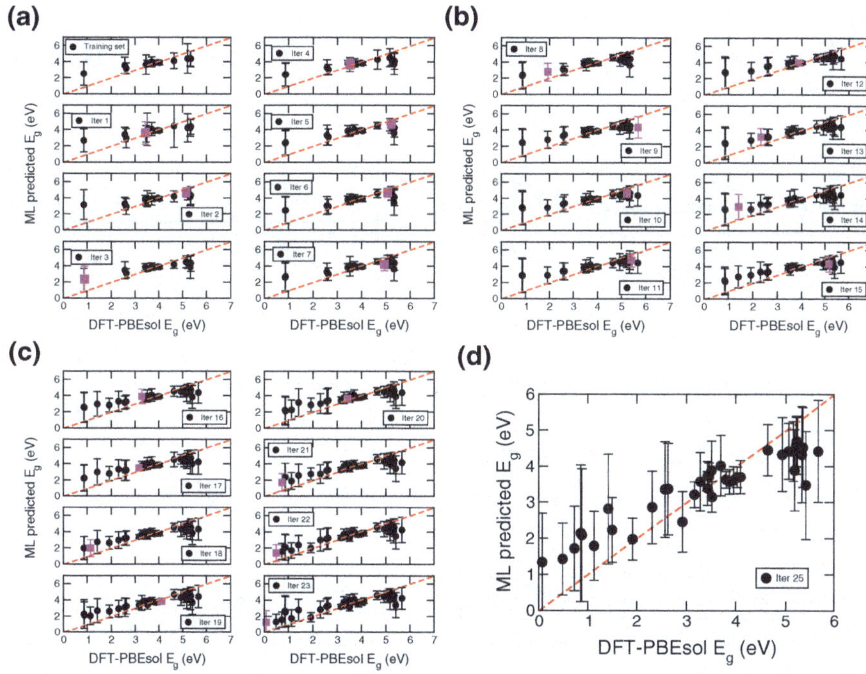

Fig. 3.6 **a–C** Evolution of our ML models that were trained on the ionic radii feature sets at the end of first 23 iterations of our design feedback loop. The calculated E_g from DFT-PBEsol is given in the x-axis and the predicted E_g from machine learning (ML) is shown in the y-axis. Error bars represent the standard deviation of the predicted E_g from 100 ML models. Red dashed line indicate the x = y line, where the predictions from ML and DFT E_g data exactly coincide. **d** Comparison between the ML predicted E_g and DFT-PBEsol calculated E_g at the end of the 25th iteration

consistently sample compositions at or near the large E_g regime ($E_g > 5$ eV). Rather, the algorithm suggested data points in the feature-E_g landscape that has the greatest potential to improve the ML models (i.e. reduce ML uncertainties). This can also be seen from surveying the chemical compositions listed in Table 3.1, where only three out of nine times the recommendations from both *max* μ-ML and *max* E(I) agree with respect to one another (before the algorithm found the optimal composition with the largest E_g in the 9th iteration). In Fig. 3.6d, we also show the performance of our final ML model at the end of the 25th iteration, where we have now trained the ML models using 38 data points. We identify two important characteristics in Fig. 3.6d, relative to Fig. 3.2a: (i) we have surveyed a substantial range of DFT-PBEsol E_g values and (ii) the uncertainties are still large.

In Fig. 3.7a–c, we also show the performance of our ML models that used the electronegativity feature set for the first 23 iterations. The quality of these ML models have lower error than the ML models trained on the ionic radii feature set. In Fig. 3.7d, we also show the performance of the ML models at the end of the 25th iteration.

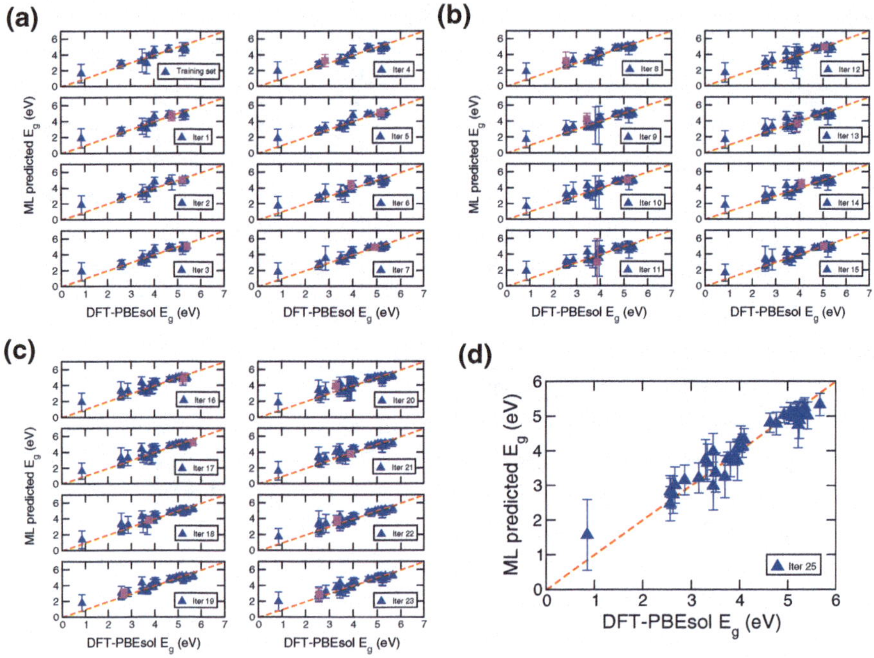

Fig. 3.7 **a–c** Evolution of our ML models that were trained on the electronegativity feature sets at the end of first 23 iterations of our design feedback loop. The calculated E_g from DFT-PBEsol is given in the x-axis and the predicted E_g from machine learning (ML) is shown in the y-axis. Error bars represent the standard deviation of the predicted E_g from 100 ML models. Red dashed line indicate the x = y line, where the predictions from ML and DFT E_g data exactly coincide. **d** Comparison between the ML predicted E_g and DFT-PBEsol calculated E_g at the end of the 25th iteration

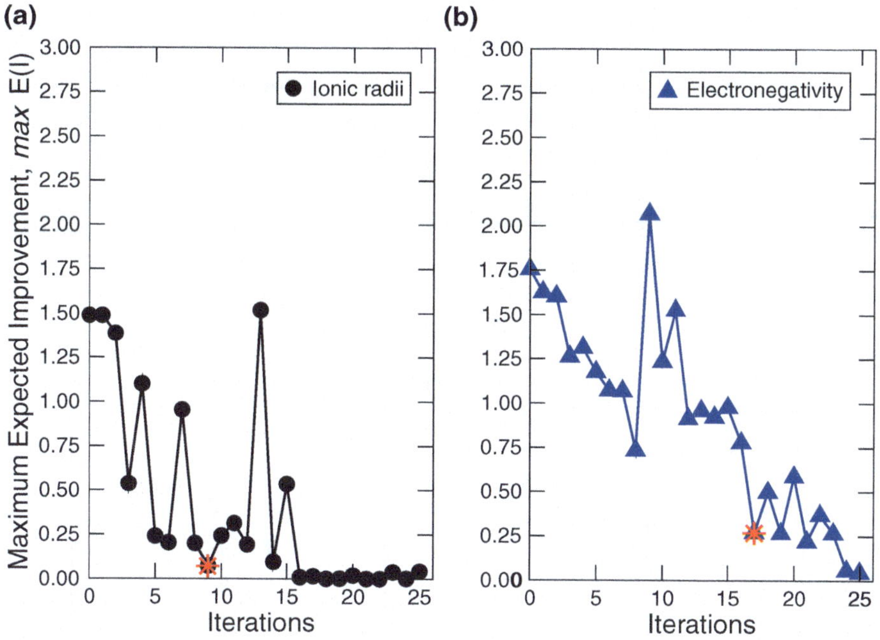

Fig. 3.8 The variation of the maximum expected improvement, *max* E(I), as a function of number of iterations for the **a** Ionic radii and **b** Electronegativity feature sets. The *max* E(I) corresponding to the optimal composition with the largest E_g is highlighted with a red star. Iteration 0 is the training set

In sharp contrast to Fig. 3.6d, the search trajectory associated with the electronegativity feature set has focussed on compositions with $E_g \geq 2.5$ eV. Furthermore, the uncertainties are also smaller. Although the electronegativity feature set appears to have possessed all desired characteristics in terms of superior model quality relative to the ionic radii feature set, intriguingly the optimal composition was found only at the end of the 17th iteration (requiring approximately twice the total number of iterations than that needed by the ionic radii feature set). This happens because the electronegativity feature was not able to clearly distinguish between P-apatites and As-apatites, due to the similarity in the electronegativity values between P and As atoms. This consequently led the design to explore relatively more compositions, before it found the optimal one.

We also follow the search process by systematically tracking the *max* E(I) at the end of each iteration (c.f. Computational Details). Ideally, we anticipate *max* E(I) to monotonically decrease as the number of iterations increases. However, as shown in Fig. 3.8, we find that the *max* E(I) fluctuates and do not decrease smoothly. One of the unanswered questions in adaptive design for accelerated materials design is the stopping criterion and we note that tracking *max* E(I) is a natural step in addressing this question. With the ionic radii feature set (Fig. 3.8a), we find that the *max* E(I) decreased towards zero from the 16th iteration onwards. In contrast, for the ML

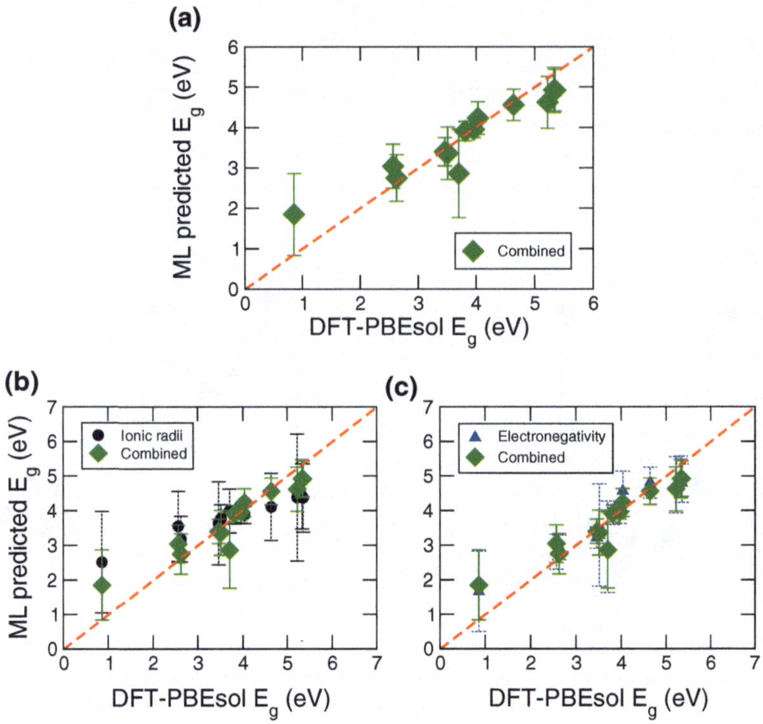

Fig. 3.9 **a** DFT-PBEsol E_g (x-axis) versus ML predicted E_g (y-axis) for a third feature set (referred to as "Combined", filled green diamonds), where we combined both ionic radii and electronegativity feature sets into one super set. In the x- and y-axis we plot the DFT-PBEsol band gap (in eV) and machine learning (ML) predicted band gap (in eV), respectively. The uncertainties (error bars) correspond to the standard deviation from the 100 bootstrap models. Red dashed line indicate the x = y line, where the predictions from ML and DFT E_g data exactly coincide. In **b** and **c** we directly compare the performance of Combined versus Ionic radii, and Combined versus Electronegativity feature sets, respectively, in reproducing the DFT-PBEsol E_g data

model that was trained using the electronegativity feature set it took 24 iterations (Fig. 3.8b) for the *max* E(I) to approach zero. Thus, with respect to the stopping criterion, we do not recommend stopping the feedback cycle immediately after *max* E(I) has reached a value of zero. Instead, we suggest running the iterative feedback loop a couple of additional iterations to confirm that the *max* E(I) is consistently zero and does not increase. An alternative criterion would be to stop the iterative cycles when a material with the desired response is found, even when the *max* E(I) did not reach zero.

Finally, we also considered a third feature set where we combined both ionic radii and electronegativity. In Fig. 3.9a, we show the DFT-PBEsol E_g versus ML predicted E_g for the combined feature set on the initial training set that contains 13 compositions (same as that used in Fig. 3.2). The performance of the ML model with seven features is comparable to the electronegativity feature set (Fig. 3.2b), but is superior to the ionic radii feature set (Fig. 3.2a). We estimate a mean squared error value of 0.21 eV/composition. We show this in Fig. 3.9b and c, where we overlay the results from combined-ionic radii and combined-electronegativity feature sets, respectively. Intriguingly, the combined feature set has more similarity with the electronegativity feature set compared to the ionic radii set. In terms of uncertainties, the major differences (between combined and electronegativity feature sets) appear for $Pb_{10}(PO_4)_6F_2$ (PbPF) and $Pb_{10}(PO_4)_6OH_2$ (PbPOH) compositions, whose DFT-PBEsol E_g is 3.7 and 3.51 eV, respectively. The ML predicted E_g with uncertainties for PbPF and PbPOH compositions using combined feature set is 2.87 ± 1.1 and 3.37 ± 0.65 eV, respectively, whereas for the electronegativity feature set it is 2.95 ± 1.32 and 3.29 ± 1.47 eV, respectively. Thus, the combined feature set has relatively smaller uncertainties compared to the electronegativity or ionic radii feature set. In addition, the top three compositions with the largest E_g of 5.35, 5.33, and 5.22 eV in the training set were SrPF, $Ca_{10}(PO_4)_6(OH)_2$ (CaPOH), and $Mg_{10}(PO_4)_6(OH)_2$ (MgPOH), respectively. The mean (μ) value of the ML predicted E_g trend for the three compositions from the three feature sets can be described as follows:

- Ionic radii: $E_g^{CaPOH} > E_g^{MgPOH} \approx E_g^{SrPF}$
- Electronegativity: $E_g^{CaPOH} > E_g^{SrPF} > E_g^{MgPOH}$
- Combined: $E_g^{CaPOH} \approx E_g^{SrPF} > E_g^{MgPOH}$

Thus, the combined feature set performs better than the ionic radii and electronegativity feature sets in reproducing the DFT-PBEsol E_g trend. We then used these ML models for adaptive design. Both *max* μ-ML and *max* E(I) recommended CaPF in the first iteration, which (as noted earlier) also has the largest E_g in our chemical space. Thus, the combined feature set has remarkably found the optimal composition in the first iteration itself. Our naive intuition informs that the combined feature set had more information about the apatites, which enabled us to fit a good ML model to the data.

3.4 Discussion

We showed that the choice of feature sets has an important role in the search for new materials and in the trajectory along which they guide the new computations. Our ML models built on ionic radii feature set, despite their relatively large uncertainties, succeeded in efficiently guiding the DFT towards promising regions (P-containing apatites) in the composition space. They also found the optimal composition [CaPF with DFT-PBEsol $E_g = 5.67$ eV] in 8 fewer iterations compared to the electronegativity feature set. After running a total of 25 iterations with feedback, the ionic radii feature set has sampled a fairly significant span of E_g space, whereas the electronegativity feature set sampled mainly $E_g > 2.5$ eV (Fig. 3.4). The best performance, however, came from the combined feature set, which gave the optimal composition in merely one iteration.

Thus, one of the insights that we uncovered is that the quality of the ML models (in terms of mean squared error) is not a sufficient indicator for achieving accelerated search. It is also important to incorporate essential features that capture the physical and/or chemical trend associated with the target property. This is reflected in our results for the combined feature set ML model: Despite a relatively poor ML model fit to the ionic radii feature set, it was efficient in finding the optimal composition in fewer iterations mainly because it carried the essential feature (i.e., B-site ionic radius). From Table 3.1, we infer that the ionic radii feature set only recommended P-containing apatites for validation and feedback. Our electronic structure calculations (DOS and partial DOS data) revealed that P-containing apatites, in general, have large E_g compared to the V- or As-containing apatites. As a consequence, the key challenge for the ionic radii feature set was to identify the optimal A- and X-atoms (only two degrees of freedom) and it took 9 iterations to find the optimal composition. In sharp contrast, the electronegativity feature set did not contain the essential feature for capturing the E_g trends of the B-site atoms. From Table 3.1, we also infer that it had to explore all three chemical degrees of freedom (A-, B-, and X-atoms), which eventually took 17 iterations to find the optimal composition. To further confirm the importance of r_B (B-site ionic radius) feature to our problem, we performed two additional tests. We augmented the electronegativity feature set in two ways, (i) added r_B as a new feature vector and (ii) added r_A and r_X as two new feature vectors, and repeated our iterative feedback look. The electronegativity plus r_B feature set found CaPF in its second iteration. On the other hand, the electronegativity plus r_A and r_X feature set did not find CaPF even after five iterations.

In the context of the cheminformatics and drug discovery literature, it is common to discuss the feature-activity (property) relationships from the viewpoint of the activity (property) landscape [41]. One of the common approaches is to visualize or analyze the relationship using similarity maps to uncover trends underlying the feature-property data [42, 43]. We borrow some of these ideas and adapt them to interpret the results reported in this work. Especially, we are interested in understanding how the ML model outcome changes as we progress from one iteration to the next. To this end, we calculate the pairwise similarity between the 96 apatite compositions using the following equation,

$$\mathrm{E}_g^{\mathrm{ML}}\ \mathrm{Closeness}_{i,j} = \frac{dist(\mathrm{E}_{g,i}, \mathrm{E}_{g,j})}{dist(i, j)} \tag{3.1}$$

where $dist(\mathrm{E}_{g,i}, \mathrm{E}_{g,j})$ is the Euclidean distance between ML predicted E_g's for compositions i and j, and $dist(i, j)$ is the Euclidean distance between the same two compositions in the feature space (ionic radii or electronegativity). The resulting outcome is a 96×96 matrix, which we refer to as $\mathrm{E}_g^{\mathrm{ML}}$ Closeness matrix. We calculated a total of 25 such matrices (one for each iteration) for both ionic radii and electronegativity feature sets. Our interest is in estimating the correlation between these matrices for the nth and $(n+1)$th iteration. In (3.1), we note that the denominator is the same for the nth or $(n+1)$th iteration, because the feature space entries remain fixed from one iteration to the next. Only the numerator changes, due to the iterative nature of our adaptive design and ML model update.

An important detail about the numerator of (3.1) between the nth and $(n+1)$th iteration is the following: At the end of the nth iteration, we have a ML model that was then used to predict the E_g's for 96 compositions. From the model, we calculate the E_g^{ML} Closeness matrix for the nth iteration. We then used the model for design, which recommends a new composition that we subsequently validate using our DFT-PBEsol calculation. We augment our dataset with this new composition and retrain our ML models. We now use the retrained ML models to predict the E_g for all 96 compositions. These updated predictions are used to calculate the E_g^{ML} Closeness matrix for the $(n+1)$th iteration.

Our hypothesis is that if the correlation between the E_g^{ML} Closeness matrices for the nth and $(n+1)$th iteration is high, then the ML model has not undergone significant change between the nth and $(n+1)$th iteration as a result of our design. On the other hand, if the correlation between the two matrices is low, then we infer that the addition of $(n+1)$th composition has affected the outcome of the ML model predictions. We utilize the Mantel test [44, 45], which is a well-known statistical method, to quantify the strength of the linear relationship between the two matrices. The standardized Mantel correlation statistic (MCS) is calculated using (3.2) given below,

$$MCS = \frac{1}{r-1}\sum_{p=1}^{r}\sum_{q=1}^{r}\frac{x_{pq}-\bar{x}}{s_x}\cdot\frac{y_{pq}-\bar{y}}{s_y} \tag{3.2}$$

where, r is the number of elements in the matrix, p and q are indices of the matrix elements, x and y are the variables associated with matrix 1 and matrix 2, respectively, $\bar{x}$ and $\bar{y}$ are the mean values for variables x and y, respectively, and s_x and s_y are the standard deviations for variables x and y, respectively.

In Fig. 3.10, we show the Mantel test results for the ionic radii and electronegativity feature sets. We find that the MCS value fluctuated substantially in the first few iterations for the ML model that was built using ionic radii feature set. In contrast, the ML model that used electronegativity did not show large variation in the MCS value during those initial iterations. After about 10 iterations and until the end of 23 iterations, we find very little change in the MCS value for both feature sets. However, the MCS value changed substantially between the E_g^{ML} Closeness matrices that represented iterations 23 and 24 for the ionic radii feature set. We can understand the reason for this behavior from Fig. 3.4a. Notice that at the end of the 23rd iteration, we found a composition with a very low DFT-PBEsol E_g (0.07 eV). When this composition was augmented to our training set and after retraining the ML models, the matrices for the 23rd and 24th iteration were affected, which is reflected in the MCS analysis. In contrast, the electronegativity ML model appears to have converged.

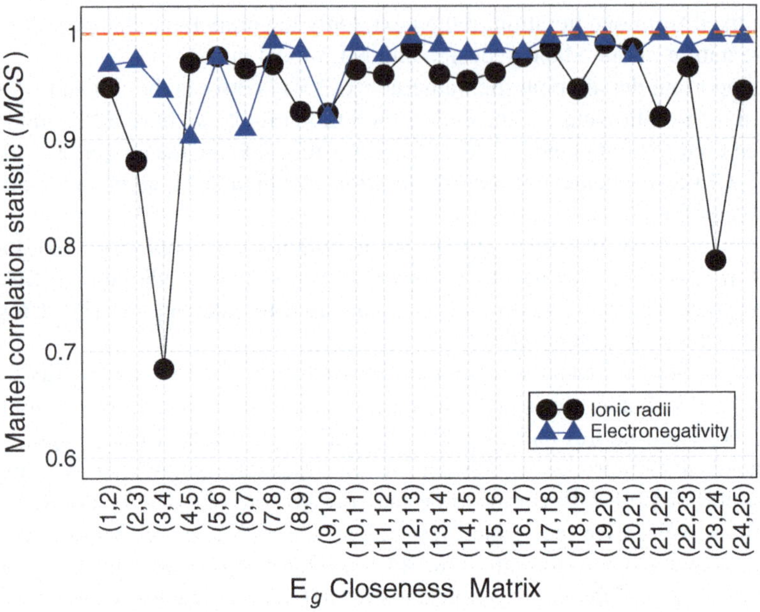

Fig. 3.10 Variation of the Mantel correlation statistic (MCS) with respect to the changes in the feature-E_g landscape between two E_g Closeness matrices estimated at the end of the nth and $(n+1)$th iteration of our adaptive design. The x-axis indicates the pair of iteration numbers for the E_g Closeness matrices that were calculated from (3.1). The results for ionic radii and electronegativity feature sets are depicted in black circles and blue triangles, respectively. Each data point indicates the strength or correlation (linear model) between two E_g Closeness matrices at the nth and $(n+1)$th iteration. Red dashed line represents $MCS = 1$, indicating perfect correlation between the two matrices

3.5 Summary

We have uncovered insights into the role of feature-property relationships within the adaptive design strategy for accelerated search using computational data. We have shown that the feature-property landscape has an intriguing and non-trivial role. The average error of the ML model in itself is not sufficient for achieving accelerated search, and we have shown that it is also important to incorporate key features that capture the underlying physical and/or chemical trends of the associated property. More studies using diverse datasets are required to validate the generality of these findings.

What are the implications of our results for the adaptive design of materials in practice? First, the adaptive design approach presented is most suitable to cases where high-throughput theory or experiment is not feasible; that is, for cases where the object of interest, here it was the band gap, is a quantity expensive to obtain accurately, if at all, from theoretical calculations or is available only from time-consuming or expensive experimental measurements. Our results illustrate that feature selection can affect the convergence of the process more strongly than the quality of the ML model. Typically, in executing the process, it is easy to start with a large list of features, and a good fit to the data will result. Reducing the size of the list generally unveils which features are key in enhancing the merit of the property of interest but can reduce fit quality.

How does then one perform feature reduction and selection? For the present study, we had prior knowledge about good choices of features. For the size of the feature sets considered, the cost of building the support vector machine regression models was insignificant. Consequently, exploring the consequences of different sets of features on the adaptive design performance was relatively computationally less demanding. In cases where such prior knowledge is absent, we could have more mechanically used regression models based on decision trees that return estimates of the relative importance of each feature in producing the fit. Hence, these methods provide a means for selecting the important features and subsequently a basis for set reduction. ML also offers other techniques. For example, principal component analysis is popular, but it often obscures which specific feature is the most important.

The dependency of the rate of the convergence of the adaptive design on the features chosen is likely our most significant observation. Its significance is not about reducing the cost of building the ML models but about reducing the cost of validating the model by subsequent theoretical calculations or experimental measurements. In this part of the process, we illustrated the fidelity of the model can be less important than its attempted validation as the validation adds to the dataset a new entry that refines the new ML model for the next prediction. The distinctive feature of the adaptive design approach is the construction of a larger dataset from a smaller one in a consistent and controlled manner.

Acknowledgements The authors acknowledge funding support from the Los Alamos National Laboratory (LANL) Laboratory Directed Research and Development (LDRD) DR (#20140013DR) on Materials Informatics. PVB and TL are grateful for the support from the Center for Non-Linear Studies (CNLS) at LANL. The authors also thank the Institutional Computing (IC) resources at LANL for providing support for running the DFT calculations.

References

1. W. Kohn, L.J. Sham, Self-consistent equations including exchange and correlation effects. Phys. Rev. **140**, A1133–A1138 (1965)
2. H.C. Andersen, Molecular dynamics simulations at constant pressure and/or temperature. J. Chem. Phys. **72**(4), 2384–2393 (1980)
3. I. Steinbach, Phase-field models in materials science. Modell. Simul. Mater. Sci. Eng. **17**(7), 073001 (2009)
4. T. Lookman, P.V. Balachandran, D. Xue, J. Hogden, J. Theiler, Statistical inference and adaptive design for materials discovery. Curr. Opin. Solid State Mater. Sci. **21**(3), 121–128 (2017)
5. D. Xue, P.V. Balachandran, J. Hogden, J. Theiler, D. Xue, T. Lookman, Accelerated search for materials with targeted properties by adaptive design. Nat. Commun. **7**, 11241 (2016)

6. T.K. Patra, V. Meenakshisundaram, J.-H. Hung, D.S. Simmons, Neural-network-biased genetic algorithms for materials design: evolutionary algorithms that learn. ACS Comb. Sci. **19**(2), 96–107 (2017)
7. R. Dehghannasiri, D. Xue, P.V. Balachandran, M.R. Yousefi, L.A. Dalton, T. Lookman, E.R. Dougherty, Optimal experimental design for materials discovery. Comput. Mater. Sci. **129**, 311–322 (2017)
8. T. Ueno, T.D. Rhone, Z. Hou, T. Mizoguchi, K. Tsuda, COMBO: an efficient Bayesian optimization library for materials science. Mater. Discov. **4**, 18–21 (2016)
9. P.V. Balachandran, D. Xue, J. Theiler, J. Hogden, T. Lookman, Adaptive strategies for materials design using uncertainties. Sci. Rep. **6**, 19660 (2016)
10. P.V. Balachandran, D. Xue, T. Lookman, Structure-Curie temperature relationships in $BaTiO_3$-based ferroelectric perovskites: anomalous behavior of (Ba, Cd)TiO_3 from DFT, statistical inference, and experiments. Phys. Rev. B **93**, 144111 (2016)
11. D. Xue, D. Xue, R. Yuan, Y. Zhou, P.V. Balachandran, X. Ding, J. Sun, T. Lookman, An informatics approach to transformation temperatures of NiTi-based shape memory alloys. Acta Materialia **125**, 532–541 (2017)
12. D. Xue, P. V. Balachandran, R. Yuan, T. Hu, X. Qian, E. R. Dougherty, and T. Lookman, "Accelerated search for $BaTiO_3$-based piezoelectrics with vertical morphotropic phase boundary using Bayesian learning," *Proceedings of the National Academy of Sciences*, vol. 113, no. 47, pp. 13 301–13 306, 2016
13. C. Kim, G. Pilania, R. Ramprasad, From organized high-throughput data to phenomenological theory using machine learning: the example of dielectric breakdown. Chem. Mater. **28**(5), 1304–1311 (2016)
14. G. Pilania, K.R. Whittle, C. Jiang, R.W. Grimes, C.R. Stanek, K.E. Sickafus, B.P. Uberuaga, Using machine learning to identify factors that govern amorphization of irradiated pyrochlores. Chem. Mater. **29**(6), 2574–2583 (2017)
15. O. Isayev, D. Fourches, E.N. Muratov, C. Oses, K. Rasch, A. Tropsha, S. Curtarolo, Materials cartography: representing and mining materials space using structural and electronic fingerprints. Chem. Mater. **27**(3), 735–743 (2015)
16. L.M. Ghiringhelli, J. Vybiral, S.V. Levchenko, C. Draxl, M. Scheffler, Big data of materials science: critical role of the descriptor. Phys. Rev. Lett. **114**, 105503 (2015)
17. P.V. Balachandran, J. Theiler, J.M. Rondinelli, T. Lookman, Materials prediction via classification learning. Sci. Rep. **5**, 13285 (2015)
18. A. Seko, H. Hayashi, K. Nakayama, A. Takahashi, I. Tanaka, Representation of compounds for machine-learning prediction of physical properties. Phys. Rev. B **95**, 144110 (2017)
19. Y. Bengio, A. Courville, P. Vincent, Representation learning: a review and new perspectives. IEEE Trans. Pattern Anal. Mach. Intell. **35**(8), 1798–1828 (2013)
20. T.J. White, D. ZhiLi, Structural derivation and crystal chemistry of apatites. Acta Crystallogr. Sect. B **59**(1), 1–16 (2003)
21. P.H.J. Mercier, Y. Le Page, P.S. Whitfield, L.D. Mitchell, I.J. Davidson, T.J. White, Geometrical parameterization of the crystal chemistry of $P6_3/m$ apatites: comparison with experimental data and ab initio results. Acta Crystallogr. Sect. B **61**(6), 635–655 (2005)
22. P.V. Balachandran, K. Rajan, Structure maps for $A^{I}_{4}A^{II}_{6}(BO_4)_6X_2$ apatite compounds via data mining. Acta Crystallogr. Sect. B **68**(1), 24–33 (2012)
23. T. White, C. Ferraris, J. Kim, S. Madhavi, Apatite—an adaptive framework structure. Rev. Mineral. Geochem. **57**(1), 307–401 (2005)
24. P.V. Balachandran, S.R. Broderick, K. Rajan, Identifying the "inorganic gene" for high-temperature piezoelectric perovskites through statistical learning. Proc. R. Soc. Lond. A: Math. Phys. Eng. Sci. **467**(2132), 2271–2290 (2011)
25. P.V. Balachandran, J. Young, T. Lookman, J.M. Rondinelli, Learning from data to design functional materials without inversion symmetry. Nat. Commun. **8**, 14282 (2017)
26. R.D. Shannon, Revised effective ionic radii and systematic studies of interatomic distances in halides and chalcogenides. Acta. Cryst. A **32**, 751–767 (1976)

27. L. Pauling, The nature of the chemical bond. IV. The energy of single bonds and the relative electronegativity of atoms. J. Am. Chem. Soc. **54**(9), 3570–3582 (1932)
28. P. Giannozzi, S. Baroni, N. Bonini, M. Calandra, R. Car, C. Cavazzoni, D. Ceresoli, G.L. Chiarotti, M. Cococcioni, I. Dabo, A. Dal Corso, S. de Gironcoli, S. Fabris, G. Fratesi, R. Gebauer, U. Gerstmann, C. Gougoussis, A. Kokalj, M. Lazzeri, L. Martin-Samos, N. Marzari, F. Mauri, R. Mazzarello, S. Paolini, A. Pasquarello, L. Paulatto, C. Sbraccia, S. Scandolo, G. Sclauzero, A.P. Seitsonen, A. Smogunov, P. Umari, R.M. Wentzcovitch, QUANTUM ESPRESSO: a modular and open-source software project for quantum simulations of materials. J. Phys.: Condens. Matter **21**(39), 395502 (2009)
29. J.P. Perdew, A. Ruzsinszky, G.I. Csonka, O.A. Vydrov, G.E. Scuseria, L.A. Constantin, X. Zhou, K. Burke, Restoring the density-gradient expansion for exchange in solids and surfaces. Phys. Rev. Lett. **100**, 136406 (2008)
30. D. Vanderbilt, Soft self-consistent pseudopotentials in a generalized eigenvalue formalism. Phys. Rev. B **41**, 7892–7895 (1990)
31. H.J. Monkhorst, J.D. Pack, Special points for brillouin-zone integrations. Phys. Rev. B **13**, 5188–5192 (1976)
32. A.D. Corso, Pseudopotentials periodic table: from H to Pu. Comput. Mater. Sci. **95**, 337–350 (2014)
33. P.V. Balachandran, K. Rajan, J.M. Rondinelli, Electronically driven structural transitions in $A_{10}(BO_4)_6F_2$ apatites (A = Ca, Sr, Pb, Cd and Hg). Acta Crystallogr. Sect. B **70**(3), 612–615 (2014)
34. H.T. Stokes, D.M. Hatch, FINDSYM: program for identifying the space-group symmetry of a crystal. J. Appl. Crystallogr. **38**(1), 237–238 (2005)
35. K. Momma, F. Izumi, VESTA: a three-dimensional visualization system for electronic and structural analysis. J. Appl. Crystallogr. **41**(3), 653–658 (2008)
36. D. Meyer, E. Dimitriadou, K. Hornik, A. Weingessel, F. Leisch, e1071: Misc Functions of the Department of Statistics, Probability Theory Group (Formerly: E1071), TU Wien, 2015, R package version 1.6-7. http://CRAN.R-project.org/package=e1071
37. R Core Team, *R: A Language and Environment for Statistical Computing* (R Foundation for Statistical Computing, Vienna, Austria, 2012). ISBN 3-900051-07-0. http://www.R-project.org/
38. D.P. MacKinnon, C.M. Lockwood, J. Williams, Confidence limits for the indirect effect: distribution of the product and resampling methods. Multivar. Behav. Res. **39**(1), 99–128 (2004)
39. D.R. Jones, M. Schonlau, W.J. Welch, Efficient global optimization of expensive black-box functions. J. Glob. Optim. **13**(4), 455–492 (1998)
40. Y. Zheng, T. Gao, Y. Gong, S. Ma, M. Yang, P. Chen, Electronic, vibrational and thermodynamic properties of $Ca_{10}(AsO_4)_6(OH)_2$: first principles study. Eur. Phys. J. Appl. Phys. **72**(3), 31201 (2015)
41. M. Cruz-Monteagudo, J.L. Medina-Franco, Y. Pérez-Castillo, O. Nicolotti, M.N.D. Cordeiro, F. Borges, Activity cliffs in drug discovery: Dr. Jekyll or Mr. Hyde? Drug Discov. Today **19**(8), 1069–1080 (2014)
42. R. Guha, J.H. Van Drie, Structure-activity landscape index: identifying and quantifying activity cliffs. J. Chem. Inf. Model. **48**(3), 646–658 (2008)
43. J.L. Medina-Franco, Scanning structure-activity relationships with structure-activity similarity and related maps: from consensus activity cliffs to selectivity switches. J. Chem. Inf. Model. **52**(10), 2485–2493 (2012)
44. N. Mantel, The detection of disease clustering and a generalized regression approach. Cancer Res. **27** (2, Part 1), 209–220 (1967)
45. J. Oksanen, F.G. Blanchet, M. Friendly, R. Kindt, P. Legendre, D. McGlinn, P.R. Minchin, R.B. O'Hara, G.L. Simpson, P. Solymos, M.H.H. Stevens, E. Szoecs, H. Wagner, vegan: Community Ecology Package, 2017, r package version 2.4-2. https://CRAN.R-project.org/package=vegan

Chapter 4
Bayesian Approaches to Uncertainty Quantification and Structure Refinement from X-Ray Diffraction

Alisa R. Paterson, Brian J. Reich, Ralph C. Smith, Alyson G. Wilson and Jacob L. Jones

Abstract This chapter introduces classical frequentist and Bayesian inference applied to analyzing diffraction profiles, and the methods are compared and contrasted. The methods are applied to both the modelling of single diffraction profiles and the full profile refinement of crystallographic structures. In the Bayesian method, Markov chain Monte Carlo algorithms are used to sample the distribution of model parameters, allowing for the construction of posterior probability distributions, which provide both parameter estimates and quantifiable uncertainties. We present the application of this method to single peak fitting in lead zirconate titanate, and the crystal structure refinement of a National Institute of Standards and Technology silicon standard reference material.

4.1 Introduction

Researchers and engineers are continually working to design new materials with enhanced or desired properties. Understanding the structure of materials is key in order to do this successfully. Diffraction, and more specifically X-ray diffraction

A. R. Paterson · J. L. Jones (✉)
Department of Materials Science and Engineering, North Carolina State University, Raleigh, NC 27695, USA
e-mail: jacobjones@ncsu.edu

B. J. Reich · A. G. Wilson
Department of Statistics, North Carolina State University, Raleigh, NC 27695, USA

R. C. Smith
Department of Mathematics, North Carolina State University, Raleigh, NC 27695, USA

T. Lookman et al. (eds.), *Materials Discovery and Design*, Springer Series in Materials Science 280, https://doi.org/10.1007/978-3-319-99465-9_4

(XRD) is a powerful tool that is commonly used by materials scientists to determine atomic structure. Simply stated, XRD relies on the constructive interference of X-rays scattered from planes of atoms. The scattered intensity I as a function of the scattering angle 2θ can be represented by

$$I(\theta) = f(2\theta|\alpha), \tag{4.1}$$

where f is a profile shape function and α is a set of parameters that determine the intensity. The X-ray scattering results in Bragg peaks, as shown in Fig. 4.1. More precisely, these are referred to as reflections; however, many reflections can overlap in one experimentally measured peak so we choose peak as the term that more reflects the experimental data.

Bragg's law describes the conditions necessary for constructive interference of the scattered X-rays, and is given by

$$2d_{hkl} \sin\theta = n\lambda, \tag{4.2}$$

where d_{hkl} is the interplanar spacing between crystal planes (hkl) (the d-spacing), θ is the Bragg angle or angle of incidence, n is the order, and λ is the wavelength of the X-rays. There are many solutions to Bragg's law for a given set of planes at different values of n, but it is customary to set n to 1 [1]. For example, when $n = 2$, the d-spacing is instead halved to keep $n = 1$.

Statistical inference deduces structural parameters of materials through analysis of the data. Inference tell us about parameters that we cannot directly observe. Crystallographers refer to this as the "inverse problem": we start with the results (the XRD pattern) and then calculate the cause (the underlying structural parameters). For example, structural details such as the d_{hkl} can be extracted from XRD patterns of intensity versus 2θ by using Bragg's law, and full profile structure refinement provides even more detailed information about the crystal structure and instrumental contributions to the profile. Uncertainty quantification, or the science of quantifying and reducing uncertainties in both computational and real-world systems [2], is also very important in structure determination. Researchers need to know how precisely the mathematical model describes the true atomic structure.

There are several different paradigms of statistical inference [4]. In this chapter, we briefly introduce classical, frequentist inference and classical methods of peak fitting and structure refinement. This is followed by an introduction to Bayesian inference and a detailed discussion of Bayesian inference applied to modelling diffraction profiles and crystallographic structure refinement. We demonstrate that Bayesian inference has several advantages over the classical methods, due to its ability to provide quantifiable uncertainty.

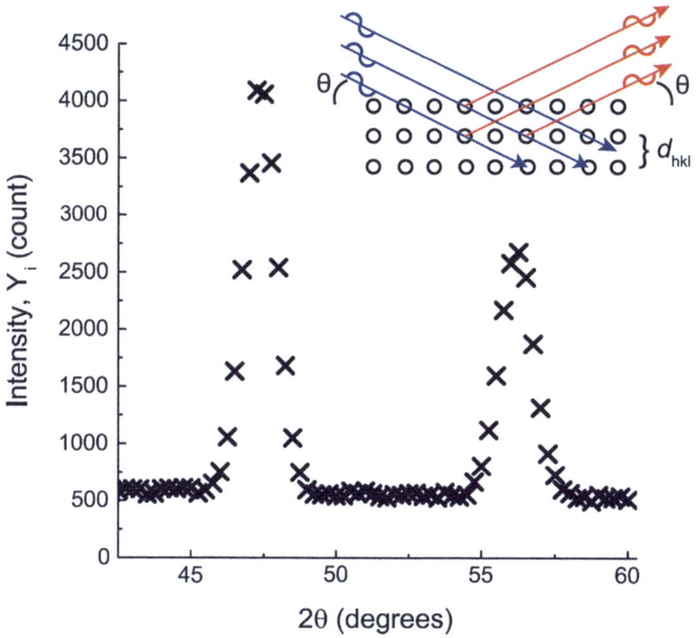

Fig. 4.1 Example of peaks observed at angles 2θ in an X-ray diffraction pattern. Peaks arise from the constructive interference of X-rays scattered from planes of atoms. The inset shows a schematic illustration of X-rays scattering from a periodic array of atoms, with X-rays incident at an angle θ, and the resulting constructive interference at an angle θ from the plane of atoms. (Reproduced from [3]). This figure is licensed under a Creative Commons Attribution 4.0 International License https://creativecommons.org/licenses/by/4.0/

4.2 Classical Methods of Structure Refinement

In this section, we briefly introduce frequentist inference and discuss the predominant classical method of structure refinement, the Rietveld method. The limitations of these approaches are also outlined.

4.2.1 Classical Single Peak Fitting

Fitting of XRD data can be accomplished by fitting a single diffraction peak or multiple peaks, up to the entire pattern. An entire pattern may contain on the order of 100 single diffraction peaks. To fit a single peak, a model is first selected to describe the profile shape. Common profile shape functions include the Gaussian, Lorentzian, and pseudo-Voigt. Classical single peak fitting uses a least-squares fitting method to minimize the difference between the intensity of the experimental diffraction peak and the model diffraction peak. Specific values for the model parameters are an output of this process and, together with the profile function, can be used to simulate the peak.

Let's consider the Gaussian model as an example. The Gaussian is a function with the form

$$f(x) = ae^{-\frac{(x-b)^2}{2c^2}}, \tag{4.3}$$

where a is the peak height, b is the centremost point of the curve, and c controls the peak width. Collectively, we will refer to these parameters as α. Changing α will produce different Gaussian curves. The goal of peak fitting is to determine the α values for the data set of interest that will minimize the sum S of the squared residuals r_i, given in (4.4) and (4.5). For a diffraction peak, the residual is defined as the difference between the observed experimental intensity (I_{data}) and the intensity predicted by the model (I_{model}), as in (4.5).

$$S = \sum_{i=1}^{n} r_i^2, \tag{4.4}$$

$$r_i = I_{data} - I_{model}. \tag{4.5}$$

By minimizing S, the "best fit curve" is produced. An example of a Gaussian fit is given in Fig. 4.2. In a plot of intensity as a function of 2θ, the centre of the fitting curve can be used to calculate the interplanar lattice spacing d_{hkl}, using Bragg's law (4.2). This parameter can help us understand the properties of the material of interest. For example, single peak fitting of the 222 peak from bismuth zinc niobate (BZN) was used by Nino et al. to examine the lattice strain in BZN cubic pyrochlore thin films deposited on Si, sapphire, MgO, and Vycor glass substrates [5]. This work demonstrates one of many applications of single peak fitting.

4.2.2 The Rietveld Method

Fitting of the whole diffraction pattern provides rich information about materials. The Rietveld method is a popular crystal structure refinement method that was developed by H. Rietveld in 1967 [6, 7]. It is a frequentist method that uses a least-squares approach to minimize the difference between a theoretical, calculated XRD pattern and an experimental XRD pattern that contains many reflections. Given a model for the crystal structure, the theoretical pattern is calculated, and model parameters such as the lattice parameters and atomic positions are adjusted to minimize the difference until a satisfactory solution is obtained. This method yields a set of specific values for all model parameters (α). An example of a Rietveld refinement is shown in Fig. 4.3 (left) for the silicon crystal structure (right). The experimental data (x) is fit with the calculated pattern (solid line), and the difference is plotted below. At least several

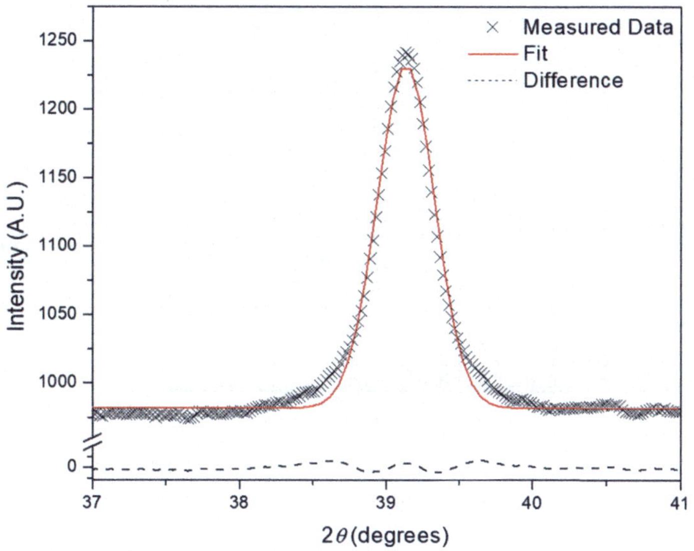

Fig. 4.2 A representative Gaussian fit of a single diffraction peak, showing the measured data, the fit, and the difference. Note that this is an imperfect fit near the peak and shoulders because the Gaussian model function cannot model these features

hundred papers a year reference this refinement method, evidencing its status as a powerful tool in crystallography [8]. Many software packages, such as General Structure Analysis Software-II (GSAS-II) [9] and TOPAS, have been developed to implement Rietveld analysis.

The parameters from Rietveld analysis have an associated uncertainty. The precision in the Rietveld refinement method is reported as the standard uncertainty (standard error). The standard uncertainty is the standard deviation of the estimator's sampling for each parameter. For example, the lattice parameter a may be reported as 3.998(5) Å. The (5) indicates the precision in the last digit of 3.998. A 95% confidence interval for this lattice parameter is $3.998 \pm (2 \times 0.005)$. Confidence intervals are discussed further in Sect. 4.2.3.

Unfortunately, studies have shown that the standard uncertainty is often incorrect or unreliable [10–13]. Moreover, the least squares method is susceptible to false minima solutions [14, 15]. False minima trap these methods and cause them to fail to find the best solution. The convergence of the refinement to a global minimum is necessary for estimated uncertainties to reflect the real uncertainty, so false minima are problematic [13]. In addition, the correct standard deviations cannot be calculated if the model does not sufficiently reproduce all the features in the diffraction pattern [16].

Another limitation of the uncertainty quantification of a standard Rietveld analysis is that the sampling distribution of the α estimator is assumed to be approximately

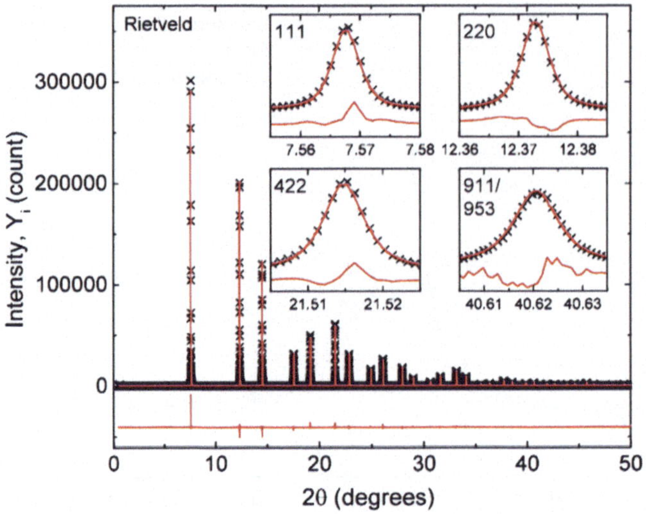

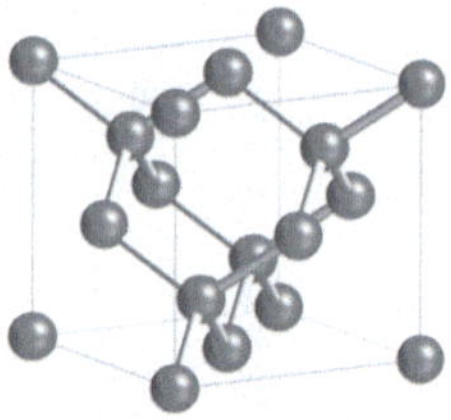

Fig. 4.3 Left: a representative Rietveld refinement of silicon X-ray diffraction data [3]. The calculated fit (solid line) is plotted with the experimental data (x) and the difference curve is shown below. The insets show the fit for specific peaks. Right: silicon crystal structure. (Left figure reproduced from [3]).

Gaussian with covariance derived from the Fisher information matrix. This assumption is tenuous for such a highly non-linear problem. Violation of this assumption could lead to poor statistical inference including under-coverage of confidence intervals.

4.2.3 Frequentist Inference and Its Limitations

The methods discussed above fall under the category of frequentist statistics. Frequentist inference is based on a frequency view of probability, where a given experiment is considered to be a random sample of an infinite number of possible repetitions of the same experiment [17]. While useful for characterizing quality of a continuous manufacturing process, this description is not well-suited to characterize uncertainty of a single experiment. For example, crystallographers often only have one set of XRD data that is used for crystal structure determination, and the frequentist perspective does not readily apply. The repeated experiments considered by the frequentist are merely hypothetical.

It is important to report the uncertainty associated with the materials parameter values obtained by frequentist inference. Uncertainty in these point estimates is often summarized using confidence intervals. However, confidence intervals are complicated, and one recent study suggests that researchers do not understand how to correctly interpret them [18]. A confidence interval is a range of values that is believed (with a particular percentage) to contain the "true" parameter value [17]. Many researchers misinterpret a 95% confidence interval to mean that there is a 95% probability that the interval includes the true value. In reality, a 95% confidence interval implies that 95% of the calculated confidence intervals from subsequent repeat experiments are expected to contain the "true" parameter value. Thus, the confidence interval does not provide any probability for the current measurement of interest, but only for future measurements, and each future measurement would lead to a different confidence interval.

Once parameter values and corresponding confidence intervals are obtained for a set of experimental data, the results are used for model selection. With frequentist approaches, the decision is often sharp. For example, a Rietveld refinement may be performed using two different space groups, *Cc* and *R*3*c*. The frequentist would accept as truth only one of these solutions, likely based on a quality of fit calculation such as a weighted residual. A Bayesian perspective, however, may be constructed such that different solutions have associated probabilities.

4.3 Bayesian Inference

Bayesian inference methods use subjective probability to quantify uncertainty [19]. These methods are based on Bayes' theorem, which describes the probability of an event, using prior knowledge of potentially related conditions. Bayes' theorem is given by

$$P(A|B) = \frac{P(B|A) \times P(A)}{P(B)}, \tag{4.6}$$

where $P(A)$ and $P(B)$ are the probabilities of observing A and B, respectively, and $P(A|B)$ is the probability of observing A given that B is true, and $P(B|A)$ is the probability of observing B given that A is true.

If we rewrite Bayes' theorem for the application to X-ray diffraction data, it is given by

$$P(\alpha|\text{data}) = \frac{P(\text{data}|\alpha) \times P(\alpha)}{P(\text{data})} \tag{4.7}$$

where α is the parameter of interest, $P(\alpha|\text{data})$ is the posterior probability distribution, $P(\text{data}|\alpha)$ is the likelihood (the probability of the data given the parameter values), $P(\alpha)$ is the prior distribution, and $P(\text{data})$ is the marginal likelihood (the probability of the data without assuming parameter values) [20]. Posterior distributions represent the probability after new evidence has been considered, whereas the prior distribution gives the probability of the parameters before considering the data.

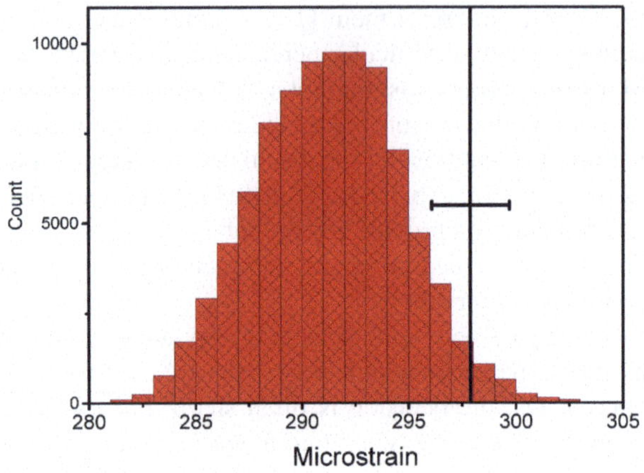

Fig. 4.4 A posterior probability distribution from Bayesian inference compared to a point estimate (vertical line) and standard uncertainty from the Rietveld method. (Reproduced from [3]). This figure is licensed under a Creative Commons Attribution 4.0 International License https://creativecommons.org/licenses/by/4.0/

Unlike classical frequentist methods, which treat data as a repeatable, random sample from an infinitely large population of data, data are fixed and known in Bayesian methods. Based on the data, Bayesian analysis asks the question "which values of α are the most plausible?" Instead of a single value for each parameter, Bayesian inference provides a probability distribution such as that shown in Fig. 4.4. This approach provides an output that has quantifiable certainty in terms of probability. For example, in a polycrystalline sample, one would not expect all crystals to have the same value for crystal size, so a distribution is more realistic. Bayesian inference also yields more quantifiable uncertainties than the frequentist approach, providing credible intervals instead of confidence intervals. A 95% credible interval means that there is a 95% probability that this interval contains the true value; this is a simply interpreted probability on whether the credible interval covers the true value. The posterior probability distribution obtained from Bayesian inference is an entirely different way of quantifying uncertainty than that available in frequentist approaches.

Selecting appropriate prior probability distributions, or priors, for Bayesian analysis is very important because the priors influence the results. If the priors selected do not contain the true value, the analysis cannot find the correct solution. Moreover, knowledge about the parameters that is known before the experiment may be used to select the priors. For instance, it is known that X-ray intensity will be positive. Knowledge for the prior may also be obtained from previous experiments. For example, electron microscopy analysis could provide information about the range of crystallite size in the sample, which can be used to select an appropriate prior distribution. The incorporation of prior knowledge is an advantage of Bayesian inference.

4.3.1 Sampling Algorithms

Many approximate inference algorithms have been proposed for Bayesian inference [21]. The studies we present in this chapter utilize Markov chain Monte Carlo (MCMC) algorithms, which are a subclass of stochastic sampling methods. Unlike least-squares minimization, which can become trapped in a region of parameter space due to false minima, the MCMC algorithm has the ability to escape from local minima due to its stochastic aspect [3].

MCMC is an iterative, general-purpose algorithm to indirectly simulate random observations from complex, high-dimensional probability distributions [17]. MCMC explores the parameter space by sampling multiple combinations of model parameters [22]. Random-walk Metropolis sampling is a versatile MCMC algorithm. Figure 4.5 shows a flowchart of this algorithm. First, a set of parameters are chosen, which can be based on prior knowledge of the material of interest if it is available. To begin, one parameter is selected, while the other parameters are fixed. The starting parameter, α_{st}, is compared with a new value of the parameter, α_{new}, obtained by randomly drawing from a proposed Gaussian distribution with mean α_{st}. The intensity is calculated based on the starting and new parameter values, and then the likelihood of these parameters is calculated to give $P(\text{data}|\alpha_{st})$ and $P(\text{data}|\alpha_{new})$. Based on these likelihoods, an acceptance criteria r, given by

$$r = \min\left(\frac{P(\text{data}|\alpha_{new})P(\alpha_{new})}{P(\text{data}|\alpha_{st})P(\alpha_{st})}, 1\right), \tag{4.8}$$

is used to decide whether to accept or reject the new parameter value α_{new}. α_{new} is accepted if $r \geq 1$. If α_{new} is accepted, it is used for the next iteration, but if it is rejected, the next iteration continues to use α_{st}. Acceptable answers are stored in a set. This process is repeated for thousands of iterations for one parameter at a time. For example, if we are applying this sampling process to a structure refinement, this process may be repeated for one model parameter at a time for 10^5 iterations to refine the crystallographic structure.

To further clarify MCMC sampling, let's consider a simple example. Suppose that we have two parameters α_1 and α_2 and have observed a set of data. Assume that α_1 and α_2 can only take on two values, 0 and 1. We have specified the model for how our data occurs given we know the values of the parameters (as in (4.1)), and also our prior distributions for α_1 and α_2, which means in this case the probability they equal 0 or 1 before observing the data. The MCMC algorithm iteratively chooses values for the parameters, with the ith values denoted as $\alpha_1^{(i)}$ and $\alpha_2^{(i)}$. The value for $\alpha_1^{(i)}$ is chosen randomly using $\text{P}(\alpha_1 = 1|\alpha_2^{(i-1)}, \text{data})$, where the probability is specified by the choices of model for the data, the prior distributions, and the most recent value for α_2. One can think of this value selection process as a coin toss that is weighted by the data's likelihood and prior probabilities.

The sampling algorithms described can also be applied to single peak fitting. The parameters that describe a Gaussian distribution, discussed in Sect. 4.2.1, can

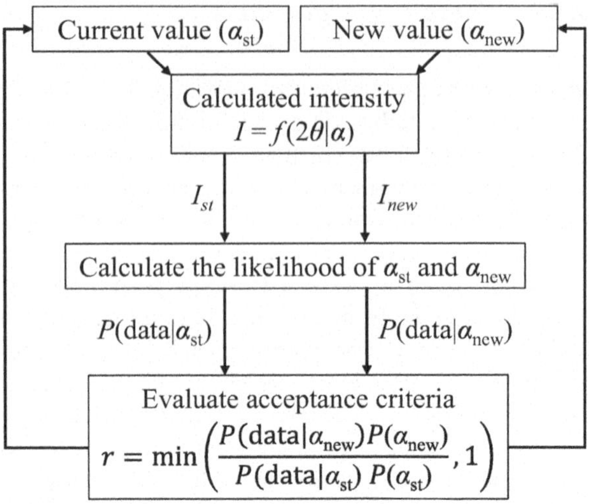

Fig. 4.5 A flowchart of a random-walk Metropolis sampling algorithm. Reproduced with permission of the International Union of Crystallography [20]

be sampled and a posterior probability distribution obtained for each parameter. A YouTube video about "Statistical Methods for Peak Fitting" was produced by the Data-Enabled Science and Engineering of Atomic Structure program at North Carolina State University, and provides a video introduction to both least-squares and Bayesian approaches to peak fitting [23].

4.4 Application of Bayesian Inference to Single Peak Fitting: A Case Study in Ferroelectric Materials

Bayesian inference has been applied to several areas of crystallography [24–34]. For example, Gagin and Levin developed a Bayesian approach to describe the systematic errors that affect Rietveld refinements, and obtained more accurate estimates of structural parameters and corresponding uncertainties than those determined from the existing Rietveld software packages [35]. Moreover, Mikhalychev and Ulyanenkov used a Bayesian approach to calculate the posterior probability distributions for the presence of each phase in a sample, allowing for phase identification comparable to existing methods [31]. This section focuses on the application of Bayesian inference to single peak fitting in a ferroelectric material.

Recently, Iamsasri et al. utilized Bayesian inference and an MCMC sampling algorithm to model single peaks [20]. The peak width may be associated with crystallite size and/or microstrain, the intensity may be affected by preferred orientation and/or the scattering factors of the crystals, and the peak position may be associated with the interatomic spacing [20]. Thus, single peak fitting can provide a great deal of information about these ferroelectric materials. Two different ferroelectric materials

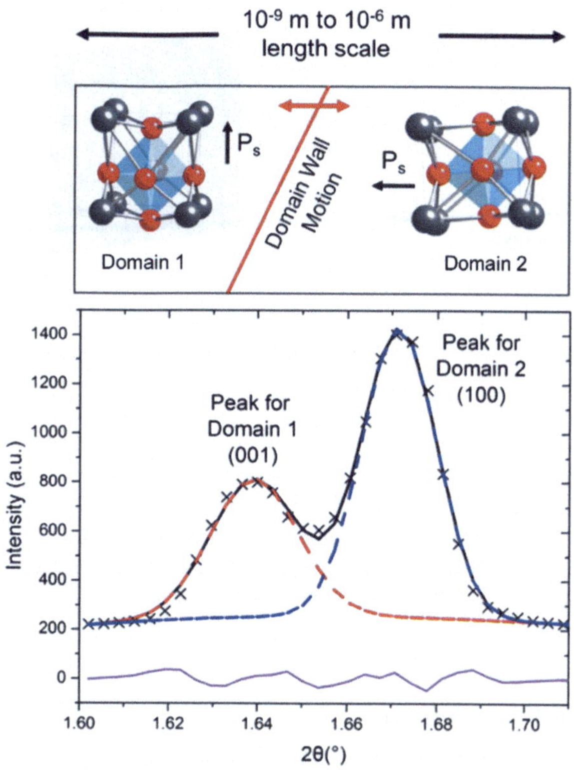

Fig. 4.6 Top: a schematic diagram of different crystalline orientations (domains) in a ferroelectric material. Bottom: a representative X-ray diffraction profile for a tetragonal ferroelectric material. The peak intensities are proportional to the fraction of each domain. Reproduced with permission of the International Union of Crystallography [20]

were studied: thin-film lead zirconate titanate (PZT) of composition $PbZr_{0.3}Ti_{0.7}O_3$ and a bulk commercial PZT polycrystalline sample.

Ferroelectric materials are materials that possess a spontaneous polarization that can be switched in direction through the application of an external electric field. The least-squares methods described above have been used extensively for peak fitting in ferroelectric materials, as shown in many reviews and studies, for example [36–41]. Peak intensities have been used to determine preferred crystallographic or ferroelectric domain orientations. Analysis is typically done on diffraction peaks that split from single peaks as a function of temperature and/or composition. For example, the $00h$ and $h00$ peaks are fit for a tetragonal perovskite, where h represents an integer. These reflections are particularly interesting because they represent crystal directions that are parallel and perpendicular to the ferroelectric polarization direction in a tetragonal perovskite. The intensities of these peaks can therefore reflect the volume fraction of different domains in a particular direction in the sample. Representative X-ray diffraction peaks for a tetragonal perovskite are shown in Fig. 4.6 (bottom). By tracking the change in intensities of these peaks as a function of applied electric field, researchers can characterize the domain wall motion (Fig. 4.6 (top)), which is a characteristic phenomenon in these materials.

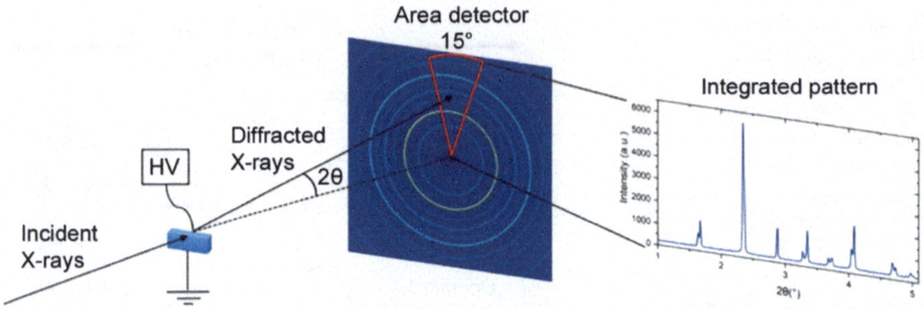

Fig. 4.7 A schematic diagram of the experimental set up on the APS beamline 11-ID-C. HV is high voltage. Reproduced with permission of the International Union of Crystallography [20]

Iamsasri et al. demonstrate in their work that Bayesian approaches can be applied to the fitting of peaks to calculate the degree of domain reorientation in the ferroelectric PZT samples under applied electric field [20]. When subjected to an external electric field, PZT exhibits a large degree of domain reorientation, which has been studied extensively [36–42]. This work is reviewed here in order to demonstrate the value of the Bayesian inference methods.

4.4.1 Methods

X-ray diffraction peaks were measured at the Advanced Photon Source (APS) at Argonne National Laboratory in Illinois, USA. A schematic of the experimental set-up for the bulk sample is shown in Fig. 4.7, and details of the set-up and experimental method for both samples can be found in [20]. The vertical direction of the two-dimensional XRD image was integrated over a 15° azimuthal range, as shown in Fig. 4.7, to select diffraction data with scattering vectors that are approximately parallel to the applied electric field and obtain intensity versus 2θ diffraction patterns. In this Chapter, we review the results on the ferroelectric PZT thin films.

The diffraction peaks of interest were first fit using a least-squares method. The $h00$ and $00h$ reflections in the diffraction pattern were fit using two pseudo-Voigt profiles for the PZT thin film. Integrated intensities were extracted from the fit peak profiles. The volume fractions were calculated using the following equation:

$$v_{00h} = \frac{I_{00h}/I'_{00h}}{I_{00h}/I'_{00h} + 2I_{h00}/I'_{h00}}, \tag{4.9}$$

where v_{00h} is the volume fraction of the $00h$-oriented domains in a particular direction of the sample, I_{00h} and I_{h00} are the integrated intensities of the $00h$ and $h00$ reflections, respectively, and I'_{00h} and I'_{h00} are reference intensities of the $00h$ and $h00$ reflections, respectively [20]. The reference intensities are obtained from the Powder Diffraction

File (card No. 01-070-4261; International Centre for Diffraction Data, Newtown Square, Pennsylvania, USA).

The domain switching fraction, η_{00h}, can be determined by calculating the difference between the volume fraction of the 00*h* reflection at voltage V and the reference value, as shown in the following equation

$$\eta_{00h} = v_{00h}^{\mathrm{V}} - v_{00h}^{\mathrm{ref}}, \tag{4.10}$$

where v_{00h}^{V} and v_{00h}^{ref} are the volume fractions of the 00*h* reflection at voltage V and for the reference, respectively. A confidence interval was acquired using an adapted variance equation (see supporting information for [20]).

In contrast, the Bayesian method employs an MCMC algorithm known as the Metropolis-in-Gibbs algorithm. A sampling process similar to the flowchart in Fig. 4.5 was followed and repeated for 10^5 iterations. A sequence of parameters is drawn from a suitable proposal distribution, and the parameters are accepted or rejected based on a probability specified by the algorithm. The parameters obtained from the first 10^3 cycles (the burn-in period) were discarded because they may be influenced by the starting parameters chosen. After convergence, histograms for each parameter can be constructed by counting the frequency of the accepted parameters in the specified ranges [20]. This fitting was repeated for all measured voltages.

The intensities from the iterations after the burn-in period are calculated from the posterior distribution of parameter values and the average intensities are plotted to obtain the peak fit typically used by crystallographers. The credible intervals in the parameter values can be propagated into credible intervals for the calculated intensities. A 95% credible interval was constructed from these calculated intensities. The domain switching fraction, η_{00h}, can then be calculated from the posterior distribution of intensities.

4.4.2 Prediction Intervals

Representative fits of the 00*h* and *h*00 reflections at 0 V for the PZT thin film are shown in Fig. 4.8a and b for least-squares and Bayesian approaches, respectively. One interesting feature of the Bayesian fit is the ability to draw the 95% credible interval, shown with a grey outline, which indicates that 95% of the calculated solutions are in this range. This demonstrates confidence in the solution, because nearly all data points fall in this 95% interval.

The domain reorientation parameter, η_{00h}, can be calculated from the parameter values of I, determined either from the least-squares or Bayesian methods. Since η_{00h} is a calculated quantity, error propagation is necessary. For the least-squares method, this involves an adapted variance equation. For the Bayesian method, subsequent distributions with associated probability density functions, can be calculated from the posterior distribution of parameters values obtained. For example, the parameters obtained from the single peak fitting can be used to calculate distributions for the

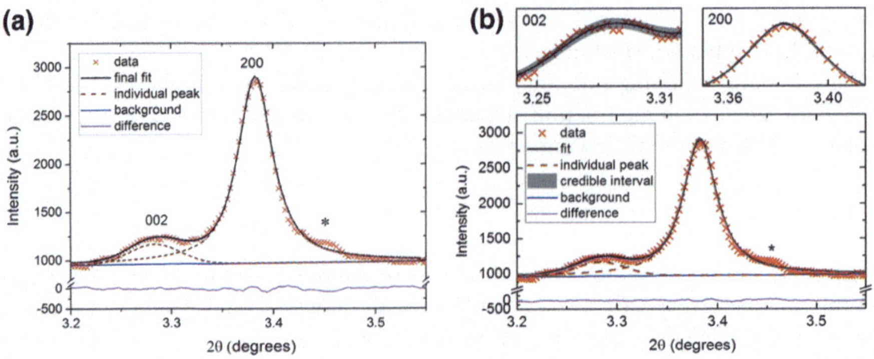

Fig. 4.8 Representative fits of the 00h and h00 reflections at 0 V for PZT as a thin film using **a** the least-squares method, and **b** the Bayesian inference method. The asterisk (*) represents an additional reflection due to a $PbTiO_3$ seed layer that is used for orientation of the sample. This seed layer was not modelled in this analysis. Reproduced with permission of the International Union of Crystallography [20]

degree of ferroelectric domain reorientation η_{00h}. For the Bayesian method, the value of η_{00h} was calculated for each iteration.

Figure 4.9 shows a comparison of the calculated η_{00h} at various electric fields for the thin film sample, with a confidence interval for the least-squares method, and a posterior distribution for the Bayesian method. This plot illustrates that the least-squares method yields a single value, while the Bayesian method gives a probability distribution of the domain reorientation values. While the values obtained from each method are similar, it is clear that the Bayesian method provides a richer description of the possible solutions. The calculated uncertainties from each method cover approximately the same range; they are comparable in amplitude, but not equal. These results suggest that uncertainty quantification from the Bayesian method is a reliable alternative to the least-squares method, and that errors can be propagated dependably from the initial results.

4.5 Application of Bayesian Inference to Full Pattern Crystallographic Structure Refinement: A Case Study

We use the previous work of Fancher et al. [3] to illustrate the application of Bayesian methods to full pattern refinement. Their work introduces a Bayesian statistical approach to refining crystallographic structures, and compares the results obtained to those determined by the classical method of Rietveld refinements [3]. An MCMC algorithm [22] is used to explore the parameter space and sample combinations of model parameters. Similar to Rietveld refinements, a theoretical model unit cell is used to calculate a diffraction pattern, but instead of obtaining single point value

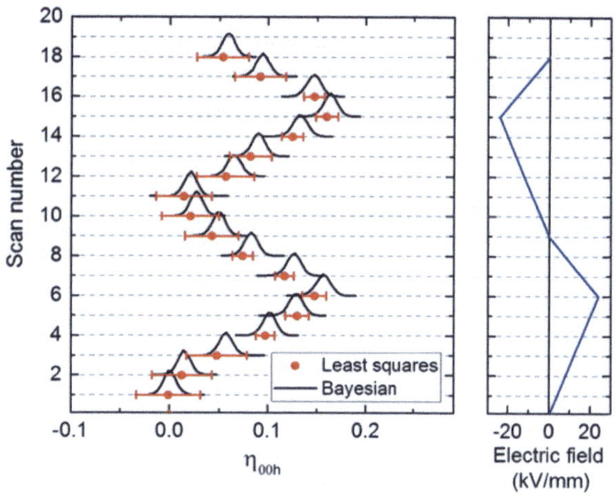

Fig. 4.9 The domain switching fraction, η_{00h}, for the PZT thin film. Reproduced with permission of the International Union of Crystallography [20]

estimates of model parameters, a posterior probability distribution of all modelled parameters is obtained. This yields estimates of the parameters with quantifiable uncertainty. The method described in this article is applied to a NIST silicon standard, and is readily adopted for use with other materials and data from other radiation sources such as neutrons.

4.5.1 Data Collection and the Rietveld Analysis

The high-resolution synchrotron XRD pattern for the NIST silicon standard (SRM 640d) was measured at 22.5 °C at the 11-BM-B beamline at the APS at Argonne National Laboratory. The Rietveld method was applied to the data first, to allow for a comparison to the Bayesian inference results, and to provide a starting point for model parameters for the MCMC algorithm. Rietveld refinements were performed using the software package GSAS-II [9]. To reduce the risk of nonconvergence of the least-squares approach, the sequence for parameter refinement suggested by Young was followed [14].

The Rietveld refinement result was previously shown in Fig. 4.3, and the refined parameters are presented in Table 4.1. The crystallite size (1.0006(6) μm) and microstrain (0.0298(2)%) for the NIST SRM640D standard differ from the values of 0.6 μm and 0, respectively, reported on the data sheet [43]. Fancher et al. suggest this may be due to differences in resolution in the synchrotron versus X-ray diffractometers, or the implementation of the method in GSAS-II versus TOPAS [44].

Table 4.1 Summary of fixed and refined structural parameters, including atomic positions and occupancies, and goodness of fit values for the Rietveld refinement of the Si standard. All refined parameters are shown with their respective standard uncertainty; remaining values are fixed

a (Å)	Crystal size (μm)	Microstrain (%*100)	λ (Å)	Profile fit
				R_p = 5.85%
5.43123	1.0006(6)	2.98(2)	0.4138490(5)	R_w = 8.28%
				χ^2 = 2.02
Site positions				
	x	y	z	U_{iso} (Å^2)
Si	0.125	0.125	0.125	0.00551(2)
Peak shape parameters				
	U	V	W	
	0.702(19)	−0.242(6)	0.0322(5)	

4.5.2 *Importance of Modelling the Variance and Correlation of Residuals*

Fancher et al. performed a simulation study to show the importance of properly modelling the variance and correlation of residuals [3]. Synthetic data was generated from a model with known parameter values and was fit using Bayesian inference to obtain estimates and credible intervals for the model parameters. These estimated values were compared to the true model values. The process was repeated several times to allow for the estimation of the accuracy and coverage of the credible intervals.

The results show that models that do not account for heteroskedastic and correlated residuals give poor estimates of the model parameters. Heteroskedasticity refers to data where the variability of a variable is unequal across a set of second, predictor variables. The poor estimates are evidenced in the 90% credible intervals. Fewer samples contain the true parameter values in the 90% interval in simpler models with independent residuals and constant variance. Therefore, it is necessary to adequately model the residual distribution of the error to obtain valid statistical inference. The Rietveld method uses independent residuals and variable variance, while the Bayesian model proposed by Fancher et al. uses a more complex residual structure which, is argued, improves the estimates and credible intervals.

4.5.3 Bayesian Analysis of the NIST Silicon Standard

To begin a Bayesian analysis, knowledge about the material or instrument parameters that we have preceding the experiment is used to specify a prior probability distribution. This is another advantage this approach has over the Rietveld method: there is no framework for incorporating prior knowledge into a Rietveld analysis. Bayesian inference allows for incorporation of knowledge such as crystallite size obtained by electron microscopy, or instrumental parameters such as wavelength. Posterior distributions for instrumental parameters could even be obtained through Bayesian analysis on a NIST standard reference material, and then utilized in subsequent analysis of more complex materials.

The observed data is used to calculate posterior probability distributions that reflect the uncertainty in the parameters. In their original analyses, Fancher et al. found that the computational expense is a bottleneck for the MCMC analysis. The reason is that GSAS-II calculates a model diffraction pattern at each MCMC iteration for each of the model parameters. An Intel CoreTM i5-3750 took 900 s, on average, to do 1000 iterations. The MCMC algorithm was run for 100,000 iterations, to be conservative. The authors are presently implementing a solution to this challenge.

Figure 4.10 compares the posterior probability distributions obtained from the Bayesian analysis to the point estimate and standard error obtained from the Rietveld method for four different parameters. Similar posterior probability distributions were obtained for all model parameters and show reasonable agreement with the point estimates determined by the Rietveld method; additional distributions can be viewed in the supplementary material for [3]. The uncertainties in the model parameters for both methods are comparable, except in the case of the 2θ offset, which has a much greater standard uncertainty in the model parameters from Rietveld than from Bayesian inference.

4.5.4 Comparison of the Structure Refinement Approaches

A great advantage to the Bayesian inference approach to full profile refinement is that the posterior probability distribution yields a much richer set of information about the uncertainty in the model parameters than the standard uncertainty obtained from least-squares minimization. For example, the posterior distributions can show asymmetric distributions of values, such as that seen in Fig. 4.4 for microstrain. It is important to note that other refinement approaches can be used to obtain distributions of certain parameters. Distributions of crystallite size have also been modelled by the Whole Powder Pattern Modelling (WPPM) method, assuming distributional forms such as normal or lognormal [45]. The advantage of the Bayesian inference approach is that these assumptions do not need to be made a priori.

The modelled diffraction pattern is typically plotted with the experimental pattern for comparison to determine the quality of a fit. Since Bayesian inference does not

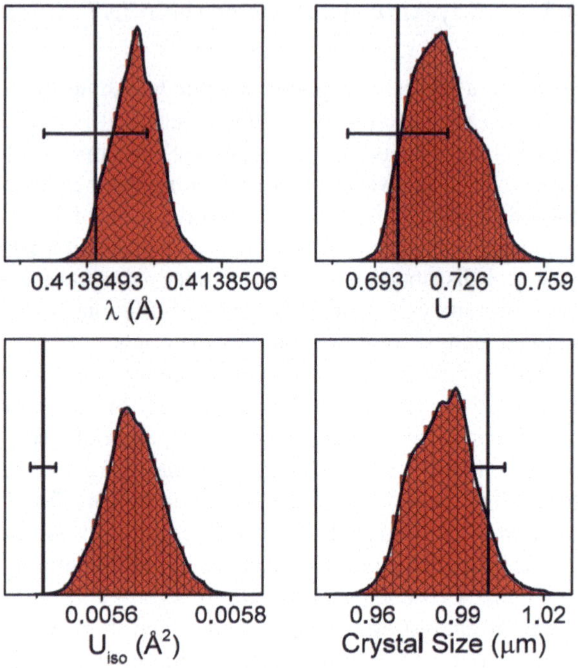

Fig. 4.10 Posterior probability distributions from Bayesian inference and point estimates (vertical lines) with corresponding standard uncertainty from Rietveld refinements. λ is the wavelength, U is an instrument parameter related to peak broadening, and U_{iso} is the isotropic atomic displacement parameter. (Reproduced from [3]). This figure is licensed under a Creative Commons Attribution 4.0 International License https://creativecommons.org/licenses/by/4.0/

yield single values for model parameters, this comparison is not as straightforward. A diffraction profile generated from MCMC data should be considered as a single observation from a collection. A representative pattern can be selected by choosing one single pattern, or by averaging many calculated patterns from MCMC samples. In this work, Fancher et al. chose the latter method, and plot a single pattern resulting from the average of the parameters obtained from the final 1000 MCMC samples, shown in Fig. 4.11. The average considers correlations, asymmetries, and uncertainties in the parameter distributions, but this representation overly simplifies the result. It does, however, demonstrate that modelled patterns fit experimental data well.

Figures 4.3 and 4.11 show the fit of the calculated model pattern for the Rietveld and Bayesian approaches, respectively, but it is difficult to see the subtle differences in the fit quality in these figures. Figure 4.12 makes these differences clearer and demonstrates that the Bayesian inference method better reproduces the experimental data. For the 111, 220, 422, and 911/953 reflections, the difference curves are positive when the Rietveld method underestimates the peak intensity, and negative when the observed intensity is overestimated. The improved estimate of peak position through Bayesian inference is evidenced in the 911/953 reflection: the Rietveld method underestimates the peak position, and while Bayesian inference estimates a peak position closer to the observed peak position.

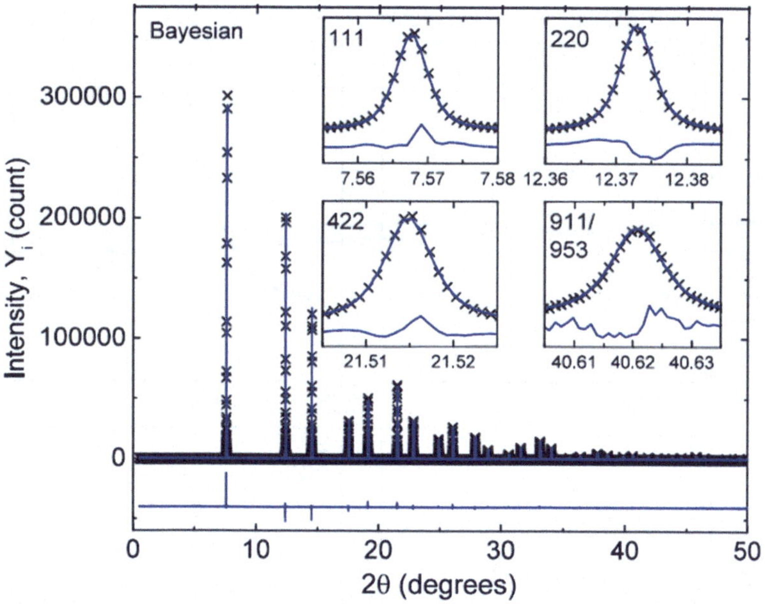

Fig. 4.11 The fitting results of the silicon powder diffraction data from Bayesian inference [3]. The results are an average of the final 1000 MCMC samples. The insets of characteristic reflections show that similar fits to the observed diffraction data are obtained by both Bayesian and Rietveld (see Fig. 4.3 insets) methods. (Reproduced from [3]). This figure is licensed under a Creative Commons Attribution 4.0 International License https://creativecommons.org/licenses/by/4.0/

4.5.5 Programs

Lesniewski et al. [46] published further applications of the Bayesian method to full pattern refinement. Most importantly, they presented a new software package, the Bayesian library for analyzing neutron diffraction data (*BLAND*). They demonstrate that even with limited knowledge of only the space group, composition, and site symmetries, adequate solutions can still be found through use of an automated Bayesian algorithm.

Esteves, Ramos, Fancher, and Jones have made available a program that implements Bayesian inference for single profile fitting [47]. The software package, Line Profile Analysis Software (LIPRAS), includes an option for Bayesian uncertainty quantification using the methods outlined in this chapter.

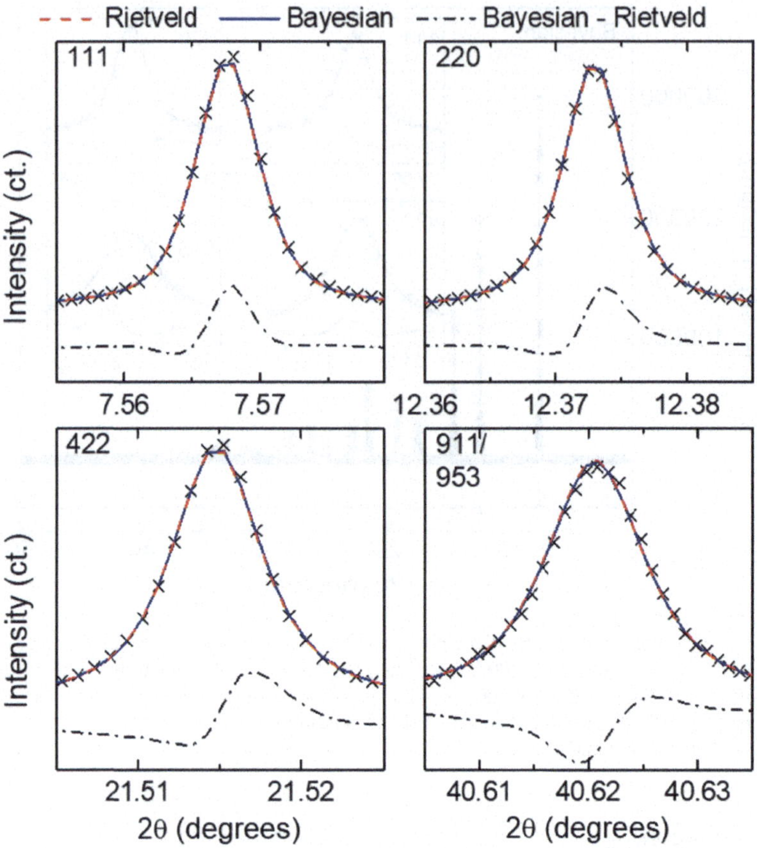

Fig. 4.12 Experimental diffraction data (x) plotted with the Rietveld (dotted) and Bayesian (solid line) analysis results for the 111, 220, 422, and 911/953 reflections in Si. The difference (Bayesian–Rietveld) is shown at the bottom, and demonstrates that the Bayesian results better model the experimental data. (Reproduced from [3]).

4.6 Conclusion

Both the least-squares frequentist approach and the Bayesian inference approach to structure refinement yield useful information about the structure of materials. Bayesian statistics were shown to provide many advantages over the current techniques, such as the ability to escape from false minima, to incorporate prior knowledge of the material into the analysis, and to provide quantifiable uncertainty through credible intervals. The application of both methods was shown for single peak fitting and full diffraction pattern fitting, and generally revealed that a better fit is obtained when using Bayesian inference. These results show that this new method is an attractive alternative to the classical least-squares methods applied to crystal structure determination and provides a richer description of the models and uncertainties than that previously available.

Acknowledgements The authors acknowledge the support from the National Science Foundation under awards DMR-1409399 and DGE-1633587.

References

1. A.R. West, *Basic Solid State Chemistry*, 2nd edn. (Wiley, West Sussex, England, 1999)
2. R.C. Smith, *Uncertainty Quantification: Theory, Implementation, and Applications* (Society for Industrial and Applied Mathematics, Philadelphia, 2014)
3. C.M. Fancher, Z. Han, I. Levin, K. Page, B.J. Reich, R.C. Smith, A.G. Wilson, J.L. Jones, Sci. Rep. **6**, 31625 (2016)
4. P.S. Bandyopadhyay, M.R. Forster, *Philosophy of Statistics*, vol. 7 (Elsevier B.V., Oxford, 2011)
5. J.C. Nino, W. Qiu, J.L. Jones, Thin Solid Films **517**, 4325 (2009)
6. H.M. Rietveld, J. Appl. Crystallogr. **2**, 65 (1969)
7. H.M. Rietveld, Acta Crystallogr. **22**, 151 (1967)
8. H.M. Rietveld, Zeitschrift Für Krist. **225**, 545 (2010)
9. B.H. Toby, R.B. Von Dreele, J. Appl. Crystallogr. **46**, 544 (2013)
10. M.J. Cooper, Acta Crystallogr. Sect. A **38**, 264 (1982)
11. M. Sakata, M.J. Cooper, J. Appl. Crystallogr. **12**, 554 (1979)
12. H.G. Scott, J. Appl. Crystallogr. **16**, 159 (1983)
13. P. Tian, S.J.L. Billinge, Zeitschrift Fur Krist. **226**, 898 (2011)
14. R.A. Young, *The Rietveld Method* (Oxford University Press, 1995)
15. G. Will, *Powder Diffraction: The Rietveld Method and the Two Stage Method to Determine and Refine Crystal Structures from Powder Diffraction Data* (Springer, Berlin/Heidelberg, 2006)
16. E. Prince, J. Appl. Crystallogr. **14**, 157 (1981)
17. B.S. Everitt, *The Cambridge Dictionary of Statistics*, 2nd edn. (Cambridge University Press, Cambridge, 2002)
18. R. Hoekstra, R.D. Morey, J.N. Rouder, E.-J. Wagenmakers, Psychon. Bull. Rev. **21**, 1157 (2014)
19. A. Gelman, J.B. Carlin, H.S. Stern, D.B. Dunson, A. Vehtari, D.B. Rubin, *Bayesian Data Analysis*, 3rd edn. (CRC Press, 2014)
20. T. Iamsasri, J. Guerrier, G. Esteves, C.M. Fancher, A.G. Wilson, R.C. Smith, E.A. Paisley, R. Johnson-Wilke, J.F. Ihlefeld, N. Bassiri-Gharb, J.L. Jones, J. Appl. Crystallogr. **50**, 211 (2017)
21. C. Yuan, M.J. Druzdzel, Math. Comput. Model. **43**, 1189 (2006)
22. S. Chib, E. Greenberg, Am. Stat. **49**, 327 (1995)
23. Data-Enabled Science and Engineering of Atomic Structure at North Carolina State University. https://youtu.be/S_ItC4ytT60 (2016)
24. S. French, Acta Crystallogr. Sect. A **34**, 728 (1978)
25. C.R. Hogg III, K. Mullen, I. Levin, J. Appl. Crystallogr. **45**, 471 (2012)
26. N. Armstrong, W. Kalceff, J.P. Cline, J.E. Bonevich, J. Res. Natl. Inst. Stand. Technol. **109**, 155 (2004)
27. C.J. Gilmore, Acta Crystallogr. Sect. A: Found. Crystallogr. **52**, 561 (1996)
28. G.P. Bourenkov, A.N. Popov, H.D. Bartunik, Acta Crystallogr. Sect. A: Found. Crystallogr. **52**, 797 (1996)
29. J. Bergmann, T. Monecke, J. Appl. Crystallogr. **44**, 13 (2011)
30. W.I.F. David, D.S. Sivia, J. Appl. Crystallogr. **34**, 318 (2001)
31. A. Mikhalychev, A. Ulyanenkov, J. Appl. Crystallogr. **50**, 776 (2017)

32. J. Clérouin, N. Desbiens, V. Dubois, P. Arnault, Phys. Rev. E **94**, 61202 (2016)
33. A. Altomare, R. Caliandro, M. Camalli, C. Cuocci, I. Da Silva, C. Giacovazzo, A.G. Giuseppina Moliterni, R. Spagna, J. Appl. Crystallogr. **37**, 957 (2004)
34. M. Wiessner, P. Angerer, J. Appl. Crystallogr. **47**, 1819 (2014)
35. A. Gagin, I. Levin, J. Appl. Crystallogr. **48**, 1201 (2015)
36. M. Wallace, R.L. Johnson-Wilke, G. Esteves, C.M. Fancher, R.H.T. Wilke, J.L. Jones, S. Trolier-McKinstry, J. Appl. Phys. **117**, 54103 (2015)
37. J.L. Jones, E.B. Slamovich, K.J. Bowman, J. Appl. Phys. **97**, 34113 (2005)
38. G. Esteves, C.M. Fancher, J.L. Jones, J. Mater. Res. **30**, 340 (2015)
39. G. Tutuncu, D. Damjanovic, J. Chen, J.L. Jones, Phys. Rev. Lett. **108**, 177601 (2012)
40. D.A. Hall, A. Steuwer, B. Cherdhirunkorn, T. Mori, P.J. Withers, J. Appl. Phys. **96**, 4245 (2004)
41. V. Anbusathaiah, D. Kan, F.C. Kartawidjaja, R. Mahjoub, M.A. Arredondo, S. Wicks, I. Takeuchi, J. Wang, V. Nagarajan, Adv. Mater. **21**, 3497 (2009)
42. P. Muralt, R.G. Polcawich, S. Trolier-McKinstry, MRS Bull. **34**, 658 (2009)
43. D.R. Black, D. Windover, A. Henins, D. Gil, J. Filliben, J.P. Cline, Powder Diffr. **25**, 187 (2010)
44. D. Balzar, N. Audebrand, M.R. Daymond, A. Fitch, A. Hewat, J.I. Langford, A. Le Bail, D. Louër, O. Masson, C.N. McCowan, N.C. Popa, P.W. Stephens, B.H. Toby, J. Appl. Crystallogr. **37**, 911 (2004)
45. J.I. Langford, D. Louër, P. Scardi, J. Appl. Crystallogr. **33**, 964 (2000)
46. J.E. Lesniewski, S.M. Disseler, D.J. Quintana, P.A. Kienzle, W.D. Ratcliff, J. Appl. Crystallogr. **49**, 2201 (2016)
47. G. Esteves, K. Ramos, C.M. Fancher, J.L. Jones. https://github.com/SneakySnail/LIPRAS (2017)

Chapter 5
Deep Data Analytics in Structural and Functional Imaging of Nanoscale Materials

Maxim Ziatdinov, Artem Maksov and Sergei V. Kalinin

Abstract Recent advances in scanning probe microscopy and scanning transmission electron microscopy have opened unprecedented opportunities in probing the materials structural parameters and electronic properties in real space on a picometre-scale. At the same time, the ability of modern day microscopes to quickly produce large, high-resolution datasets has created a challenge for rapid physics-guided analysis of data that typically contain several hundreds to several thousand atomic or molecular units per image. Here it is demonstrated how the advanced statistical analysis and machine learning techniques can be used for extracting relevant physical and chemical information from microscope data on multiple functional materials. Specifically, the following three case studies are discussed (i) application of a combination of convolutional neural network and Markov model for analyzing positional and orientational order in molecular self-assembly; (ii) a combination of sliding window fast Fourier transform, Pearson correlation matrix and canonical correlation analysis methods to study the relationships between lattice distortions and electron scattering patterns in graphene; (iii) application of a non-negative matrix factorization with physics-based constraints and Moran's analysis of spatial associations to extracting electronic responses linked to different types of structural domains from multi-modal imaging datasets on iron-based superconductors. The approaches demonstrated here are universal in nature and can be applied to a variety of microscopic measurements on different materials.

M. Ziatdinov (✉) · A. Maksov · S. V. Kalinin
Oak Ridge National Laboratory, Institute for Functional Imaging of Materials, Oak Ridge, TN 37831, USA
e-mail: ziatdinovma@ornl.gov

M. Ziatdinov · A. Maksov · S. V. Kalinin
Oak Ridge National Laboratory, Center for Nanophase Materials Sciences, Oak Ridge, TN 37831, USA

A. Maksov
Bredesen Center for Interdisciplinary Research, University of Tennessee, Knoxville, TN 37996, USA

T. Lookman et al. (eds.), *Materials Discovery and Design*, Springer Series in Materials Science 280, https://doi.org/10.1007/978-3-319-99465-9_5

5.1 Introduction

According to the established paradigm of structure-property relationship, there is a direct link between materials atomic structure and their optical, mechanical, electronic, and magnetic functionalities [1, 2]. This allows scenarios in which relatively small changes in the material structural and chemical compositions may have a decisive impact on the physical properties of the system. Examples include ultra-high piezoelectric response of relaxor ferroelectrics due to interaction between nanopolar domains and acouwestic phonon mode [3], filamentary superconductivity associated with nonuniform distribution of Pr dopants in iron arsenides [4], high critical-current density due to clustering of oxygen vacancies in cuprates [5–7], reduced mobility of Dirac electrons in graphene transistor devices due to formation of charge nano-puddles [8, 9], fluctuating superconducting state above a transition temperature (Tc) in high-Tc cuprates associated with emergence of nanometre-sized electron pairing regions [10], and emergence of glassy mixed-phases state in manganites linked to a quenched chemical disorder [11].

The advances in scanning transmission electron and scanning probe microscopies (STEM and SPM) have opened an unprecedented path towards simultaneously probing the material structural parameters (e.g. bond lengths) and its functional properties (e.g. electronic polarization or superconducting gap) in real space with a nanometer precision, making them the perfect tools for studying nanoscale inhomogeneities and their role in bulk crystalline behavior [12, 13]. Examples in SPM include direct imaging of chemical bonds in molecules [14], visualizing atomic collapse in artificial nuclei on graphene [15], and inferring mechanisms behind fundamental physical phenomena, such as high-Tc superconductivity, from single atom defect induced scattering patterns [6]. Meanwhile, STEM experiments can produce picometer-resolved images of ferroelectric polarization [16, 17], octahedral tilts [18], and chemical expansion strains [19]. Furthermore, combination of STEM and SPM with different spectroscopic techniques, such as optical and Raman spectroscopy, electron energy loss spectroscopy and mass spectroscopy have led to a rise of new multi-modal imaging capabilities that now allow a simultaneous capturing of materials structural, electronic, chemical, and optical properties at the nano and meso-scales. Such experimental capabilities allow, in principle, constructing combinatorial libraries of lattice configurations and functionalities at the single-defect level. This, however, requires first a development of methods for extracting all the experimentally accessible (spatially-dependent) information on structure and function variables and for cross-correlating the information from different "channels" in physically-meaningful and statistically-meaningful ways.

We illustrate several frameworks based on machine learning and multivariate analysis that allow automated and highly accurate extraction and mapping of different structural and functional descriptors from experimental datasets as well as studying their local correlations. The approach for a two-channel microscopic imaging experiment is schematically outlined in Fig. 5.1. It starts with recording 'structure' and 'function' information over the same sample area via two different acquisition

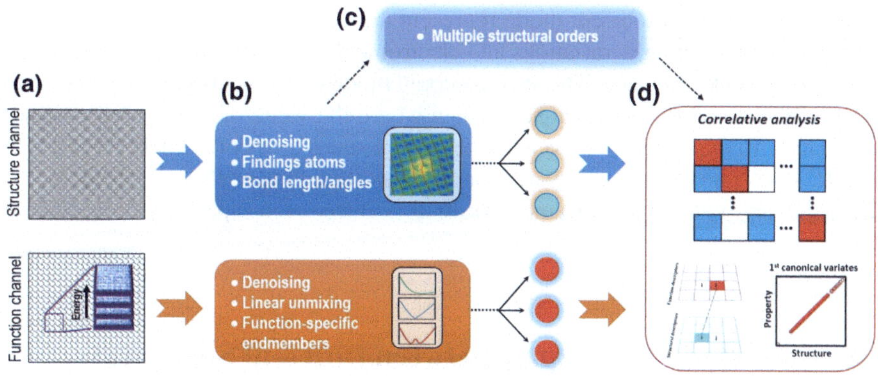

Fig. 5.1 Schematic workflow for structure-property relationships analysis. **a** 2-channel ('structure' and 'function') data acquisition. **b** Processing data from both channels to extract relevant structure and function descriptors. **d** Mining the combinatorial library of lattice configurations and functionalities. For systems with multiple structural orders one can apply correlative analysis 'toolbox' directly to the processed structural data (**c**–**d**)

channels (Fig. 5.1a). In this case, the first channel corresponds to 2D images in which Z is a 'structural' variable used to calculate lattice parameters, such as inter-atomic (or atomic columns) distances and apparent heights. The second channel represents 3D dataset in which G is a 'function' variable, for example, differential conductance or electron energy loss. After performing an image alignment such that, the data from both channels is cleaned from spurious noise features and outliers in a way that minimizes the information loss (e.g., using principal component analysis). The next step is constructing structural and functional descriptors. For structure channel, one may adapt various pattern recognition techniques from a field of computer vision, such as sliding window Fast Fourier Transform, deep neural networks and Markov random field. For function channel, blind source un-mixing/decomposition methods such as Bayesian linear unmixing and non-negative matrix factorization performed on hyperspectral "functional" data can generally provide a physically meaningful separation of spectral information when multiple 'phases' are present in the dataset (Fig. 5.1b, c). Once completed, one proceeds to performing direct data mining of structure-property relationships from correlative analysis of the derived structural and functional descriptors (Fig. 5.1d). The correlation analysis 'toolbox' typically includes methods such as Pearson correlation matrix, global and local Moran's correlative analysis, and linear and kernel canonical correlation analysis. Note well that for systems with multiple order parameters and/or systems where both structural and electronic information can be effectively extracted from a single image, the correlation analysis can be performed directly on variables extracted from the structure channel.

In the following, we analyze structure-property relationship on different molecular and solid state systems using data obtained from constant-current mode and spectroscopic mode of scanning tunneling microscope [20]. The STM topographic images obtained in a constant-current mode represent a 2-dimensional dataset where $Z(\mathbf{R})$ is a convolution of height variations and electronic density of states in each $\boldsymbol{R}(X, Y)$ point (pixel) on the surface. The spectroscopic mode of STM (usually referred to as STS) produces a 3-dimensional set of data where the value of differential conductance $G(\boldsymbol{R}', V)$ is proportional to local density of states at specific energy $E = eV$ at each $\boldsymbol{R}(X', Y')$ point on the surface. For all cases studied here $\boldsymbol{R}(X', Y') = \boldsymbol{R}(X, Y)$. The necessary mathematical frameworks will be introduced separately for each case study.

5.2 Case Study 1. Interplay Between Different Structural Order Parameters in Molecular Self-assembly

5.2.1 Model System and Problem Overview

To demonstrate an application of advanced data science tools to molecular resolved STM images, a self-assembly of $C_{21}H_{12}$ molecules [21, 22] is chosen as a model system (Fig. 5.2a). Each individual molecular unit in the self-assembly can be viewed as a fragment bowl of buckminsterfullerene (hereafter, buckybowl). A buckybowl in the self-assembly can reside in two different structural conformations (bowl-up and bowl-down) as well as in multiple lateral orientations with respect to the substrate. In the absence of external perturbation and/or substrate disorder the molecular monolayer forms a long-range superperiodic pattern, in which each bowl-down state is surrounded by six bowl-up states. In the following, this superstructure is referred to as **2U1D**, where **U** and **D** stand for bowl-up and bowl-down states, respectively. At the low tip-sample separation distances in the constant current STM experiment (typically achieved at sample bias $U_s \lesssim 0.1$ V) it is usually possible to induce a switching between different molecular degrees of freedom via mechanochemistry effects, whereas at large separation distances (at $U_s \gtrsim 1$ V) the switching events, particularly those involving structural changes, are minimized [22]. Thus one can interpret the scans at low and high bias voltages as "writing" (albeit randomly) and "reading" molecular patterns, respectively. The representative STM image of buckybowl self-assembly is shown in Fig. 5.2c. The STM data used as an input in the current analysis was acquired in the reading regime; prior to acquisition of the image of interest, several STM scans were performed over the same area at the lower tip-surface distances (switching regime) producing additional "excitations", that is, enhancing a disorder, in the initial molecular structure. A global 2-dimensional Fast Fourier Transform (2D FFT) obtained from image in Fig. 5.2c shows a strong suppression of peaks associated with **2U1D** structure (compared to peaks in the outer hexagon associated with positional order in molecular lattice) indicating a presence of disorder in the molecular film. In the following, an approach based on a synergy of ab-initio simulations, Markov random field model and convolutional neural network

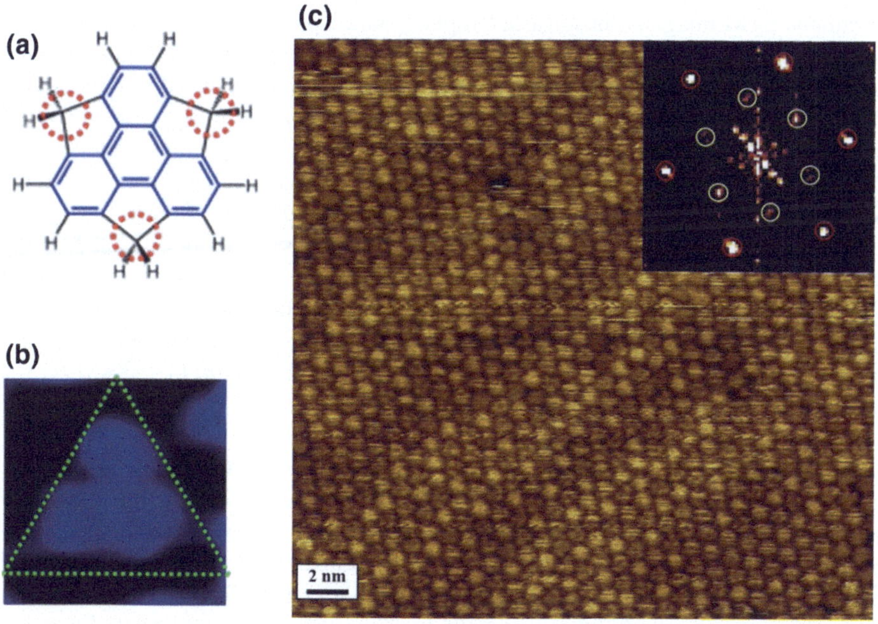

Fig. 5.2 Self-assembly of sumanene molecules (buckybowls) on gold substrate. **a** Chemical structure of sumanene. **b** Experimental STM image of individual buckybowl. Adapted with permission from [22]. Copyright 2018 American Chemical Society.**c** Large-scale STM image over field of view with approximately 1000 molecules. The inset shows FFT transform of data in (**c**). The yellow circles denote FFT spots associated with a formation of **2U1D** superlattice. Adapted from [23]

is introduced for "reading out" complex molecular patterns of buckybowls on gold substrate from molecule-resolved STM images [23].

5.2.2 How to Find Positions of All Molecules in the Image?

The first crucial step in analyzing the STM data on complex surface molecular structures is the identification and extraction of positions of all molecules for each image. Simple visual examination of STM image in Fig. 5.2c suggests that it contains up to about 1000 individual molecules. The normalized cross-correlation is performed to obtain correlation surfaces defined as

$$\gamma(\mathrm{u},\mathrm{v}) = \frac{\sum_{x,y}[f(x,y)-\overline{f}_{u,v}][t(x-u,y-v)-\overline{t}]}{\{\sum_{x,y}[f(x,y)-\overline{f}_{u,v}]^2\sum_{x,y}[t(x-u,y-v)-\overline{t}]^2\}^{0.5}} \tag{5.1}$$

where f is the original image, t is the template, $\overline{f}_{u,v}$ is the mean of $f(x,y)$ in the region under the template, $\overline{t}$ is the mean of the template. The bowl-up DFT-simulated STM image is chosen as a template, which produced the highest accuracy in determination the positions of molecular centers. The uniform threshold is applied to the generated correlation surface γ, with cutoff set to 0.35, in order to maximize the number of extracted molecules. This results in a binary image, for which the connected components are identified and their centers are assigned as centers of the corresponding molecules. The apparent height I_m of each molecule, which represents a convolution of an actual geometric height and local density of electronic states, is calculated as $I_m = \sum_{x=1}^{15} \sum_{y=1}^{15} i_{x,y}$ where $i_{x,y}$ is the intensity of pixel at position x, y in the extracted image patch for molecule m. The summation is performed for 15×15 pixel patches around the center of each molecule. To remove outliers due to possible contaminations on a surface which may not directly associate with molecules, a maximum intensity value defined as $I_{max} = mean(I) + 3 * std(I)$ is introduced such that all intensities that exceed the maximum value are scaled back set to I_{max}.

Once all positions and intensities are identified a principal component analysis is performed on the stack of images of individual molecules. The aim of the principal component analysis (PCA) can be interpreted as finding a lower dimensional representation of data with a minimum loss of important (relevant) information [24]. Specifically, in PCA one performs an orthogonal linear transformation that maps the data into a new coordinate system such that the greatest variance comes to lie on the first coordinate called the first principal component, the second greatest variance on the second coordinate, and so forth. Hence, the most relevant information (including information on the orientation/rotation of molecules) can be represented by a small number of principal components with the largest variance, whereas the rest of the (low-variance) components correspond to 'noise'. The PCA analysis suggests that suggests that a likely number of rotational classes needed to be considered for this dataset is four.

5.2.3 Identifying Molecular Structural Degrees of Freedom via Computer Vision

Convolutional neural networks. The identification of molecular "shapes" (different orientation with respect to substrate) is performed using a technique from a field of computer vision known as convolutional neural networks. Convolutional neural networks (cNN) represent one of the key examples of a successful application of neuroscientific principles to the field of machine leaning. The cNNs are used for processing data which is characterized by a known, grid-like topology such as 2-dimensional grid of pixels obtained in the STM constant current experiments [25]. The architecture of the convolutional network used in the current work is shown in Fig. 5.3a and it includes convolutional layers, pooling layers, as well as a fully connected "dense" layer. The convolution layer is formed by running *learnable* kernels ('filters') of the selected size over the input image (or image in the previous layer).

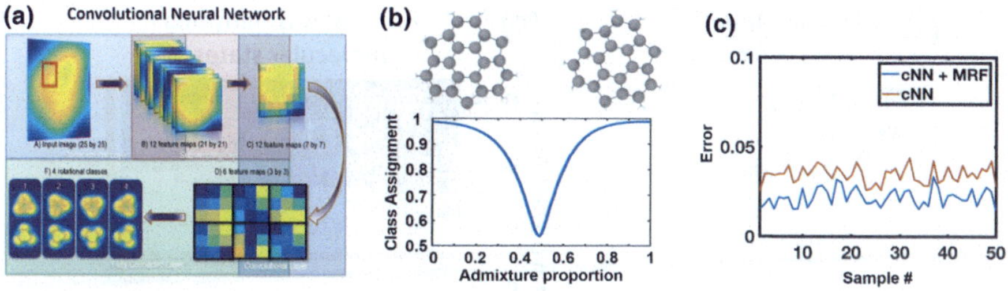

Fig. 5.3 Deep learning of molecular features. **a** Schematic graph of convolutional neural network (cNN) architecture for determining of molecular lateral degrees of freedom on the substrate. **b** Role of dynamical averaging (admixture of a different rotational class) in probability of the correct class assignment. **c** Error rate for cNN only and for cNN refined with Markov random field model. Adapted from [23]

The pooling layers produce downsampled versions of the input maps. The i-th feature map in layer l, denoted as V_i^l can be expressed as [26]

$$V_i^l = \sum_{i \in M_i} V_j^{(l-1)} * K_{i,j}^l + B_i^l \tag{5.2}$$

Here K is a kernel connecting the i-th feature map in layer l and the j-th feature map layer $(l - 1)$, B_i^l describes the bias, and M_i corresponds to a selection of input maps. The output Z_i^l is a fully connected ("dense") layer that takes as input the "flattened" feature maps of the layer below it:

$$Z_i^l = \sum_{i \in M_i} \sum_{m \in M_i} \sum_{n \in M_i} (V_j^{(l-1)})_{m,n} W_{i,j,m,n}^l \tag{5.3}$$

where $W_{i,j,m,n}^l$ connects i-th unit at position m, n in the feature map of layer $(l - 1)$ to the j-th unit in layer l. The cNN is trained on a set of synthetic STM images (25,000 samples) obtained from DFT simulations of different rotational classes.

Markov random field. The unique aspect of the present approach is that the cNN is followed by Markov random field model [27] which takes into account probabilities of neighboring molecules to be in the same lateral orientation on the substrate. This allows us to "refine" the results learned by neural network in a fashion that takes into account physics of the problem. The MRF model makes use of an undirected graph $G = (V, E)$, in which the nodes V are associated with random variables $(X_v)_{v \in V}$, and E is a set of edges joining pairs of nodes. The underlying assumption of Markov property is that each random variable depends on other random variables only through its neighbors:

$$X_v \perp X_{V \setminus v \cup N(v)} | X_{N(v)}, \tag{5.4}$$

for $N(v)$ = neighbors of v. Importantly, the explicit Markov structure implicitly carries longer-range dependencies. These priors are directly linked to the underlying

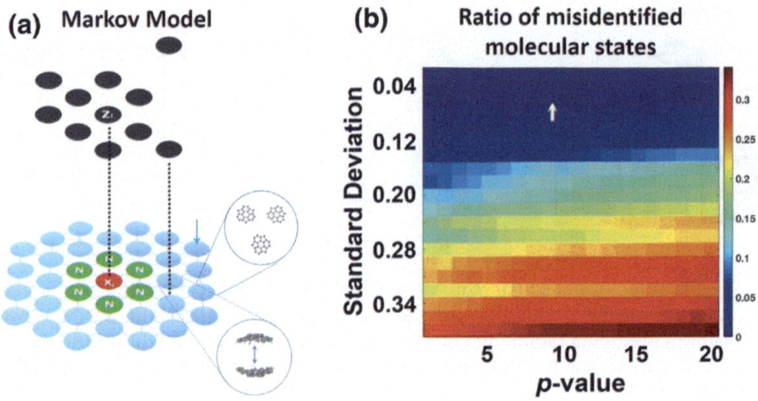

Fig. 5.4 Molecular self-assembly as Markov random field model (MRF). **a** Graphical Markov model structure used for analysis of a molecular self-assembly. **b** Error rate as a function of standard deviation of normalized STM intensity distributions and an optimization parameter (p-value). The arrow shows the value of these parameters for the analysis of the synthetic data. Adapted from [23]

physics of the system, that is, the presence of short-range interactions in molecular assembly which are now explicitly taken into account during image analysis. The experimental STM data on buckybowls is mapped on to a graph such that each molecule is represented as a node, and edges are connections to each molecule's nearest neighbors (Fig. 5.4a). The posterior distribution of an MRF can be factorized over individual molecules such that

$$P(x|z) = \frac{1}{Z} \prod_{<ij>} \Psi_{ij}(x_i, x_j) \prod_{i} \Psi_i(x_i, z_i) \tag{5.5}$$

where Z is the partition function, and $\Psi_i(x_i, z_i)$ and $\Psi_{ij}(x_i, x_j)$ are unary and pairwise potentials, respectively. These potentials are defined based on the knowledge about physical and chemical processes in the molecular system, such as a subtle interplay between a difference in adsorption energy for **U** and **D** molecules, molecular interactions different molecular configurations, and imperfection of the substrate. Finding an exact solution to MRF model is intractable in such a case as it would require examining all 2^n combinations of state assignments, where n is the number of molecules, that is, about 1000 for examined images. However, one can obtain a close approximate solution by using a max-product loopy belief propagation method [28], which is a message-passing algorithm for performing inference on MRF graphs, with unary and pairwise potentials as an input. Briefly, from initial configuration, nodes propagate message containing their beliefs about state of the neighboring nodes given all other neighboring nodes messages. This results in an iterative algorithm. All messages start at 1, and are further updated as max-product of potentials and incoming messages:

$$msg(x_j)_{i\to j} = max_l[\sum_{x_i} \Psi_{ij}(x_i, x_j)\Psi_i(x_i, z_i) * \prod_{k=neighbors\ of\ i\neq j} msg(X)_{k\to i}] \tag{5.6}$$

At each iteration belief is calculated for each node and the state with highest belief is selected, until message update converges:

$$Belief(x_i) = \Psi_i(x_i, z_i) * \prod_{j=neighbors\ of\ i} msg(x_i)_{j\to i} \tag{5.7}$$

According to theoretical modeling, it is unlikely that two neighboring molecules can have the same rotational state [29]. Therefore assign probability of each class to have a neighbor of its own class is considered to be 1% and probabilities to have a neighbor of other 3 rotational classes is considered to be 33%. Finally, the decoding using loopy belief propagation is performed in order to acquire a more precise solution. Note well that by tuning a graph structure and/or form of the potentials one can easily apply Markov random field approach to other molecular order parameters or even different molecular architectures. Indeed, one can also apply MRF to decoding different conformational states of molecules (note that an application of the cNN to a problem of determining different conformational states typically returns relatively poor results). For MRF modelling of bowl-up and bowl-down states, the unary potentials $\Psi_i(x_i, z_i)$ over molecular states are assigned based on the proximity of a particular molecule's intensity in the STM image to the threshold value between the states T. The node probabilities are calculated as two logistic functions:

$$\Psi_i(x_i = 1, z_i = I_i) = \frac{1}{1 + Exp[S * (T - I_i)]} \tag{5.8a}$$

$$\Psi_i(x_i = 2, z_i = I_i) = 1 - \Psi_i(x_i = 1, z_i = I_i) \tag{5.8b}$$

where $I_i \in [0, 1]$ is the intensity of a given molecule i, and S is a parameter that controls the growth rate of the logistic function. The logistic functions allow us to assign molecular intensities sufficiently far from the threshold as belonging to their corresponding class with probability of $\sim$1, while also providing more flexibility in the region around the threshold value itself. Next, the pairwise potentials $\Psi_{ij}(x_i, x_j)$ for the molecular system are determined. The optimal **2U1D** configuration proposed above is characterized by six **U** molecules surrounding one **D** molecule, such that **D** molecule is never allowed to have the nearest neighbor in the same bowl conformation. As we are interested in the distortion of an ideal **2U1D** structure (six bowl-up molecules surrounding one bowl-down molecule), a disorder parameter p is introduced such that a probability of **D** and **U** molecules having their neighbor in the same conformational state becomes p and $1 - p$, respectively.

Testing on synthetic data. Prior to analyzing real experimental data, a validity of the described approach is tested on synthetic dataset(s). Specifically, the DFT-based calculations of the STM signal associated with an individual molecule for each configuration are combined with Markov Chain Monte Carlo sampler to generate synthetic images of molecular self-assembly containing a large number ($\succcurlyeq$1000) of molecules.

Additionally, the synthesized data is "distorted" by addition of blurring associated with a convolution with the STM tip probe function, Poisson noise associated with tunnelling statistics, and dynamical averaging due to potential admixture of another azimuthal rotational state to a given structural configuration. Since the exact distribution of molecular states in synthetic data is known for each sample, one can evaluate an error rate for this method. It was found that the proposed approach results in a remarkably accurate identification of different molecular conformational and rotational states in scenarios where the distribution of the STM intensities in the synthetic data closely resembles the typical experimental data. The MRF approach allowed to identify accurately distributions of bowl-up and bowl-down configurations in the large scale synthetic STM images, even when no estimations regarding the p-value is available apriori (Fig. 5.4b), while its addition to cNN helped to improve the decoding results by reducing number of misclassified states (Fig. 5.3c). It was also found that the cNN framework allows to obtain a reliable classification of molecules rotational states even in the presence of relatively strong dynamical averaging between proximate rotational states of the molecule (Fig. 5.3b) which is relatively common in the STM experiments [30, 31].

5.2.4 *Application to Real Experimental Data: From Imaging to Physics and Chemistry*

Having confirmed that the introduced approach works on synthetic data we proceed to analysis of real experimental data. The results if full decoding of rotational (via cNN+MRF) and conformational (via MRF) states are presented in Fig. 5.5. Once a full decoding is performed, it becomes possible to explore a nature of disorder in the molecular self-assembly by searching for *local* correlations between different molecular degrees of freedom. Of the specific interest is a potential interplay between molecule bowl inversion and azimuthal rotation of the neighboring molecules. To obtain such an insight, method based on calculation the so-called Moran's I is adopted that can measure a spatial association between the distributions of two variables at nearby locations on the *lattice* [32]. The 'correlation coefficient' for global Moran's I is given by

$$I = \frac{N}{\sum_i \sum_j w_{ij}} \frac{\sum_i \sum_j w_{ij}(X_i - \overline{X})(Y_j - \overline{Y})}{\sum_i (Y_i - \overline{Y})^2} \tag{5.9}$$

where N is the number of spatial units, X and Y are variables, $\overline{X}$ and $\overline{Y}$ are corresponding means, and w is the weight matrix defining neighbor interactions. It is worth noting that the presence of the spatial weight matrix in the definition of Moran's I allows us to impose constrains on the number of neighbors to be considered. For highly inhomogeneous system, one may use the so-called local indicators of spatial association which can evaluate the correlation between two orders at the neighboring points on the lattice for each individual coordination sphere. This is achieved through

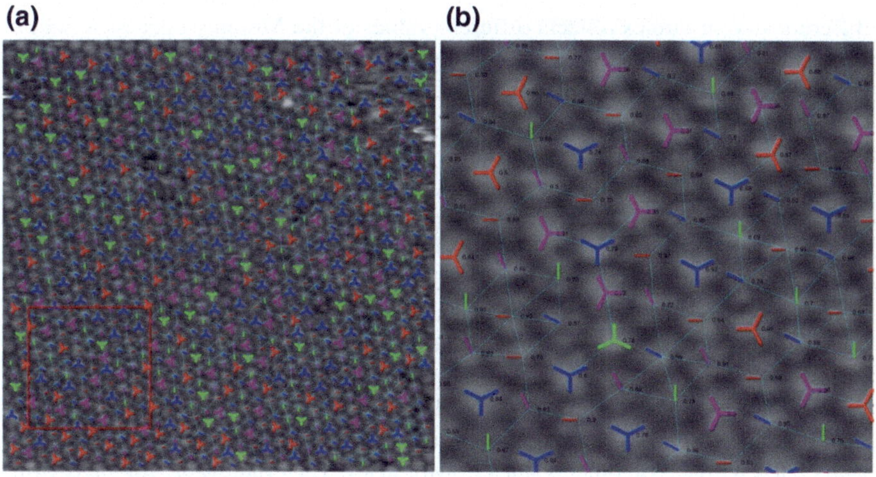

Fig. 5.5 Application of the current method to experimental data of buckybowls on gold (111). Decoding of rotational states (cNN+MRF) and bowl-up/down states (MRF, $p = 7$) for the experimental image from Fig. 5.1c. **b** Zoomed-in area from *red rectangle* in **a** where numbers denote an accuracy of state determination. Adapted from [23]

calculating *local* bivariate Moran's I for each spatial unit such as

$$I_{xy} = \frac{\sum_i \sum_{j \neq i} w_{ij} x_i y_i}{W} \tag{5.10}$$

where x and y are standardized to zero mean and variance of 1.

The results for spatial correlation between bowl-up/down configuration and different rotational classes for the first 'coordination sphere' is shown in Fig. 5.6a where

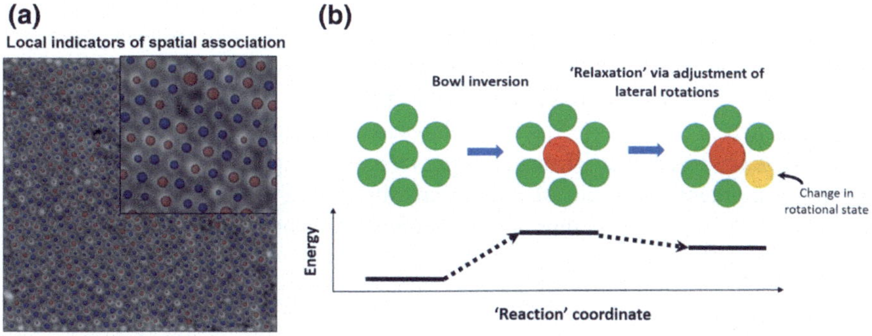

Fig. 5.6 From imaging to physics. **a** Local indicators of spatial associations based on the Moran's I calculated for the first coordination "sphere". **b** Proposed reaction mechanism involving change in molecular rotational state(s) after bowl inversion. Adapted from [23]

a different size of circles reflects different values of the Moran's I across a field of view. Generally, the map in Fig. 5.6a implies a spatial variation in coupling between the two associated order parameters, which could also be sensitive to presence of defects. The average value of Moran's I for the first 'coordination sphere' is 0.310, whereas the average value for correlation of rotational classes with bowl-up and bowl-down molecular conformations are 0.246 and 0.426 respectively. This result can be interpreted as that a bowl-up-to-bowl-down inversion of a molecule that creates an 'additional' molecule in the **D** state requires a larger change in a rotational state of the neighboring molecules in order to compensate for a formation of energetically unfavorable, "extra" bowl-down state (as compared to a reversed, bowl-down-to-bowl-up inversion). Based on these findings, it is possible to propose a two-stage "reaction" mechanism, where in the first stage an excitation of a new bowl-down state elevates the energy of the system, which is then relaxed in the second stage of the proposed reaction through adjustment of rotational states of the nearby molecule(s). The latter is associated with the obtained values of Moran's I. The crude value for energy difference between different rotational states induced by bowl inversion, and calculated by estimating Boltzmann factor directly from the ratio of two different correlation values, is $\approx$0.015 eV.

Unlike previous studies which only considered a bowl inversion process for an *isolated* single molecule, the presented analysis based on synergy of convolutional neural networks, Markov random field model and ab-initio simulations allowed to obtain a deeper knowledge of local interactions that accompany a switching of conformational state of neighboring molecules in the self-assembled layer. This new advanced understanding of local degrees of freedom in the molecular adlayer could lead to a controllable formation of various molecular architectures on surfaces which in turn could result in a realization of multi-level information storage molecular device or systems for molecular level mechanical transduction. As far as future directions of applying machine learning and pattern recognition towards molecular structures are concerned, it should be noted that the physical priors used for input in cNN and MRF could be also in principle extracted from state-of-the-art ab-initio analysis and molecular dynamics (MD) simulations. This could potentially provide more accurate decoding results. In addition, a choice of the optimization parameter in MRF analysis could be optimized in future using a statistical distance approach [33]. Finally, we envision an adaption of deep learning technique called domain-adversarial neural networks [35] which allows to alter theoretically predicted classes based on the observed data. The underlying idea of this approach is that the theoretical and experimental datasets are similar yet different in such a way that traditional neural networks may not capture correct features just from the labeled data.

5.3 Case Study 2. Role of Lattice Strain in Formation of Electron Scattering Patterns in Graphene

5.3.1 *Model System and Problem Overview*

Graphene, a two-dimensional honeycomb lattice of sp^2-carbon atoms, has attracted enormous research interest mostly due to its unique electronic properties, such as anomalous quantum Hall effect and Klein tunneling, which are a consequence of massless Dirac fermions with linear energy dispersion in the electronic band structure. Presence of a disorder in graphene lattice, such as substitutional dopants, vacancies and adatoms, as well as nanoscale variations in bond lengths (due to in-plane and out-of-plane surface deformations), can have a major impact on the material electronic (and magnetic) structure. Below we describe the study on a relationship between nanoscale modulations of lattice strain and parameters of electron scattering induced by point defects in graphene [34]. This study was performed by applying a combination of sliding window fast Fourier transform, Pearson correlation matrix and canonical correlation analysis to low-bias atomically-resolved scanning probe microscopy images of graphene.

Two graphenic systems on different substrates with different types of defects were chosen. The first system is a topmost graphene layer of graphite peppered with hydrogen-passivated single atomic vacancies (hereafter denoted as G_H) [36]. The second system is a monolayer graphene of reduced graphene oxide on gold (111) substrate (hereafter G_O) covered with oxygen-passivated atomic defects and oxygen functional groups [37]. The representative scanning probe microscopy images for G_H and G_O samples are shown in Fig. 5.7a and b, respectively. Both images were obtained in a low-bias regime ($U_s \leq 0.1$ V) where the current is proportional to the density of states at the Fermi level. The global 2D FFTs for data in Fig. 5.7 a, b shows (see insets) similar reciprocal space patterns for both systems characterized by the two hexagons rotated by 30° with respect to each other, with their lattice constants differ by a factor of $\approx \sqrt{3}$. The outer and inner hexagon is associated with lattice structure and electronic density of states, respectively. Specifically, a formation of the inner hexagon in undistorted graphene is explained as due to the constructive interference between incident and backscattered states from the electron valleys at opposite corner points of the hexagonal Brillouin zone [38, 39]. Owing to the symmetry of graphene lattice, there are three backscattering channels. For point defect that do not preserve the symmetry of graphene lattice as well as in graphene with distorted lattice the scattering probability may be different for each of the three channels. Indeed, it is possible to observe experimentally (in a real space) a fine structure of the electronic superlattice around the defects characterized by the alternation of intensities of the FFT spots in the inner hexagon (see Fig. 5.7a, d and e). The precise origin of such a modulation in graphene electronic superlattice is not yet well understood.

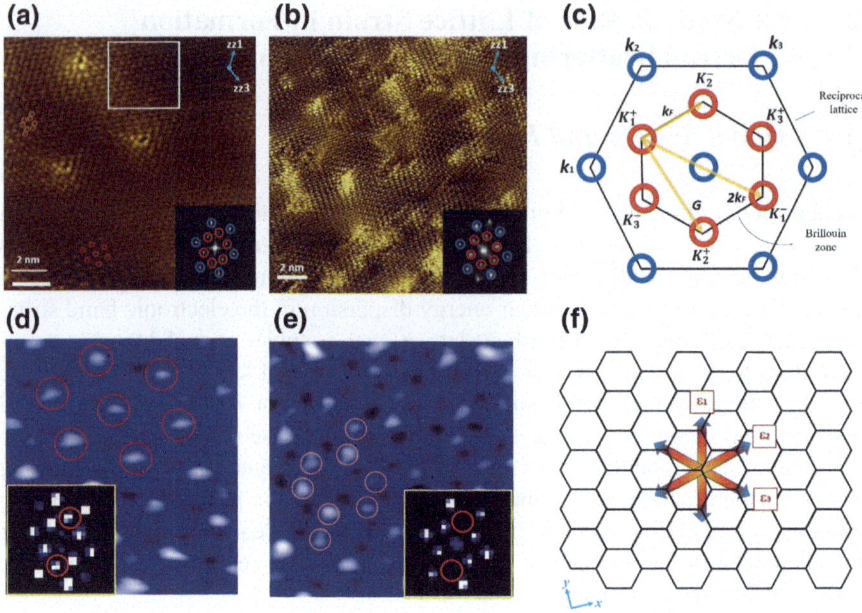

Fig. 5.7 Imaging lattice and electronic structure in graphenic samples. **a** STM image of the top graphene layer of graphite with hydrogen-passivated monoatomic vacancy. $U_s = 100$ mV, $I_{setpoint} = 0.9$ nA. The sliding window used for our analysis is overlaid with the image. **b** Low-bias (2 mV) current-mapping c-AFM image of reduced graphene oxide on gold (111) substrate. The 2D FFT data for both images is shown in the insets. **c** Schematics of graphene electron scattering in the reciprocal space. **d** Hexagonal superperiodic lattice and its 2D FFT. **e** Staggered-dimer-like electronic superlattice and it 2D FFT. Both superlattices are also marked in (**a**). **f** Schematic depiction of 3 different strain components in real space used in our analysis.

5.3.2 *How to Extract Structural and Electronic Degrees of Freedom Directly from an Image?*

Sliding FFT. The goal is to analyze a structure-property relationship in the two graphene systems by studying the correlation between local lattice distortions associated with k_l peaks and electronic features associated with K_e peaks (see Fig. 5.7c). First, a square window of size (w_x, w_y) is created and being shifted across the input image (T_x, T_y) in series of steps x_s and y_s such that the entire image is scanned. At each step, the 2D FFT is computed for the image portion that lies within the window [40]. Hanning window is used to minimize edge effects, as well as a 2 zoom combined and a 2× interpolation function for higher pixel density during the each step of this sliding FFT procedure. The amplitudes and coordinates of the selected peaks are extracted from each 2D FFT image by fitting them with 2D Gaussian distribution, defined as

$$G(q_x, q_y) = A\exp[-(\frac{(q_x - q_x^0)^2 + (q_y - q_y^0)^2}{2\sigma^2})] \tag{5.11}$$

Here A is the peak amplitude, (q_x, q_y) are the Cartesian coordinates of the peak position, and σ is the standard deviation. The unique aspect of graphene is that charge density oscillations are commensurate with the underlying atomic lattice. Therefore, the sliding FFT maps can be used to extract information on both electronic and structural properties of the material. Specifically, the values of intensity and coordinates associated with inner hexagon peaks provide information about intensity of electronic scattering and position of Dirac cone. For the outer hexagon, the coordinates of the peaks from local FFT maps give information about the nanoscale strain distribution in the sample.

The Dirac point drift and electron scattering intensities along the i-th channel are computed as ΔK, $\Delta K_i = (K_i - \overline{K}_i)/\overline{K}_i$ and $I_{K_i} = I\,(K_i^+ \rightarrow K_i^-)$, respectively. To derive a strain map, a strain ε_i is defined as a variation of the lattice vector a_i along the i-th direction, that is, $\varepsilon_i = (a_i - \overline{a}_i)/\overline{a}_i$, where $\overline{a}_i$ is the mean value of the lattice vector in the full image (Fig. 5.7f). It is assumed that for the randomly fluctuating strain fields the mean value of the lattice vector is close to the value of lattice constant in the unperturbed lattice. The a_i is calculated for each step of the sliding FFT algorithm using a standard relation between real space and reciprocal space lattices in graphene. The resolution of spatial maps of the derived structural and electronic descriptors is determined by the size of sliding FFT window and the size of step.

5.3.3 Direct Data Mining of Structure and Electronic Degrees of Freedom in Graphene

Pearson and canonical correlation analysis. Once all the structural and electronic variable of interest are extracted, it becomes possible to explore potential correlations between the corresponding descriptors. Specifically, Pearson correlation matrix analysis and canonical correlation analysis are adopted to explore how formation of various electron interference patterns can be affected by nanoscale variations in the lattice strain. The correlation parameter for each pair of variables x and y is defined as a linear Pearson correlation coefficient,

$$r_{xy} = \frac{\sum_{i=1}^{N}(x_i - \overline{x})(y_i - \overline{y})}{\sqrt{\sum_{i=1}^{N}(x_i - \overline{x})^2}\sqrt{\sum_{i=1}^{N}(y_i - \overline{y})}} \tag{5.12}$$

where $\overline{x}$ is the mean of x, $\overline{y}$ is the mean of y, and N is a number of scalar observations. While Pearson correlation matrix analysis is a useful technique for studies of bivariate correlations, it is useful to adopt a method called canonical correlation analysis (CCA) that allows grouping the variables in each multivariate dataset such that optimal

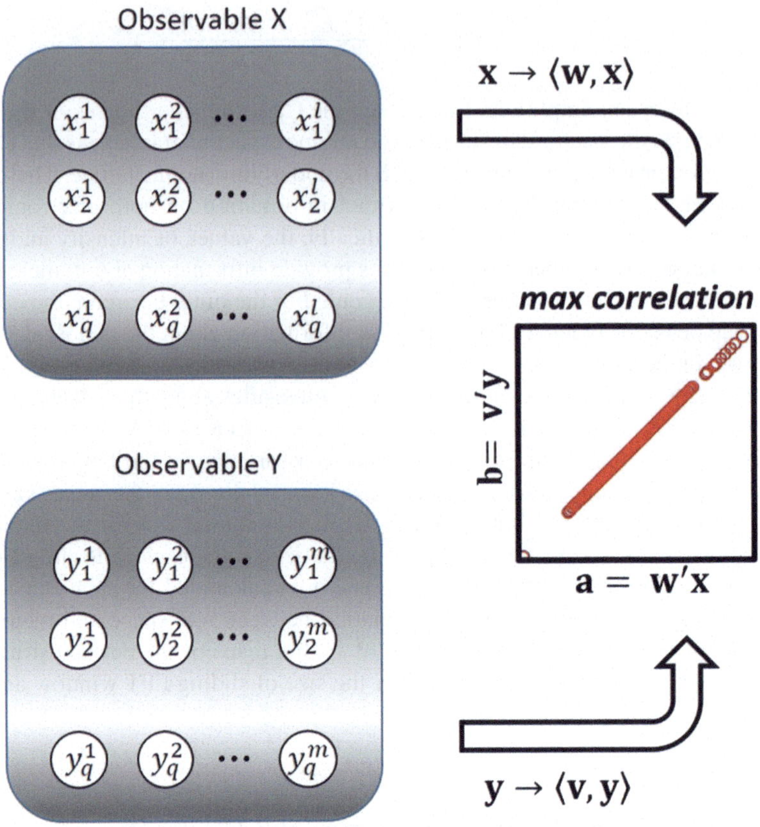

Fig. 5.8 Canonical correlation analysis (CCA). Schematics of CCA workflow.

correlation is achieved between two sets [41]. Specifically, CCA solves the problem of finding basis vectors **w** and **v** for two multi-dimensional datasets X and Y such that the correlation between their projections $x \to \langle w, x \rangle$ and $y \to \langle v, y \rangle$ onto these basis vectors is maximized. The canonical correlation coefficient ρ is expressed as

$$\rho = max_{w,v} \frac{w' C_{xy} v}{\sqrt{w' C_{xx} w v' C_{yy} v}} \tag{5.13}$$

where C_{xx}, C_{yy} are auto-covariance matrices, and C_{xy}, C_{yx} are cross-covariance matrices of **x** and **y**. The projections $a = w'x$ and $b = v'y$ represent the first pair of canonical variates (Fig. 5.8).

Application to experimental data. The results of correlation matrix and canonical correlation analysis for G_H sample are summarized are summarized in Fig. 5.9a and b, respectively. The canonical correlation coefficient is 0.62 and the associated canonical scores are given by

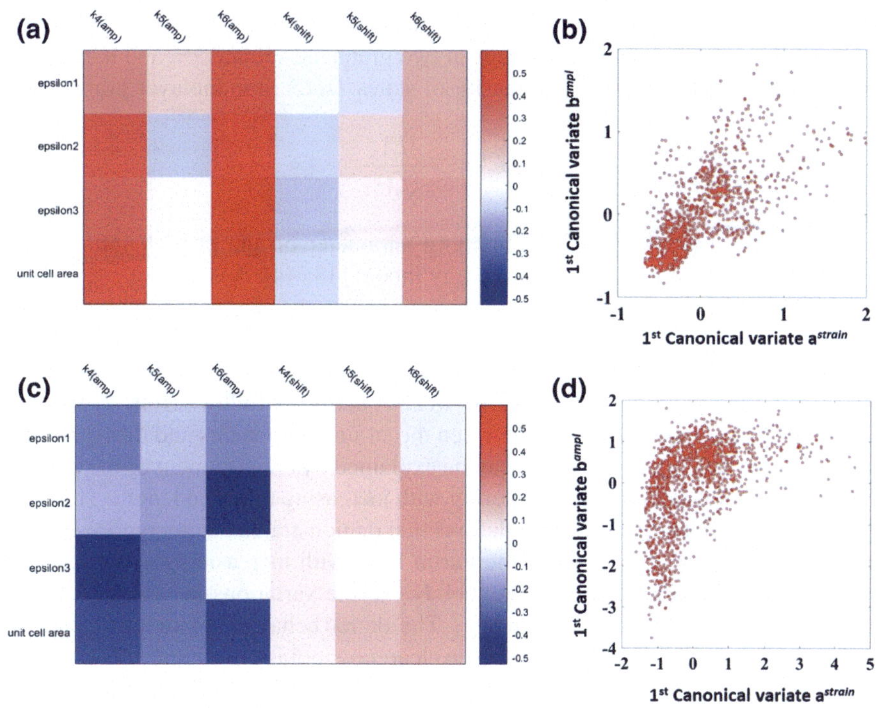

Fig. 5.9 Correlative analysis of graphene structural and electronic degrees of freedom. **a–b** Pairwise Pearson correlation matrix (**a**) and plot of the canonical variable scores for the correlation between strain components and scattering intensity for the G_H sample. **c–d** Same for G_O sample.

$$a_i^{strain} = 0.37(\varepsilon_1)_i + 0.50(\varepsilon_2)_i + 0.36(\varepsilon_3)_i \tag{5.14a}$$

$$b_t^{ampl} = 0.39(I_{k1})_i - 0.33(I_{K2})_i + 0.80(I_{K3})_i \tag{5.14b}$$

where the magnitudes of the coefficients before the variables give the optimal contributions of the individual variables to the corresponding canonical variate. Here the scattering intensities associated with two channel I_{K1} and I_{K2} show a non-negligible positive correlation with strain components in both Pearson correlation matrix and the canonical scores. A dependence of electron scattering intensity on lattice strain for G_H sample can be in principle understood within nearest-neighbor tight-binding model. Specifically, the tight-binding Hamiltonian for graphene monolayer is expressed as [42]

$$H = -\gamma \sum_{\langle i,j \rangle} (a_i^{\dagger} b_j + h.c.) \tag{5.15}$$

where γ is the nearest neighbor hopping parameter, operators $a_i^\dagger$ ($b_i^\dagger$) and a_i (b_i) create and annihilate an electron, respectively, at two graphene sublattices, and h.c. stands for the Hermitian conjugate. The density of states $D(E)$ in monolayer graphene is given by

$$D(E) = \frac{|E|}{\pi\sqrt{3}\gamma^2} \tag{5.16}$$

Further, the dependence of the hopping parameter on the bond length can be described in terms of the exponential decay model [43, 44],

$$\gamma \cong \gamma_0 exp(-\tau\varepsilon) \tag{5.17}$$

where τ is typically assigned values between 3 and 4. It follows from (5.16) and (5.17) that the positive correlation between the strain components and the scattering amplitudes in channels I_{K1} and I_{K3} can be explained by enhancement of the density of electronic states available for scattering with increasing the bond length. This also agrees with the first-principles calculations that demonstrated an emergence of new peaks in the density of states near the Fermi level with increasing the bond length [45]. Interestingly, a response of channel I_{K2} to the variations in strain is clearly different from that of channels I_{K1} and I_{K3}. The altered behavior of structure-property relationship for I_{K2} channel becomes even clearer by looking at canonical variates in (5.14) that show a negative sign of a coefficient in front of I_{K2}. Such altered behavior in one of the scattering channels may lead to the formation of observed fine structure of electronic superlattice, namely, coexistence of staggered dimer-like and hexagonal superlattices.

Unlike the G_H sample, the oxidized graphene layer G_O shows a negative correlation between lattice strain and scattering intensities for all the scattering channels (Fig. 5.9c and d). The CCA canonical variates for GO sample are

$$a_i^{strain} = 0.31(\varepsilon_1)_i + 0.73(\varepsilon_2)_i + 0.32(\varepsilon_3)_i \tag{5.18a}$$

$$b_t^{ampl} = -0.37(I_{k1})_i - 0.41(I_{K2})_i + 0.80(I_{K3})_i \tag{5.18b}$$

with CCA coefficient equal to 0.50. This indicates a presence of *apparent* lattice contraction in the 2D-projected SPM images caused by out-of-plane "rippling" of graphene lattice in the presence of oxygen functional groups on the surface. In addition to out-of-lane surface deformations [46, 47], the attached oxygen functional groups also cause an expansion of the lattice constant in their vicinity [47, 48] which, in this case, is hidden from our view "under" the rippled regions in the image. Similar to the analysis for G_H sample, the correlation between scattering intensity and lattice stain can be explained based on the nearest neighbor tight binding model, where an increased lattice constant under the curved regions leads to enhanced density of electronic states available for scattering. Interestingly, the ε_2 strain component and the scattering intensity in I_{K3} channel display the strongest contribution to their respective canonical variates indicating non-uniform strain-scattering relation at the nanoscale and their potential connection to the variations in the electronic superlattice patterns in G_O sample.

We now comment on a character of Dirac point shift. It is worth recalling that for the underformed graphene lattice the positions of electron scattering maxima ("Dirac valleys") are located at the corners of graphene Brillouin zone. Interestingly, however, only relatively small correlation between positions of Dirac point and lattice strain was found in both G_O and G_H systems. Since the position of the Brillouin zone corners in both deformed and non-deformed graphene are given by a direct linear transformation of the reciprocal lattice vectors, these results suggest that in the deformed graphene lattice the locations of electron scattering maxima do not necessarily coincide with the corners of the (new) Brillouin zone.

To summarize this section, we have demonstrated a successful approach for analyzing structure-property relationship at the nanoscale using a combination of sliding window fast Fourier transform, Pearson correlation matrix and canonical correlation analysis. A peculiar connection between variations in coupling between lattice strain components and intensity of electron scattering was found that could explain an emergence of the experimentally observed fine structure in the electronic superlattice. It is worth noting that the analysis demonstrated here was mainly limited to linear structure-property-relationships. One potential way to overcome this limitation would be to use kernelized version of CCA [49] with physics based kernels. For example, one may construct a certain function $F(x, z)$, where z is a physical parameter that determines a non-linearity, so that the resultant kernel $K(x, y) = F'(x, z) * F(y, z)$ will approximate a linear behavior in a limit of very small z, whereas for large values of z it will approximate a non-linear behavior.

5.4 Case Study 3. Correlative Analysis in Multi-mode Imaging of Strongly Correlated Electron Systems

5.4.1 Model System and Problem Overview

In our last case study, a structure-property relationship is analyzed for the case where structural and electronic information are obtained through two separate channels of scanning tunneling microscopy experiment on iron-based strongly correlated electronic system. This type of materials display a rich variety of complex physical phenomena including an unconventional superconductivity [6]. The Au-doped $BaFe_2As_2$ compound was selected which, at the dopant level of $\sim$1%, presides in the spin-density wave (SDW) regime below $T_N \approx 110$ K [50, 51]. At increased concentration of Au-dopants, the magnetic interactions associated with SDW phase become suppressed and the system turns into a superconductor ($T_c \approx 4$ K) at $\sim$3% [51]. The interactions present in SDW regime may thus provide important clues about mechanisms behind emergence of superconductivity in FeAs-based systems. Of specific interest is a region of cleaved $Ba(Fe_xAu_{1-x})_2As_2$ surface (Fig. 5.10) that

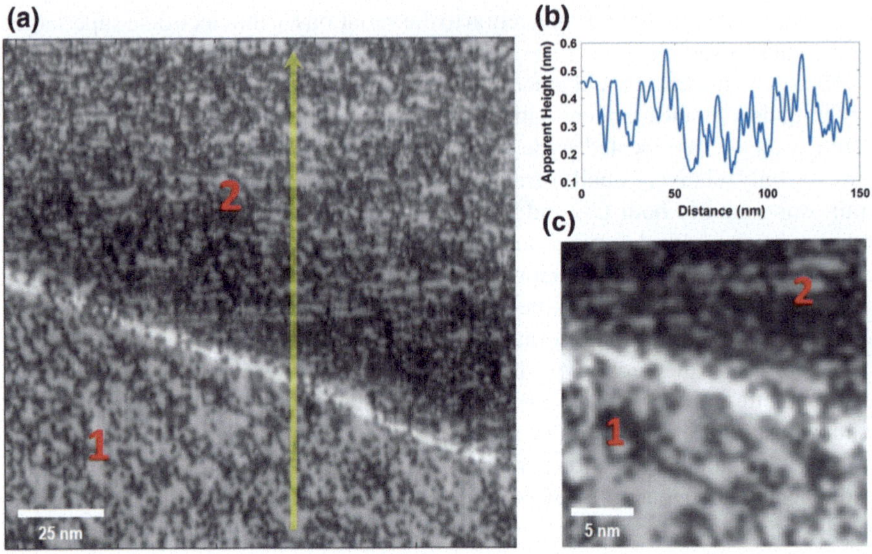

Fig. 5.10 Scanning tunneling microscopy data on Au-doped $BaFe_2As_2$. **a** STM topographic image showing domain-like structure where two different (as seemingly appears from the topography) domains are denoted as 1 and 2. **b** Topographic profile along yellow line in (**a**). **c** Smaller topographic area of a 2-domain-like structure that was used for scanning tunneling spectroscopy (STS measurements)

seemingly shows a presence of two different domains-like structures (marked 1 and 2 in Fig. 5.10a) separated by a bright linear topographic feature. Manual inspection of conductance maps at several different values of energy from such region demonstrates a spatially inhomogeneous electronic structure across the FOV, as well as potentially different *dominant* forms of electronic behavior in domain 1 and domain 2, but does not allow an accurate mapping of these electronic behaviors.

5.4.2 How to Obtain Physically Meaningful Endmembers from Hyperspectral Tunneling Conductance Data?

To gain a deeper insight into the types and spatial distribution of different electronic behaviors in this 2-domain-like structure, the non-negative matrix factorization (NMF) method is applied to a scanning tunneling spectroscopy (STS) dataset of dimensions $100 \times 100 \times 400$ pixels recorded over a portion of the structure of interest (Fig. 5.10c). NMF solves the problem of decomposing the input data represented by matrix X of size $m \times n$, where m is the number of features (m = 512 for this dataset) and n is the number of samples (n = 10,000 for this dataset), into two non-negative factors W and H such that $X \approx WH$ [52]. The k columns of W are interpreted as source signals (endmembers) whereas H defines the loading maps (abundance). Due to the non-negativity constraint, NMF can be applied to problems involving finding $k \ll min(m, n)$ physically-meaningful source signals (i.e. physically-defined phases) from the input data, such that all the data can be explained as a mixture of the k basic phases. The NMF can be formally defined as a constrained optimization problem, which can be written, according to Li and Ngom, in a general form as [53]

$$min_{W,H} f(W, H) = \frac{1}{2} \|X - WH\|_F^2 + \\ + \sum_{i=1}^{k}(\alpha_1 \|w_i\|_1 + \frac{\alpha_2}{2} \|w_i\|_2^2) + \sum_{i=1}^{k}(\lambda_1 \|h_i\|_1 + \frac{\lambda_2}{2} \|h_i\|_2^2) \tag{5.19}$$

subject to $W \geq 0,\ H \geq 0$ and where $\|\bullet\|_F$ is the Frobenius norm, w_i and h_i are the i-th columns of W and H, respectively, α_1 and α_2 are regularization parameter for sparsity and smoothness, respectively, for the endmembers domain, while λ_1 and λ_2 control sparsity and smoothness, respectively, for the loading maps (abundancies) domain.

The results on NMF based decomposition into 3 components are shown in Fig. 5.11 (no new information was obtained by increasing a number of compo-

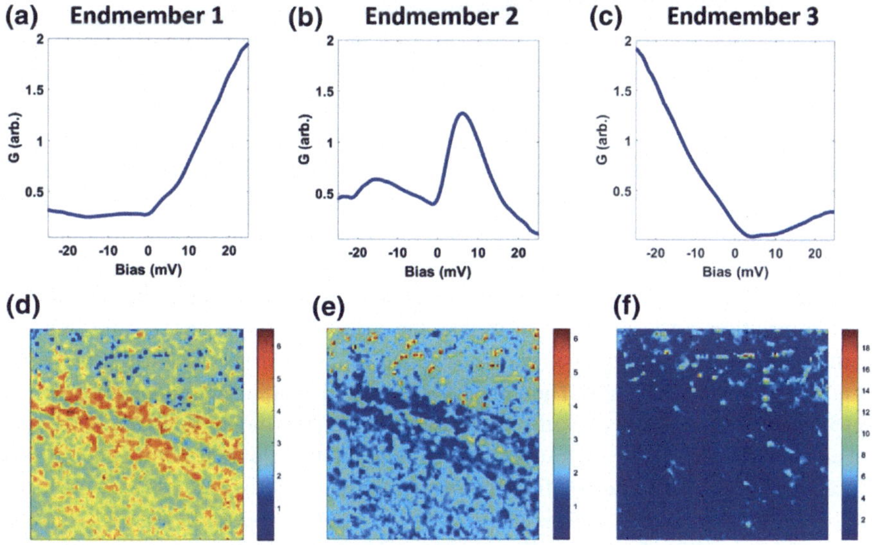

Fig. 5.11 Extraction of electronic descriptors from STS dataset on Au-doped $BaFe_2As_2$. **a–c** NMF decomposed spectral endmembers. **d–f** corresponding loading maps (the same region as shown Fig. 5.10c)

nents). Spatial weight of endmember 1 is mainly concentrated within the domain 1 (Fig. 5.11a, d). The corresponding spectral curve shows a reduced density of states at the negative energies that agrees with the theoretical and angle-resolved photoemission spectroscopy evidence for partial gap opening just below the Fermi level in the SDW regime. The endmember 2 spectral curve shows a well-defined asymmetric double-peak structure (Fig. 5.11b). Analysis of loading maps for this component (Fig. 5.11e) reveals that this type of electronic behavior is constrained to point-like features on the surface. Furthermore, these features are predominantly located in the domain 2. Therefore, they are associated with a presence of dopant states. Interestingly, the asymmetric double peak structure observed in the endmember 2 is in a good qualitative agreement with non-magnetic dopant-induced double resonance peak model in SDW phase. Analysis of loading maps for the endmember 3 suggests that it may also originate from some form of localized disorder (Fig. 5.11f). These point-like defect states are located mainly in the domain 2 although there is a diluted concentration of defect in the domain 1 as well. While there is no well-defined peak in the density of states associated with this type of defect in the low energy range of interest (Fig. 5.11c), an alternation of the local density of states around the Fermi level was still observed as compared to SDW phase (endmember 1). It is therefore concluded that endmember 2 and endmember 3 describe two distinct types of point defect/dopants that have different structural and/or chemical origin. Thus, the characteristic difference between two domain-like structures 1 and 2 is that there is a significant accumulation of point "impurities"/dopants in only one of those domains. This effectively can be interpreted a peculiar transition between "heavily-doped" and "lightly-doped" regions on the surface.

Correlative analysis of surface geometry and electronic structure. We next proceed to correlative analysis of STM topographic data and loading maps of NMF electronic components. Since no atomic lattice was resolved for this surface region, a correlative analysis is carried out in a pixel-by-pixel fashion. The global Moran's I analysis for the NMF components 1, 2, and 3 and topography returns the values of −0.472, 0.351, and −0.282, respectively. In order to derive physics from such type of structure-property cross-correlation analysis it is crucial to be able to visualize directly those regions on the surface that show higher/lower correlation values. For this purpose, the local indicators of spatial associations described earlier for analysis of correlation between different molecular orders are employed. In addition the results of local Moran's analysis can be mapped on to quadrants resulting into what is known as Moran's Q maps. The local Moran's I and Moran's Q maps are shown in Fig. 5.12. The analysis of Moran's I correlation maps for the endmember 1 (SDW) and endmember 2 (localized defect state) captures a well-defined point-like regions of positive and negative correlation, respectively, which indicates a relatively large number of impurities (characterized by localized states) residing in local dips of the topographic map (Fig. 5.12a, b). The correlative analysis also offers a unique chance to get an insight into 'coupling' of different electronic orders to the boundary between domain 1 and domain 2 (linear bright topographic feature in Fig. 5.10a, c). Particularly, a peculiar depletion of SDW phase along the domains boundary was found that

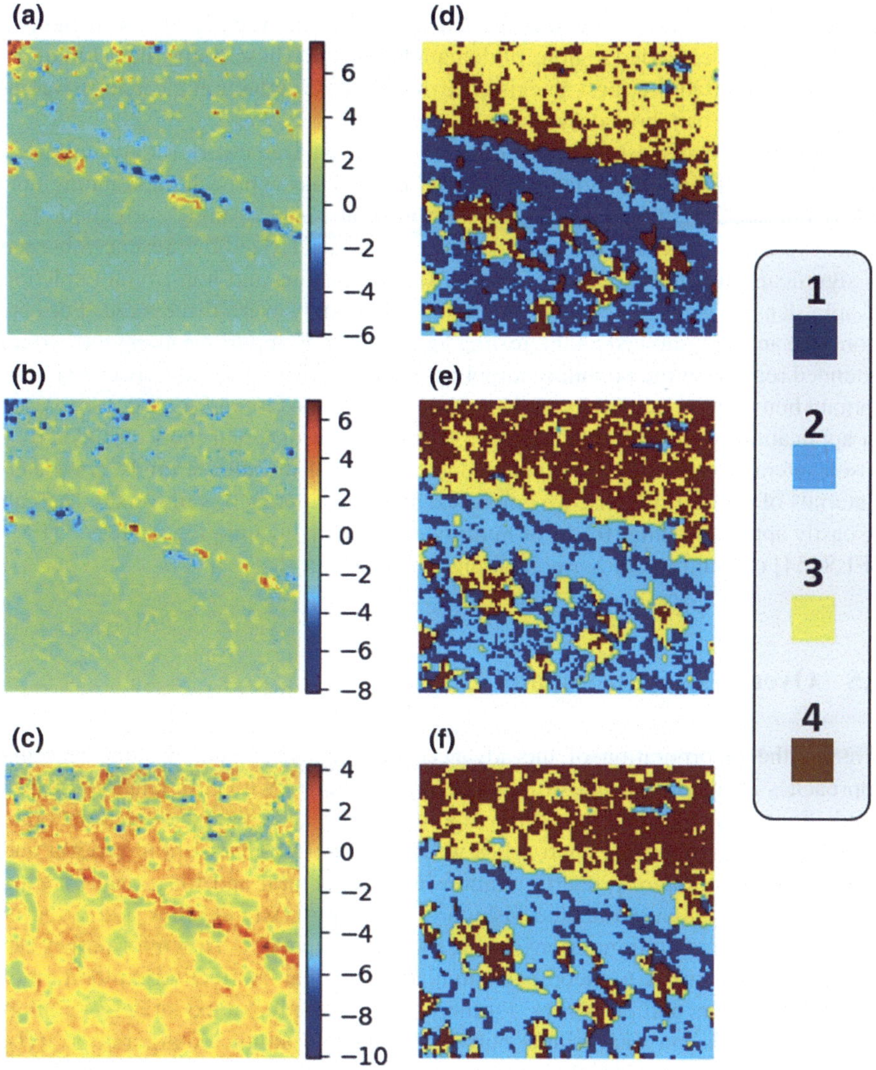

Fig. 5.12 Local indicators of spatial association. Local bivariate Moran's I and Moran's Q (quadrants) calculated for relationship between topographic data (apparent height) and endmember 1 (**a**, **d**); endmember 2 (**b**, **e**); endmember 3 (**c**, **f**). Quadrants legend: Q = 1—positive correlation between high x and high neighboring y's; Q = 2—negative, low x and high neighboring y's; Q = 3—positive, low x and low neighboring y's; Q = 4—negative, high x and low neighboring y's

is clearly evident from appearance of a well-defined linear Q = 2 feature in Moran's Q maps (Fig. 5.12d), that is, a region in which low local values of SDW component correspond to high local values of apparent height (topography). Meanwhile, a presence of Q = 1 features in Fig. 5.12e, f indicates an aggregation of localized states associated with both types of structural/chemical disorder (i.e., NMF components 2 and 3) along the extended regions of domain boundary. These chain-like formations of defects potentially suggest an existence of different conduction mechanism along the quasi-1D domain boundary.

To summarize this last section, we have developed a framework for an automated analysis of multimodal imaging data, and illustrated our approach on scanning tunneling microscopy/spectroscopy datasets from iron-based strongly correlated electronic systems. A peculiar domain-like structure characterized by presence/absence of significant dopants accumulation in different domains and non-trivial depletion of spin density wave state along the domain boundary were discovered. Furthermore, the analysis showed an interesting aggregation of impurities along the certain extended regions of the boundary implying a potential for realizing a special type of domain boundary conductivity under certain conditions. Going forward, we foresee an application of the outlined approach to analysis of different modes of electron-boson interaction in high-T_c superconductors as well in other strongly correlated materials of interest. Finally, we emphasize that this approach is universal, and can be easily applied to other forms of multimodal imaging techniques, such as STEM-EELS [54] or multimodal X-ray imaging techniques [55].

5.5 Overall Conclusion and Outlook

Overall, the incorporation of the advanced data analytics and machine learning approaches in functional and structural imaging coupled with computational-based simulations could lead to breakthroughs in the rate and quality of materials discoveries. The use of these approaches would enable full information retrieval and exploration of structure-property relationship in structural and functional imaging on atomic level in an automated fashion. This, in turn, would allow a creation of libraries of atomic configurations and associated properties. This information can be then directly linked to theoretical simulations to enable effective exploration of material behaviors and properties. Furthermore, knowledge of extant defect configurations in solids can significantly narrow the range of atomic configurations to be probed from the first-principles, thus potentially solving an issue with exponential growth of number of possible configurations with system size. These approaches can further be used to build experimental databases across imaging facilities nationwide (as well as worldwide), establish links to X-ray, neutron and other structural databases, and enable immediate in-line interpretation of information flows from microscopes, X-Ray and neutron facilities and simulations.

Acknowledgements This research was sponsored by the Division of Materials Sciences and Engineering, Office of Science, Basic Energy Sciences, US Department of Energy (MZ and SVK). Part of research was conducted at the Center for Nanophase Materials Sciences, which is a DOE Office of Science User Facility.

References

1. T. Le, V.C. Epa, F.R. Burden, D.A. Winkler, Chem. Rev. **112**(5), 2889–2919 (2012)
2. O. Isayev, D. Fourches, E.N. Muratov, C. Oses, K. Rasch, A. Tropsha, S. Curtarolo, Chem. Mater. **27**(3), 735–743 (2015)
3. G. Xu, J. Wen, C. Stock, P.M. Gehring, Nat. Mater. **7**(7), 562–566 (2008)
4. K. Gofryk, M. Pan, C. Cantoni, B. Saparov, J.E. Mitchell, A.S. Sefat, Phys. Rev. Lett. **112**(4), 047005 (2014)
5. O.M. Auslaender, L. Luan, E.W.J. Straver, J.E. Hoffman, N.C. Koshnick, E. Zeldov, D.A. Bonn, R. Liang, W.N. Hardy, K.A. Moler, Nat. Phys. **5**(1), 35–39 (2009)
6. I. Zeljkovic, J.E. Hoffman, Phys. Chem. Chem. Phys. **15**(32), 13462–13478 (2013)
7. M. Daeumling, J.M. Seuntjens, D.C. Larbalestier, Nature **346**(6282), 332–335 (1990)
8. Y. Zhang, V.W. Brar, C. Girit, A. Zettl, M.F. Crommie, Nat. Phys. **5**(10), 722–726 (2009)
9. J. Martin, N. Akerman, G. Ulbricht, T. Lohmann, J.H. Smet, K. von Klitzing, A. Yacoby, Nat. Phys. **4**(2), 144–148 (2008)
10. K.K. Gomes, A.N. Pasupathy, A. Pushp, S. Ono, Y. Ando, A. Yazdani, Nature **447**(7144), 569–572 (2007)
11. E. Dagotto, Science **309**(5732), 257 (2005)
12. S.V. Kalinin, S.J. Pennycook, Nature **515** (2014)
13. S.V. Kalinin, B.G. Sumpter, R.K. Archibald, Nat. Mater. **14**(10), 973–980 (2015)
14. D.G. de Oteyza, P. Gorman, Y.-C. Chen, S. Wickenburg, A. Riss, D.J. Mowbray, G. Etkin, Z. Pedramrazi, H.-Z. Tsai, A. Rubio, M.F. Crommie, F.R. Fischer, Science (2013)
15. Y. Wang, D. Wong, A.V. Shytov, V.W. Brar, S. Choi, Q. Wu, H.-Z. Tsai, W. Regan, A. Zettl, R.K. Kawakami, S.G. Louie, L.S. Levitov, M.F. Crommie, Science (2013)
16. C.-L. Jia, S.-B. Mi, K. Urban, I. Vrejoiu, M. Alexe, D. Hesse, Nat. Mater. **7**(1), 57–61 (2008)
17. H.J. Chang, S.V. Kalinin, A.N. Morozovska, M. Huijben, Y.-H. Chu, P. Yu, R. Ramesh, E.A. Eliseev, G.S. Svechnikov, S.J. Pennycook, A.Y. Borisevich, Adv. Mater. **23**(21), 2474–2479 (2011)
18. A. Borisevich, O.S. Ovchinnikov, H.J. Chang, M.P. Oxley, P. Yu, J. Seidel, E.A. Eliseev, A.N. Morozovska, R. Ramesh, S.J. Pennycook, S.V. Kalinin, ACS Nano **4**(10), 6071–6079 (2010)
19. Y.-M. Kim, J. He, M.D. Biegalski, H. Ambaye, V. Lauter, H.M. Christen, S.T. Pantelides, S.J. Pennycook, S.V. Kalinin, A.Y. Borisevich, Nat. Mater. **11**(10), 888–894 (2012)
20. W.J. Kaiser (ed.), *Scanning Tunneling Microscopy* (Academic Press, San Diego, 1993), p. ii
21. H. Sakurai, T. Daiko, T. Hirao, Science **301**(5641), 1878 (2003)
22. S. Fujii, M. Ziatdinov, S. Higashibayashi, H. Sakurai, M. Kiguchi, J. Am. Chem. Soc. **138**(37), 12142–12149 (2016)
23. M. Ziatdinov, A. Maksov, S.V. Kalinin, npj Computational Materials **3**, 31 (2017)
24. S. Jesse, S.V. Kalinin, Nanotechnology **20**(8), 085714 (2009)
25. I. Goodfellow, Y. Bengio, A. Courville, *Deep Learning* (MIT Press, 2016)
26. D. Stutz, *Seminar Report* (RWTH Aachen University, 2014)
27. G.R. Cross, A.K. Jain, IEEE Trans. Pattern Anal. Mach. Intell. **PAMI-5**(1), 25–39 (1983)
28. M. Schmidt, http://www.cs.ubc.ca/~schmidtm/Software/UGM.html (2007)
29. R. Jaafar, C.A. Pignedoli, G. Bussi, K. Aït-Mansour, O. Groening, T. Amaya, T. Hirao, R. Fasel, P. Ruffieux, J. Am. Chem. Soc. **136**(39), 13666–13671 (2014)
30. H. Amara, S. Latil, V. Meunier, P. Lambin, J.C. Charlier, Phys. Rev. B **76**(11), 115423 (2007)
31. A.A. El-Barbary, R.H. Telling, C.P. Ewels, M.I. Heggie, P.R. Briddon, Phys. Rev. B **68**(14), 144107 (2003)
32. L. Anselin, Geogr. Anal. **27**(2), 93–115 (1995)
33. L. Vlcek, A.A. Chialvo, J. Chem. Phys. **143**(14), 144110 (2015)
34. M. Ziatdinov et al., Nanotechnology **27**, 495703 (2016)
35. Y. Ganin, E. Ustinova, H. Ajakan, P. Germain, H. Larochelle, F. Laviolette, M. Marchand, V. Lempitsky, *ArXiv e-prints*, vol. 1505 (2015)
36. M. Ziatdinov, S. Fujii, K. Kusakabe, M. Kiguchi, T. Mori, T. Enoki, Phys. Rev. B **89**(15), 155405 (2014)

37. S. Fujii, T. Enoki, ACS Nano **7**(12), 11190–11199 (2013)
38. P. Ruffieux, M. Melle-Franco, O. Gröning, M. Bielmann, F. Zerbetto, P. Gröning, Phys. Rev. B **71**(15), 153403 (2005)
39. K.-I. Sakai, K. Takai, K.-I. Fukui, T. Nakanishi, T. Enoki, Phys. Rev. B **81**(23), 235417 (2010)
40. R.K. Vasudevan, A. Belianinov, A.G. Gianfrancesco, A.P. Baddorf, A. Tselev, S.V. Kalinin, S. Jesse, Appl. Phys. Lett. **106**(9), 091601 (2015)
41. W.J. Krzanowski, *Principles of Multivariate Analysis: A User's Perspective* (Oxford University Press, Inc., 1988)
42. P.R. Wallace, Phys. Rev. **71**(9), 622–634 (1947)
43. V.M. Pereira, A.H. Castro Neto, N.M.R. Peres, Phys. Rev. B **80**(4), 045401 (2009)
44. R.M. Ribeiro, M.P. Vitor, N.M.R. Peres, P.R. Briddon, A.H.C. Neto, New J. Phys. **11**(11), 115002 (2009)
45. V.J. Surya, K. Iyakutti, H. Mizuseki, Y. Kawazoe, Comput. Mater. Sci. **65**, 144–148 (2012)
46. S. Fujii, T. Enoki, J. Am. Chem. Soc. **132**(29), 10034–10041 (2010)
47. V.V. Shunaev, O.E. Glukhova, J. Phys. Chem. C **120**(7), 4145–4149 (2016)
48. J. Ito, J. Nakamura, A. Natori, J. Appl. Phys. **103**(11), 113712 (2008)
49. K. Fukumizu, F.R. Bach, A. Gretton, J. Mach. Learn. Res. **8**, 361–383 (2007)
50. M. Ziatdinov, A. Maksov, L. Li, A.S. Sefat, P. Maksymovych, S.V. Kalinin, Nanotechnology **27**(47), 475706 (2016)
51. L. Li, H. Cao, M.A. McGuire, J.S. Kim, G.R. Stewart, A.S. Sefat, Phys. Rev. B **92**(9), 094504 (2015)
52. D.D. Lee, H.S. Seung, Nature **401**(6755), 788–791 (1999)
53. Y. Li, A. Ngom, Source Code Biol. Med. **8**(1), 10 (2013)
54. M. Varela, J. Gazquez, S.J. Pennycook, MRS Bull. **37**(1), 29–35 (2012)
55. O. Bunk, M. Bech, T.H. Jensen, R. Feidenhans'l, T. Binderup, A. Menzel, F. Pfeiffer, New J. Phys. **11**(12), 123016 (2009)

Chapter 6
Data Challenges of In Situ X-Ray Tomography for Materials Discovery and Characterization

Brian M. Patterson, Nikolaus L. Cordes, Kevin Henderson, Xianghui Xiao and Nikhilesh Chawla

Abstract Since its development in the 1970s (Hounsfield, Br J Radiol 46(552):1016–1022, 1973) [1], X-ray tomography has been used to study the three dimensional (3D) structure of nearly every type of material of interest to science, both in the laboratory (Elliott and Dover, J Microsc 126(2):211–213, 1982) [2] and at synchrotron facilities (Thompson et al., Nucl Instrum Methods Phys Res 222(1):319–323, 1984) [3]. The ability to nondestructively image internal structures is useful in the medical community for patient diagnosis. For this same reason, it is critical for understanding material structural morphology. X-ray tomography of static materials can generate a true 3D structure to map out content and distribution within materials including voids, cracks, inclusions, microstructure, and interfacial quality. This technology is even more useful when applying a time component and studying the changes in materials as they are subjected to non-equilibrium stimulations. For example, testing mechanical properties (e.g., compressive or tensile loading), thermal properties (e.g., melting or solidification), corrosion, or electro-

Disclaimer: Commercial products are identified in this document in order to specify the experimental procedure and options adequately. Such identification is not intended to imply recommendation or endorsement by LANL or DOE, nor is it intended to imply that the products identified are necessarily the best available for the purpose.

B. M. Patterson (✉) · N. L. Cordes · K. Henderson
Materials Science and Technology Division, Engineered Materials Group, Los Alamos National Laboratory, Los Alamos, NM 87545, USA
e-mail: bpatterson@lanl.gov

X. Xiao
X-ray Photons Sciences, Argonne National Laboratory, Argonne, IL, USA

N. Chawla
4D Materials Science Center, Arizona State University, Tempe, AZ, USA

T. Lookman et al. (eds.), *Materials Discovery and Design*, Springer Series in Materials Science 280, https://doi.org/10.1007/978-3-319-99465-9_6

static responses, while simultaneously imaging the material in situ, can replicate real world conditions leading to an increase in the fundamental understanding of how materials react to these stimuli. Mechanical buckling in foams, migration of cracks in composite materials, progression of a solidification front during metal solidification, and the formation of sub-surface corrosion pits are just a few of the many applications of this technology. This chapter will outline the challenges of taking a series of radiographs while simultaneously stressing a material, and processing it to answer questions about material properties. The path is complex, highly user interactive, and the resulting quality of the processing at each step can greatly affect the accuracy and usefulness of the derived information. Understanding the current state-of-the-art is critical to informing the audience of what capabilities are available for materials studies, what the challenges are in processing these large data sets, and which developments can guide future experiments. For example, one particular challenge in this type of measurement is the need for a carefully designed experiment so that the requirements of 3D imaging are also met. Additionally, the rapid collection of many terabytes of data in just a few days leads to the required development of automated reconstruction, filtering, segmentation, visualization, and animation techniques. Finally, taking these qualitative images and acquiring quantitative metrics (e.g., morphological statistics), converting the high quality 3D images to meshes suitable for modeling, and coordinating the images to secondary measures (e.g., temperature, force response) has proven to be a significant challenge when a materials scientist 'simply' needs an understanding of how material processing affects its response to stimuli. This chapter will outline the types of in situ experiments and the large data challenges in extracting materials properties information.

6.1 Introduction

The penetration of X-rays through materials and their subsequent use for imaging dates back to the first radiograph in 1895 by Wilhelm Röntgen of his wife Anna Bertha Ludwig's hand. It was quickly used for medical diagnosis within a few months. Since then, radiographic uses have expanded to include medical (e.g., dental, mammography, skeletal imaging, and with dyes, soft tissue examination), engineering (e.g., cracks and welds), security (e.g., airport and ports), and (the focus of this chapter) materials science (e.g., polymers, metals, explosives, etc.). However, since two dimensional (2D) radiographs convolute the three dimensional (3D) information, further technique refinement led to the invention of nondestructive tomographic imaging, which deconvolutes the 3D structure. Understanding the 3D structure of newly manufactured, aged, refurbished, novel, or remanufactured materials is critical to understanding a material's property-structure-function relationship. Few techniques provide a better picture than 3D X-ray computed tomography (CT).

When the X-rays interact with matter, several processes may occur:

- the X-rays pass through with no interaction,

- the X-rays are absorbed by the atom,
- the X-rays are absorbed and a fluorescent X-ray is emitted,
- or the X-rays are diffracted or scattered.

This chapter will primarily focus on the absorption of the X-rays by the material and the processing of the resultant data. In addition to X-ray absorption radiographic and CT imaging, other X-ray imaging techniques include Extended X-ray Absorption Fine Structure (EXAFS) spectroscopy, X-ray Absorption Near Edge Structure (XANES [4]) spectroscopy, Grazing-Incidence Small-Angle X-ray Scattering (GISAXS [5]), X-ray fluorescence (XRF [6, 7]), and X-ray diffraction (XRD) [8–10] techniques. EXAFS can measure the chemical state of materials, especially low concentration elements in a bulk structure. XANES can measure the electronic transitions within materials. The fluorescence (XRF) of X-rays by atoms allows scientists to identify and quantify what elements are present and where they are located; recent advances in XRF spectroscopy have led to nondestructive 3D elemental imaging [11–13]. Scattering and high energy diffraction of X-rays by crystallographic materials (HEDM) can be reconstructed into 3D maps of the crystals within the materials. Both 3D XRF and HEDM have their own data challenges, but will not be covered in this chapter; see Chap. 7 by Pokharel for more information on HEDM.

X-ray CT is simply the radiographic imaging of a sample as a function of sample rotation. These radiographs are then mathematically reconstructed to yield a representative 3D digital image of the material. To image, a sample is placed between an X-ray source and detector. A beam of X-rays passes through the sample, where they are absorbed by matter that scales with the electronic density of the material. This photon-matter interaction typically scales with the atomic number and density of the material. From this, an absorption-contrast radiograph is collected. A series of radiographs can be collected as the sample is rotated, either ~180° or 360°, and are then reconstructed into a series of 2D slices that can be rendered in a 3D image. Many different laboratory-based X-ray tomographic systems are available to researchers. Figure 6.1 diagrams several of the geometries used for CT imaging. The top graphic shows the cone-beam geometry present in most laboratory-based X-ray CT systems. Simple laboratory-based systems (Fig. 6.1a) without optics are capable of several micrometers resolution, limited by the X-ray spot size of the source. The typical time for acquiring a high contrast 3D image can vary between 2 and 24 h depending upon the contrast of the material and the resolution needed. The field-of-view and resolution of the image is based upon the location of the X-ray source and detector (geometric magnification) as well as the pixel size of the detector, and the presence of any additional focusing optics. The researcher can specify (from a limited list) the X-ray anode used from a variety of manufacturers that offer different optics geometries.

The addition of X-ray objective optics (Fig. 6.1b) makes it possible to build transmission X-ray microscopy (X-ray microscopy, TXM) that are capable of imaging materials with a resolution of ~25–40 nm and fields-of-view of ~10's of micrometers [14, 15]. With TXM-based X-ray CT, full 3D image time is typically 12–36 h. An advantage of using laboratory-based systems is that flexibility is very high in that

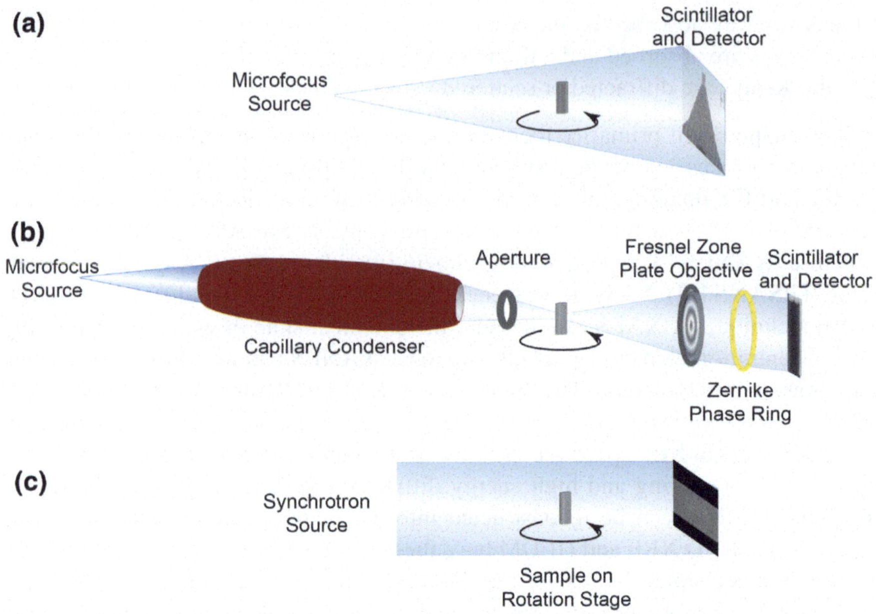

Fig. 6.1 Three graphics showing the geometry of 3D X-ray imaging instruments. **a** a typical lab-based microCT arrangement with a cone-beam X-ray source shining the X-rays through the sample and onto the detector. Magnification is governed geometrically by the positions of the source and detector as well as optics on the detector. **b** The geometry of an X-ray microscope XRM (nano-scale tomography) in which a beam of X-rays illuminates a small volume of the sample and are focused and collimated by Fresnel zone plate optics. Finally, **c** represents the geometry of a synchrotron-based setup with the collimated beam shining through the sample. X-ray focusing optics may be used to increase the flux at the sample and microscope optics may also be present after the scintillator to increase the resolution of the measurement

the instrument (e.g., field-of-view and resolution) can be designed from the ground up for the experiment; however, the relatively low photon flux in laboratory-based instruments (compared to the high photon flux of synchrotrons), usually precludes dynamic in situ measurements.

The use of a synchrotron light source opens up other areas of X-ray usage that are not possible with laboratory-based sources (Fig. 6.1c). Synchrotrons do not offer better spatial resolution than laboratory systems per se, rather, the high X-ray photon brightness available can greatly improve the temporal resolution by several orders of magnitude. Their flexibility typically means that new techniques are tested at synchrotron beamlines to better understand the commercial potential of these new techniques. Therefore, novel experimental techniques are available at synchrotrons several years before they are available commercially in the laboratory. Because synchrotrons use a parallel beam geometry, they offer more flexibility for unique, larger, in situ equipment; due to the longer working distances, opportunities also exist for phase contrast imaging and monochromatic absorption contrast radiography (due to the high photon flux). XRM in the synchrotron reduces the scan time from 12–36 h to ~15 min for a 3D image with the resolutions down to 25–40 nm resolution [16]. While fast imaging times and increased temporal resolutions are advantages of synchrotron-based imaging, the main disadvantages of these shared-user facilities are the lack of quick and easy access and the high setup times for each measurement.

With laboratory-based and synchrotron-based X-ray CT techniques, it is possible to collect a 3D image of a material, render it in 3D, extract visual information, and conduct morphological measurements. The proper 3D analysis of materials requires a multi-step process to reconstruct, process, segment, and extract statistical measures (e.g., size, shape, distribution) that can provide a wealth of information which can be correlated back to material processing and performance.

X-ray CT has impacted all areas of materials science, including cellular materials [17] (metal [18] and polymer foams [19–23], wood [24], bone [25–27]), actinides [28], fuel cells [29], high explosives [30], biological materials such as cells [31], additive manufacturing (metals [32] and polymers [33]), carbon fiber composites, geology (fossils [34], minerals [35]), batteries [36–39], works of provenance [40], superconductor [41] and catalyst development [42, 43], and metals (corrosion [44–46] and microstructure [47–51]). X-ray CT is used to image old and new formulations, aged materials, material fatigue [52–54] and degradation, as well as damaged materials [55, 56].

Beyond creating static 3D renderings of materials and extracting their morphological, dimensional, and distributional attributes, it is important to understand material behavior when exposed to real-world conditions. During a typical life cycle, a material experiences a variety of strains that it is required to respond to in order to fulfill the requirements of its service life. Often the material experiences several orthogonal strains at once. For example, bridge steels experience a dynamic, cyclical load while slowly corroding, aluminum aircraft bodies stretch and compress with changes in altitude, polymer foams present in running shoes compress with every step of the athlete's foot, metal alloys exhibit very different properties based upon the microstructure (which is controlled by the solidification conditions), and explosives hanging off an aircraft wing experience a temperature cycle that is very different in Phoenix compared to Anchorage. While many of these materials challenges have been explored with traditional 2D microscopy techniques, these example materials experience 3D stressors and are met by 3D responses. To fully understand these responses, in situ 3D X-ray imaging techniques, described below, are needed.

6.2 In Situ Techniques

To address these materials science challenges, a multitude of in situ techniques have been and continue to be developed for X-ray computed tomography. The techniques must be engineered in such a way as to allow for the collection of X-ray CT data while replicating the real-world stimulus conditions as accurately as possible, with the added caveat of not disturbing the experiment. The in situ apparatus, which supplies the external stimulus, must not block the X-ray beam during the experimental imaging. Therefore, in situ apparatus typically use either an open architecture so that the sample may be rotated or the entire in situ apparatus is rotated with a ring of uniform, X-ray transparent material to allow the X-rays to pass through. The in situ techniques developed to date include:

- mechanical loading (compression, tension),
- thermal loading (heating, cooling),
- environmental stimuli (corrosion, humidity) and,
- electromechanical (current charging).

Additionally, in situ rigs have been developed that replicate several of these conditions simultaneously (e.g., tension at elevated temperatures). Each of these techniques has its own set of special challenges in establishing, maintaining, and recording the environment during imaging.

In situ uniaxial mechanical loading of materials is the application of a mechanical stress to a material in a single direction, while measuring the load response and imaging the resultant bending, buckling, densification, cracking or other damage within the material as a result. A laboratory-based in situ loading apparatus must be small and compact to fit within the small sample chamber as well as to accommodate minimal source-sample-detector distances for optimal image magnification. However, the in situ loading apparatus must be able to cope with the large X-ray flux generated by the beamline. An example of a synchrotron in situ loading apparatus is shown in Fig. 6.2. In both cases, the apparatus must be interfaced to the instrument and remain stable over many hours. The simultaneous compressive loading and imaging of materials using synchrotron X-ray CT goes back at least to 1998 [57] by Bart-Smith et al. Al tensile specimens [58–60] are often imaged using X-ray CT for comparison to processing. In the nearly 20 years since, this technique has expanded beyond metals [61] and foams to include polymers and polymer foams [62–64], high explosives [30] (including 3D printed high explosives), textiles [65], and composite materials [66, 67], down to the nanoscale [68]. Additionally, loading of the materials has moved beyond ex situ compression due to the development of in situ load cells both in the laboratory [69] as well as at the synchrotron [62]. Tension strain in materials is now practiced both in the laboratory [69] and the synchrotron [33]. It is possible to collect images up to 20 Hz [70], in situ. Crack initiation, propagation, delamination in composite materials can be imaged at up to a demonstrated 10^{-2} s^{-1} strain rates. In situ mechanical loading coupled with XRM to investigate biological materials, organic crystal fracture, and metal microstructures has also begun to appear in the literature [69].

In situ heating is often used to understand changes in microstructure within metallic alloys. Some experiments are conducted to soften the interfacial adhesion strength [71], while experiments to higher temperatures are conducted to examine the solidification of the material upon cooling. Interfacial strength within carbon fiber [72] and SiC [73] ceramic composites weakens their mechanical properties; therefore, in situ apparatus that apply a tensile load at elevated temperatures (>1700 °C) have been developed. The solidification conditions (e.g., cooling rate) of metal materials

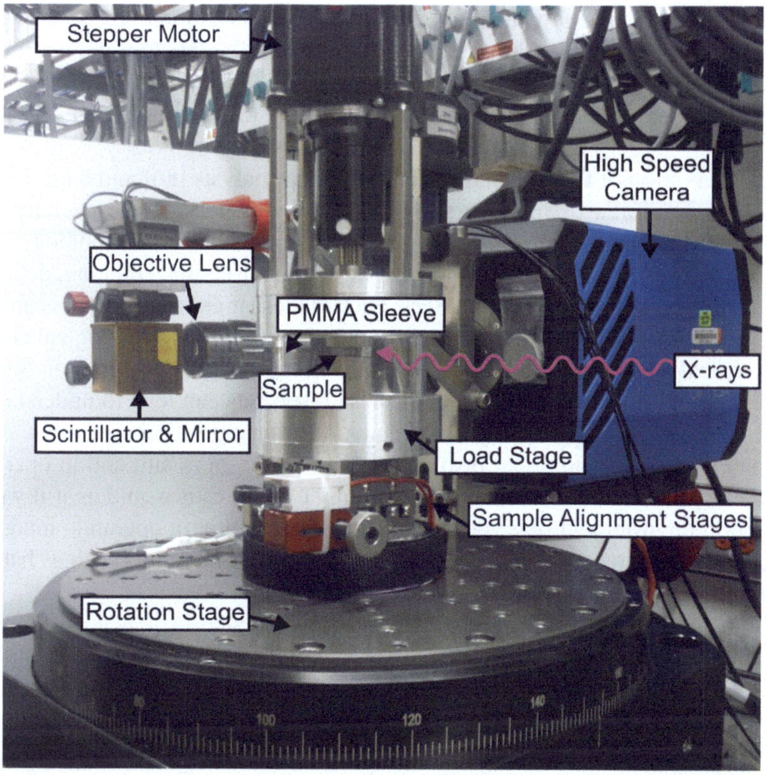

Fig. 6.2 Photograph of an in situ mechanical loading apparatus in a compression configuration at a X-ray synchrotron. The X-rays enter from the right side of the image, pass through the poly(methyl methacrylate) sleeve and the sample, illuminating the scintillator on the left (yellow). The now visible wavelength photons are collected by the objective lens and the high speed camera. The camera holds all of the radiographs in memory to be transferred to a hard drive at the completion of the experiment

governs the microstructure and the resultant properties of the materials. In situ solidification imaging is widely practiced on a variety of metal alloy systems. Most typical are high X-ray contrast Al-Cu [49, 50, 74], Al-In [75], and Al-Zn [48] bimetallic alloys. Experiments have been performed using in situ heating cells (usually graphite furnaces), laser heating [49], or high intensity lights [76]. Experiments can now be conducted that freeze the solidification front for further post experimental analysis [77]. The rate of morphological change within a material as it crosses the solidification boundary is typically quite fast (front velocities of multiple micrometers per second [78]) when compared to feature size and resolution requirements, and therefore is often practiced at the synchrotron, not in the laboratory.

In situ corrosion of metal materials is used to study intergranular defects, pitting, stress corrosion cracking [79], and hydrogen bubble formation [80]. Metals examined consist of Al [44, 81], Fe [82], and AlMgSi alloys [83]. These experiments typically are the simplest to perform in that the specimen is mounted directly into a caustic solution and the resulting corrosion is typically quite slow (hours to days), allowing for both laboratory-based and synchrotron-based experiments to be performed. Often, experiments are conducted with cyclic testing, such as in Al [84] and steels [85].

The observation of functional materials, such as catalysts [86] and batteries [39], is also an active area of in situ tomographic imaging research, especially using XRM techniques. During battery charge-discharge cycles, the morphology of the microstructure [37] can change through expansion, contraction, cracking, delamination, void formation, and coating changes. Each of these material responses can affect the lifetime of the material. Measuring the statistics of these morphological changes [38] is critical to locating fractures, especially changes that may occur on multiple size scales [43]. In operando imaging of these responses can lead to understanding how that 3D morphology changes as a function of charge-discharge rates, conditions, and cycles. An important distinction exists between in situ and in operando: the latter implies that the material is performing exactly as it would as if it were in a real-world environment (e.g., product testing). Therefore, in operando imaging is often the nomenclature when referring to the imaging of materials such as batteries [36], double-layer capacitors, catalysts, and membranes.

In situ experiments are important to materials science in that they attempt to replicate a real-world condition that a material will experience during use, and concurrently, image the morphological changes within the material. X-rays are critical to this understanding in that they usually do not affect the outcome of the experiment; however, for some soft and polymeric materials, the X-ray intensity during synchrotron experiments may affect the molecular structure. Performing preliminary measurements and understanding how the data is collected can improve the success during in situ imaging. For scientific success, the data collection must be thoroughly thought out to ensure that the acquisition parameters are optimal for quality reconstructions; the in situ processing conditions must be close to real-world conditions so that the material's response is scientifically meaningful.

6.3 Experimental Rates

Depending upon the rate at which the observed phenomena occurs (either by experimentalist decision or by the laws of physics), several 'styles' of in situ observations are practiced [87]. These include:

- ex situ tomography
- pre/post mortem in situ tomography,
- interval in situ tomography,
- interrupted in situ tomography,
- dynamic in situ tomography.

Each of these styles is shown graphically in Fig. 6.3. These styles are listed by the correlation between experimental rates to the in situ experiment. Each of the vertical red bars represents the acquisition of a 3D image. The diagonal black line represents the stimulus applied to the sample (e.g., mechanical load, heat, corrosion, electrochemical). The choice of modality is dependent upon the imaging rate and the experimental rate of progression. The critical aspect of in situ imaging is that the tomographic imaging must be significantly faster than the change in the structure of the material. Otherwise, the reconstructed 3D image will have significant image blur and loss in image resolution. In reality, during a static CT acquisition, the only motion of the sample permitted is the theta rotation. Therefore, the imaging rate must be calculated based upon the experimental rate. However, there are techniques to overcome this limitation, including iterative reconstructions [88] (but that adds another layer of complexity to the reconstruction of the images). In 'ex situ tomography' (Fig. 6.3a), a 3D image is acquired before the experiment and another 3D image is acquired after the experiment. This technique is practiced when an in situ apparatus has either not been developed or cannot be used in conjunction with the CT instrument. The lack of imaging during the progression of the experiment causes a loss of information in morphological changes that occurs between the two tomograms. For 'pre/post mortem tomography', also represented by Fig. 6.3a, the experiment is performed within the CT instrument but the imaging data is collected before and after the experiment. The progression data is still lost but registering the two images (e.g., aligning for digital volume correlation, tracking morphological feature progression, or formulating before and after comparisons) is much simpler. Figure 6.3b shows the progression of an 'interval in situ experiment'. The progression of the stimulus is so slow that a 3D data can be collected without blurring of the tomographic image [80]; therefore, the mismatch in experimental rate and imaging rate does not require the removal or stopping of the external stimulus during imaging. Figure 6.3c depicts an 'interrupted in situ' experiment [63, 64, 69, 89, 90]. The stimulus is applied and held or removed while imaging, followed by continuation or reapplication of the stimulus in an increasing pattern. A great deal of information can be collected on the progression of the change in the material; however, this technique may not provide a true picture of the behavior of the material. For example, consider an experiment in which a hyperelastic material (e.g., a soft polymer foam or a marshmallow) is subjected to an incremental compressive load. In order to image the material at 10% strain, the compressive load must be held for the duration of the imaging time. However, a hyperelastic material may continue to flow for a duration of minutes to hours and the material must relax before the image is collected. This relaxation may blur the image. This requirement leads to the loss in high quality information on the deformation of the material. This effect has been observed in the interrupted in situ imaging of a silicone foam under uniaxial compressive load in a laboratory-based X-ray microscope operating in CT mode. The stress versus time and displacement versus time of the silicone foam is shown graphically in Fig. 6.4a [62]. To collect seven CT images (i.e., tomograms), 1.5 days was required in instrument time. Due to the material relaxation, structural information is lost. The reconstructed images of the material undergoing this static compressive load shows a uniform compression, which may not be true [63].

Ideally, and especially, for fast-acting processes (e.g., high strain rate mechanical loading or solidification), the ability to collect X-ray tomograms at very high rates is critical to completely capturing the dynamic processes that occur, shown graphically in Fig. 6.3d. This imaging technique can continue throughout the dynamic process at a rate either high enough to not blur the image, or at a slightly lower rate than the experimental stimulus which would cause a slight blur in the resulting reconstructed tomograms (with advanced post processing, some of the blur can be removed). Figure 6.4b shows the stress versus time and displacement versus time curves of a silicone foam collected during a 'dynamic in situ experiment' (Fig. 6.3d). Collecting a series of tomograms during this entire experimental cycle is critical to understanding how materials deform and break. Similarly, temperature curves can be correlated to tomographic images during metal alloy solidification. The dynamic process (e.g., mechanical load, temperature, or corrosion) is being applied to the sample continuously while the 3D images are simultaneously collected. With these experimental measurements, a true picture of the changes in the material are collected [33, 70]. The advantage of this high rate imaging technique is that the experiment is not paused or slowed for the data collection. Very fast tomograms are collected and can even be parsed so that the moment of the critical event can be captured in 3D.

After collecting the in situ tomographic data (which can be gigabytes to terabytes, depending on the experiment), the data must be processed and analyzed. Processing multiple gigabytes of data in a meaningful way such that it is accurate, repeatable, and scientifically meaningful is the challenge for the experimenter. This book chapter will focus on the multi-step, multi-software package, multi-decision making process. The initial in situ experimental data collection is often the simplest and least time-consuming step. Reconstructing, processing, rendering, visualizing, and analyzing the image data requires significant computational resources and several computer programs, each requiring operator input.

Collecting the data, starting with the radiographs, then reconstructing, filtering, rendering and visualizing the 3D data, segmenting for the phases of interest (e.g., voids, material 1, material 2, cracks, phase 1, phase 1, etc.), processing to collect morphological statistics, interpreting these statistics, generating meshes of the 3D data as a starting point for modeling the performance, correlating each of these morphological measures, additionally correlating the in situ data (e.g., load, temperature, etc.) to the images as well as to orthogonal measures (e.g., nanoindentation, XRD, elemental composition, etc.) and finally drawing scientific conclusions are all much more time consuming than the actual data collection. There are approximately eight distinct steps in processing in situ tomographic data in materials science. The steps are:

1. Experimental and Image Acquisition
2. Reconstruction
3. Visualization
4. Segmentation

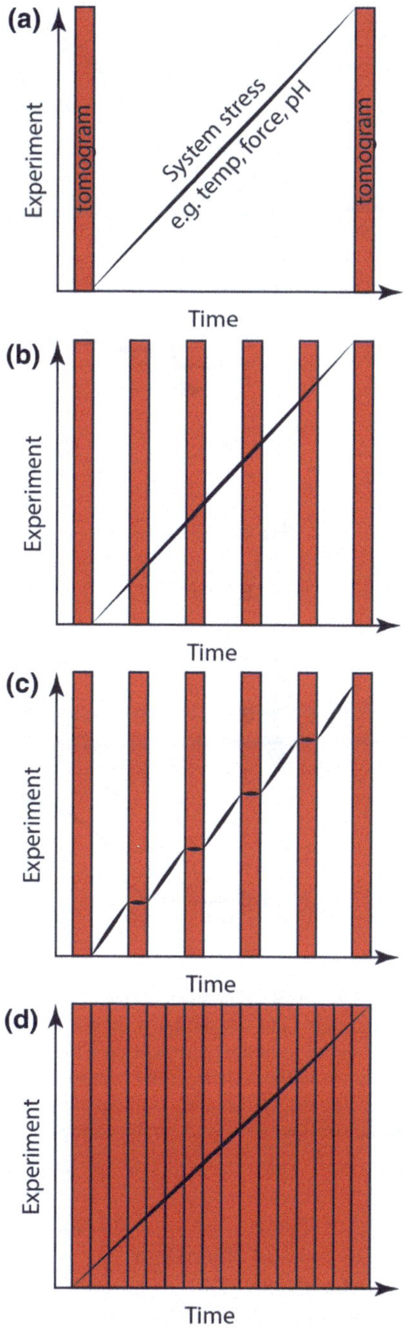

Fig. 6.3 The four types of in situ experiments. The increasing trend (black line) represents the conducted experiment. It may be increasing mechanical load (compression or tension), changing temperature, voltage, or concentration. The red bars represent the collection of the tomographic data. Graph **a** represents the collection of a series of radiographs, then some stressor applied to the material followed by a series of radiographs. Graph **b** represents the collection of CT images while the stressor is slowly applied. The third graph **c** represents the interrupted in situ collection of data with a paused experiment. Finally the graph **d** represents the dynamic in situ experiment where the CT images are rapidly collected. The method used depends upon the imaging rate available as well as the rate of change in the material

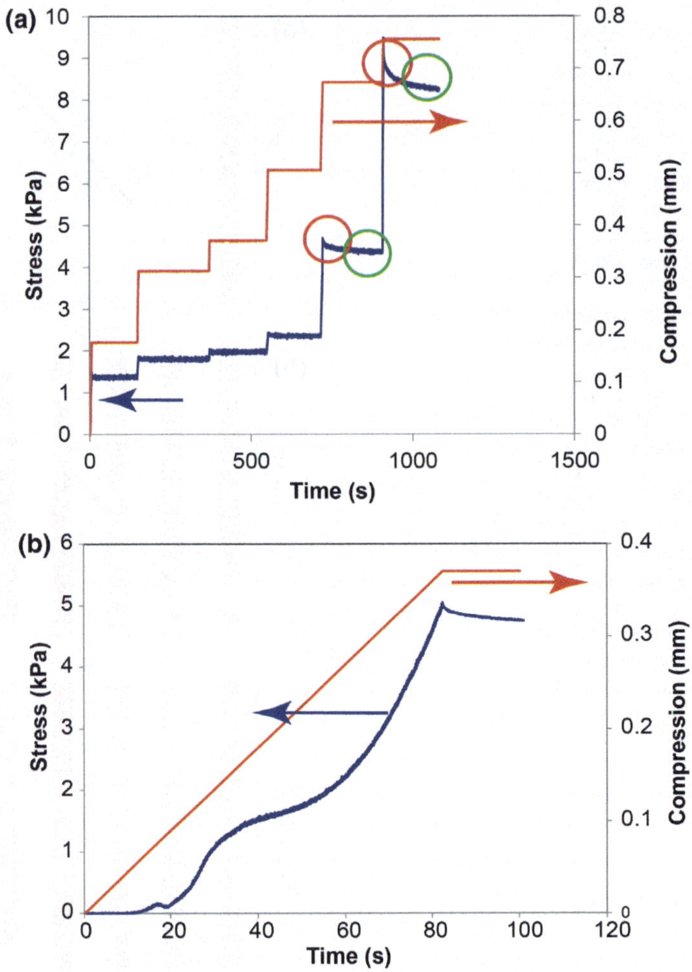

Fig. 6.4 Displacement versus time curves (red) and stress versus time (blue) curves acquired using an interrupted in situ modality (**a**) and dynamic in situ (**b**) experiments of a soft polymer foam. The interrupted in situ experiment (see Fig. 6.3c) must be paused (i.e., the application of the stress), as shown in the red circle in order to collect the 3D image (green circle). Therefore, information regarding the deformation of the material is lost. In the dynamic in situ experiment (see Fig. 6.3d), a true stress-strain curve can be collected and then correlated to each 3D image

5. Advanced Analysis and Data Processing
6. In situ and other Data
7. Modeling
8. Scientific Conclusions

Figure 6.5 graphically outlines the progression from data collection to the production of answers to the materials science challenges. The complexity not only lies in the sheer number of steps, but also in the multiple decisions that need to be made in every step of this schema. Additionally, passing the data through each of these steps may require entirely different software packages for each step! Each of these steps is an active area of research. They are active for simplifying the process itself, improving its accuracy, and understanding the processing steps boundary conditions. At the end of the chapter, future directions will be outlined in automated data processing, leaving the reader with a strong understanding of what techniques are available, which time and size scales are used, which areas of technique development are active, and which areas are needed for future growth.

6.4 Experimental and Image Acquisition

X-ray tomography begins with simple 2D X-ray radiography. The radiograph provides a 2D image of the material. The geometry of the measurement is that an object of interest is placed between a X-ray source and detector [91]. A digital radiograph is collected, which may be several megabytes in size and is often viewable as a .tiff, HDF5, or other image format. Interpretation is relatively straight forward and simple measures of the object's density and size may be obtained. However, the 3D information is convoluted into the 2D image; therefore, structural information is lost in this direction. In order to retrieve the third spatial dimension of information, a series of digital 2D radiographs are collected as either the specimen or the imaging equipment is rotated (the latter configuration is standard for medical CT). For 3D tomography, a series of radiographs is collected by shining a beam or cone of X-rays through a material while the sample is rotated. Just as in 2D imaging, the X-rays are absorbed by the material, an amount proportional to the materials electron density. The rotation angle may be anywhere between 180° to a full 360°. The number of radiographs collected are typically between a couple hundred to a few thousand. Figure 6.2 shows the geometry of an in situ loading experiment. An in situ rig, containing the sample to be tested, must be placed at the location of the sample.

As mentioned previously, the integration time for each radiograph is proportional to the brightness of the X-ray source. A variety of X-ray sources are available to researchers including fixed anode, rotating anode, liquid metal jet, and synchrotron. Fixed anode, rotating anode, liquid metal jet X-ray sources, and novel compact light sources are all available in the laboratory, whereas synchrotron X-rays are only available at national user facilities. Each of these laboratory sources produce a polychromatic beam (or cone) of X-rays to shine on the sample. By coupling with optics, it is possible to reduce the chromaticity of the beam; although, due to the brightness limitations, this is typically only performed at the synchrotron. Synchrotron X-rays offer more flexibility in the X-ray energy and flux and experimental design that may not be possible with laboratory-based systems.

Each radiograph must have sufficient exposure time so that the signal-to-noise level is high enough for proper reconstruction. This level may be governed by the reconstruction software itself. The flux of the X-ray source governs the speed at

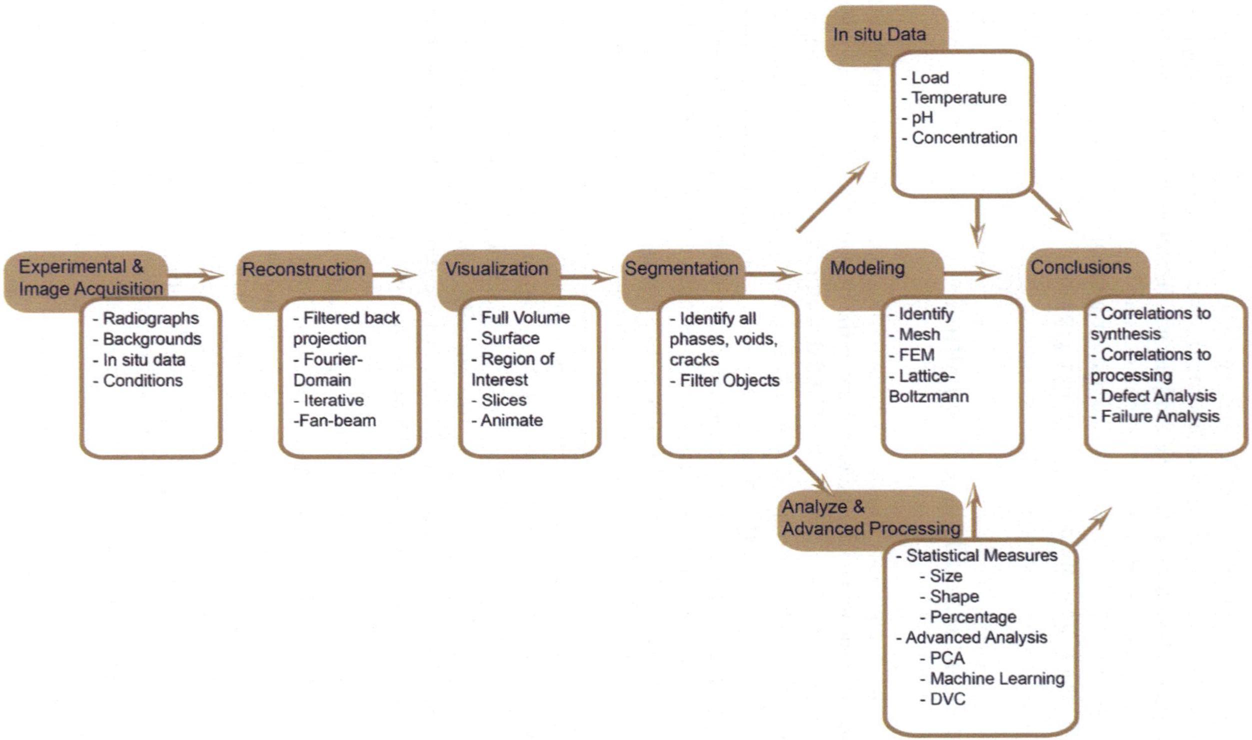

Fig. 6.5 Outline of the workflow required for in situ X-ray tomographic imaging, from collecting the in situ data to answering the scientific challenges. Often, each of these steps requires a different software package and a multitude of decisions by the user. The metrics that can be extracted are diverse, from the percent void volume, to advanced analyses with digital volume correlation or principle components analysis. The likelihood of similar decisions and methods being used by one research team to another is probably very low

which the individual radiographs may be collected. If the flux is high enough, then the scintillator and detector governs the frame rate. For laboratory-based CT systems, individual radiographic frame rates of ~0.1–0.01 s^{-1} are typical. To minimize the reconstruction artifacts, the optimal number of radiographs per tomogram collected should be ~$\pi/2$ times the number of horizontal pixels on the detector [92]. The number of radiographs times the integration time per radiograph plus some delay between images determines the approximate total time for each tomogram to be collected. This leads to full CT images collected in approximately 2–18 h. Synchrotron-based tomography systems have frame rates from ~0.01 to ~20 Hz [70]. For clear reconstructions of the 3D images, any motion by the sample must be significantly shorter than a few voxels over the imaging time of each tomogram, or special compensation techniques must be implemented to correct for this motion.

Experimentally, in situ experiments require a rig that applies the stress to the material [93], that must not obfuscate the X-ray CT measurement, that must be controllable remotely, that must operate on the timescale that is useful for the imaging rate, and that must be coordinated with the imaging technique. Figure 6.2 shows an in situ load rig or apparatus inside of a synchrotron beamline. The geometry for laboratory-based systems and synchrotron-based systems for compression or tension measurements are basically identical, although synchrotron systems have more space to build larger rigs. Common between them is the open X-ray path through the rig, the sample, and onto the detector. A ring of uniform composition (e.g., Al, plastic, carbon fiber composite) and thickness is present at the imaging plane. It must be uniform to maintain a consistent flux of X-rays through the sample as it is rotated [94]. Cabling is present to record readout signals and drive the motor. The cabling must either be loose (for single rotations of the stage) or have a slip-ring for multiple sequential rotations. In XRM-scale in situ studies (10's of micrometer fields-of-view, ~10's of nm resolution), low keV X-rays are often used (e.g., ~5–10 keV); therefore, due to low penetration energy, the rig support is often a counter arm [69]. This reduces the angles to be used for reconstruction but removes the artifacts due to absorption of the X-rays by the collar. Due to weight requirements, in many thermal solidification experiments, the sample is mounted on a rotary stage, but a furnace is mounted around and suspended above the sample, with a pair of holes for the X-rays to pass through [95].

Data acquisition consists of radiographs that are typically 1k × 1k pixels and 16 or 32 bit dynamic range. Typically, several hundred radiographs are collected for each tomographic data set. Six radiographs (Fig. 6.6a–f), out of the 10's of thousands that are collected for one in situ experiment, along with a bright (Fig. 6.6g) and dark image (Fig. 6.6h), shows just a miniscule amount of data collected during an experiment. The radiographs are the images as a polymer foam sample passes through 0° rotation at increasing compressive strains. Therefore, each tomogram is often several gigabytes in size. An in situ CT data set can be 10's of gigabytes. At an acquisition of one in situ data set per 30 min, it is possible to collect upwards of ~2 million radiographs (translating to ~7 terabytes in size) at the synchrotron per weekend. Unautomated, this may involve over 120 individual samples, subdivided into groups that each have their own acquisition parameters. Robotic automation and

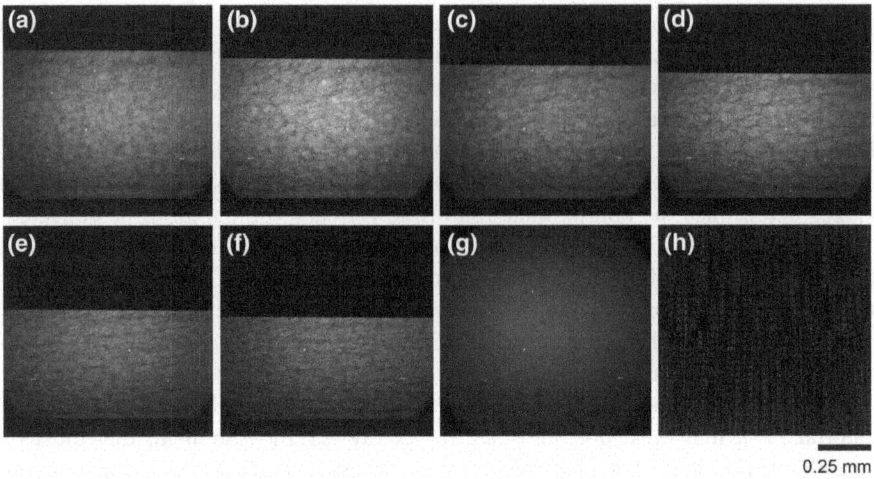

Fig. 6.6 A series of radiographs collected using synchrotron X-ray tomography as the sample passes the 0° rotation at increasing stress (**a**–**f**). A bright field (**g**) and dark field (**h**) image is shown for comparison. Each of these radiographs are interspersed with thousands of other radiographs as the sample is rotated. Depending upon the conditions and reconstruction software, every 180 or 360° of rotation are then used to reconstruct into a single 3D rendering

remote access for CT data collection allows for the changing of 100's of samples, albeit not in situ, per day [96]. Thanks to advances in hardware storage and data transfer rates, collecting and saving this data is not currently a data challenge. The challenge is the post processing.

Concurrently, during the in situ CT data collection, each experimental stimuli's data must be collected and saved in a format that can be later correlated back to the radiographic or tomographic data. Load data can be directly read out, and from a calibration equation, the stress can be directly measured. The strain can be encoded within the drive motor or measured from the radiographic images. Thermal conditions can be measured using embedded thermocouples, however, there may be some error in this measurement since the sample must be rotated during imaging. Directly placing the thermocouple on the sample is impossible.

Challenges in the data acquisition include using the appropriate in situ apparatus, choosing the correct image acquisition parameters, identifying the X-ray energy and flux for the needed imaging rate and contrast, as well as coordinating the in situ data collection from the experimental apparatus. Optimally setting these conditions may require months of preparation.

6.5 Reconstruction

Reconstruction of the collected radiographs is the mathematical calculation and conversion from the series of collected 2D radiographs into a stack of individual slices through the material (i.e., tomogram). Types of reconstruction techniques include filtered back-projection (the most common), cone beam, Fourier transform [97], fan beam, iterative [98], Radon transform, and others. All commercial XRM and X-ray CT instruments provide their own proprietary reconstruction software and at a minimum, a simple 3D rendering package. The type of reconstruction process used depends upon how the data is collected. Synchrotron facilities often use in-house or open source software for the reconstruction. One of the most common ones is Tomopy [99]. It is used at Argonne National Laboratory's Advanced Photon Source (APS) and Lawrence Berkeley National Laboratory's Advanced Light Source (ALS).

A typical reconstruction must manage a wide variety of instrument, X-ray source, sample, and in situ apparatus conditions. Therefore, several user decision-making steps are required. These include: parsing the data (e.g., determining the number of radiographs per reconstruction), image alignment, cropping, filtering, center shift, pixel range (to set the brightness and contrast), and artifact corrections (e.g., beam hardening [100], ring artifacts [101], edges [102], sample alignment (e.g., wobble)). There may be a dozen or more different decisions that are made by the researcher during the reconstruction process with regard to correcting for any issues.

Depending upon the reconstruction parameters chosen (e.g., whether or not the data was cropped or binned), the data size at this point will approximately double in storage requirements. Additionally, the data acquisition parameters must take into account any potential for image blur. For example, compressing a sample more than several voxels in the direction orthogonal to the rotational direction during a single reconstructed image may lead to image blur. Mertens et al. (see Fig. 6.3) [33] demonstrates this phenomena, in which an additively manufactured material 'snaps' back into place after tension-induced failure. The force of the recoil moves the sample faster than the imaging rate will allow for a clear reconstruction, resulting in significant image blur. For most mechanical studies, the strain rate can be chosen to minimize this; however, for some experiments, (e.g., metal solidification) the rate of morphological change within the solidification front within the material cannot be controlled. Some clever compensation methods have been developed, including Time-interlaced model based iterative reconstruction (TIMBIR) [88], in which the interlacing of frames from successive CT acquisitions are used. In general, model-based reconstruction [103] techniques and machine learning [104] can be used, especially in low dose situations. However, this adds a yet another layer of sophistication to the experimental design, data acquisition, and reconstruction, further increasing reconstruction decisions.

6.6 Visualization

Upon reconstruction of the 2D radiographs into the reconstructed slices, the scientist now has the opportunity to view the data in 3D for the first time. Ideally, this step would be semi-automated and would be available as part of the experimental time; however, due to the large data rate and the semi-manual development of the reconstruction parameters in synchrotron experiments, this step is often not reached for days or even months after data collection. It would be preferable to reach this point quickly, especially when conducting experiments so that the experiment can be assessed for data quality with a real-time feedback loop for an understanding of the success of the experiment. To visualize the in situ 3D data sets, the researcher must have access to computing systems that can load and render multiple multi-gigabyte datasets; therefore, a multi-core workstation with many gigabytes of RAM and a high-end graphics card is required (e.g., NVidia Quadro, AMD ATI, Intel HD Graphics).

Many software packages are available for visualizing 3D X-ray tomography data sets. Some of the more common open source software packages include: Chimera, ImageJ, OsiriX [105], Paraview, and Tomoviz. Additionally, proprietary software packages are available for rendering the 3D data including: Amira (Thermo Scientific), Avizo (Thermo Scientific), DragonFly (ORS), EFX-CT (Northstar), Octopus (XRE), and VGStudioMax (Volume Graphics). All instrumentation manufactures provide, at minimum, a package to render their data. Many are now beginning to include workflows for in situ data.

The challenge is in determining what types of visualization are most appropriate for conveying the scientific answer. Visualizing reconstructed slices (Fig. 6.7) gives the researcher the first clue to the data quality. Digitally cutting or 'slicing' through these reconstructed grayscale images can aid in visualizing void structures, inclusion frequency, or crack locations and constructing animated movies of these slice-throughs are useful for scientific presentations. However, this is a purely qualitative approach. Partial volume, full volume, or isosurface (Fig. 6.8) renderings of the reconstructed grayscale images begin to show the researcher the results of the experiment. Figures 6.7, 6.8, and 6.9 show the compression of a stochastically structured, gas-blown silicone foam as orthoslices, isosurfaces, and full volume renderings, respectively. The orthoslices are in the 'xz' direction, that is, the same orientation as the radiographs shown in Fig. 6.6 (the mechanical loading upon the sample is from the top of each rendering). This foam was imaged with 20 tomograms acquired within 100 s during uniaxial compression. Visualizing the deformation of the foam, whether on the bulk-scale or single ligament-scale, are possible [62].

The static 2D figures, presented in Fig. 6.7, of a dynamic process are an example of the complexity of conveying to the reader the time-scale of the sample motion during dynamic in situ 3D imaging. Fortunately, supplementary data on publisher websites is becoming more commonplace and are a great method for sharing animations of the in situ images; it is recommended that the use of supplementary data to publish animations of these processes should be used to the maximum extent

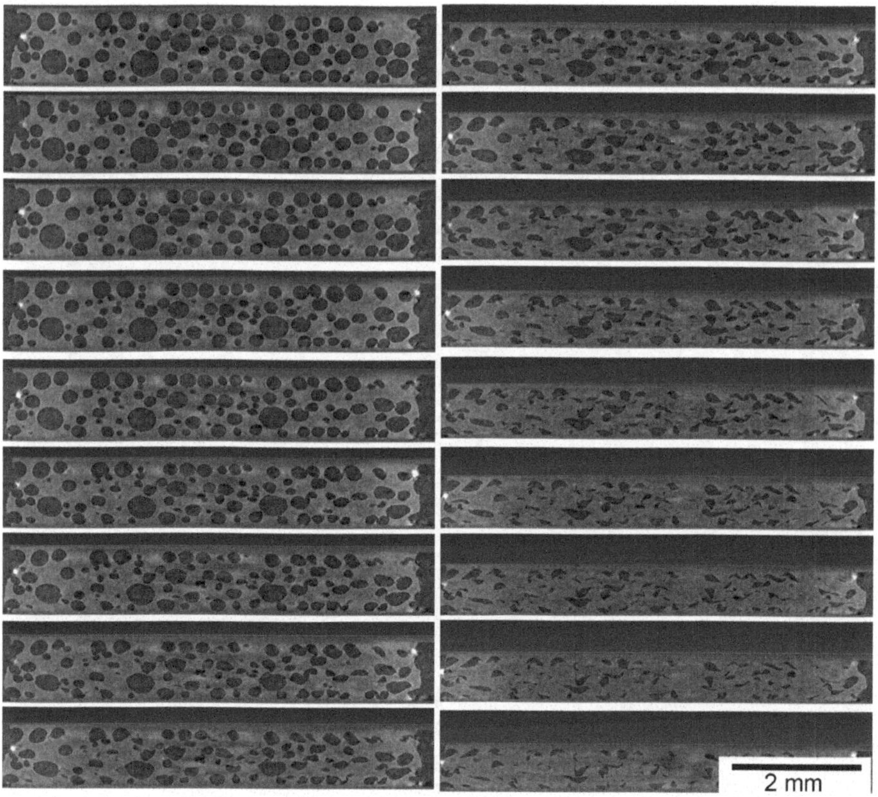

Fig. 6.7 Series of a single reconstructed slice of a polymer foam at increasing strains. Each slice represents one central slice out of approximately 1000 slices for the image. The eighteen 3D images were collected in ~5 s at a 10^{-2} s^{-1} strain rate. Finding the portion or attribute of the structure that has the largest affect upon the overall mechanical response is the challenge

possible. It is critical, when making these visualizations, to include as much information as possible. Scale bars, stress/strain values, temperatures, time-stamps, etc. can greatly improve the observers' understanding of the context and morphological change within the experiment [106]. Reporting the visualization of scientific data must also include all parameters used to render the data. Researchers must especially report the filters applied during the visualization. Additionally, researchers must keep a copy of the unprocessed data file, never adjust sub-regions of the image, manipulate all images in the series identically to improve side-by-side comparisons, and avoid the use of bad compression filters. Many of these requirements can be a challenge for in situ data. Finally, resolution and voxel counts within objects should be sufficient for accurate measurements (see the later section on analyzing 3D data).

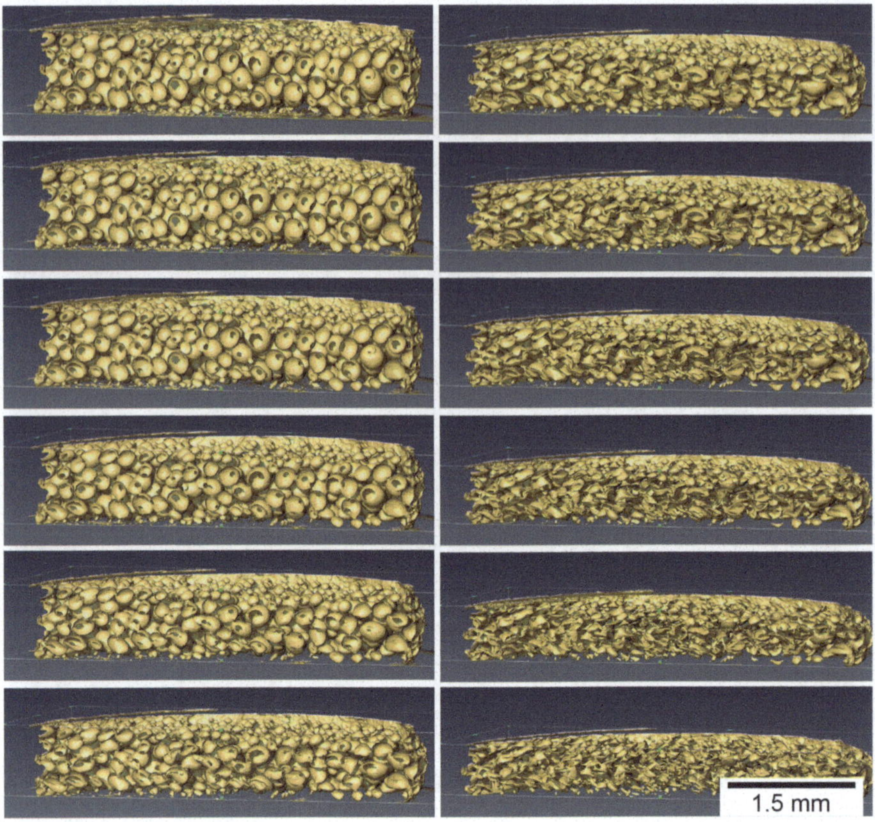

Fig. 6.8 Isosurface renderings of one half of the polymer foam shown in Fig. 6.7, at increasing strains. Flow is noted in the void collapse. Some voids are inverted during the compression

6.7 Segmentation

In situ data is constantly balancing between imaging rate and experimental stimulus rate, obtaining as many radiographs per tomogram as possible while providing enough contrast for adequate segmentation [107]. Segmentation is the act of labeling and separating the grayscale volume elements (i.e., voxels) of reconstructed image data into discrete values, thus creating groups or subgroups of voxels that constitute specific phases of the material. In order to process the data and make morphological measurements or convert to mesh surfaces for modeling, the grayscale of the reconstructed image must be segmented to reduce it down to only a few values. Typically, the data is reconstructed into 16 or 32 bit grayscale, meaning that there may be 2^{16} or 2^{32} grayscale values in an image. Ideally, the segmented images are correlated to the phases of the material, creating an image amenable for processing. Often, the segmentation of polymer foams may only contain two phases, air (i.e., voids) and

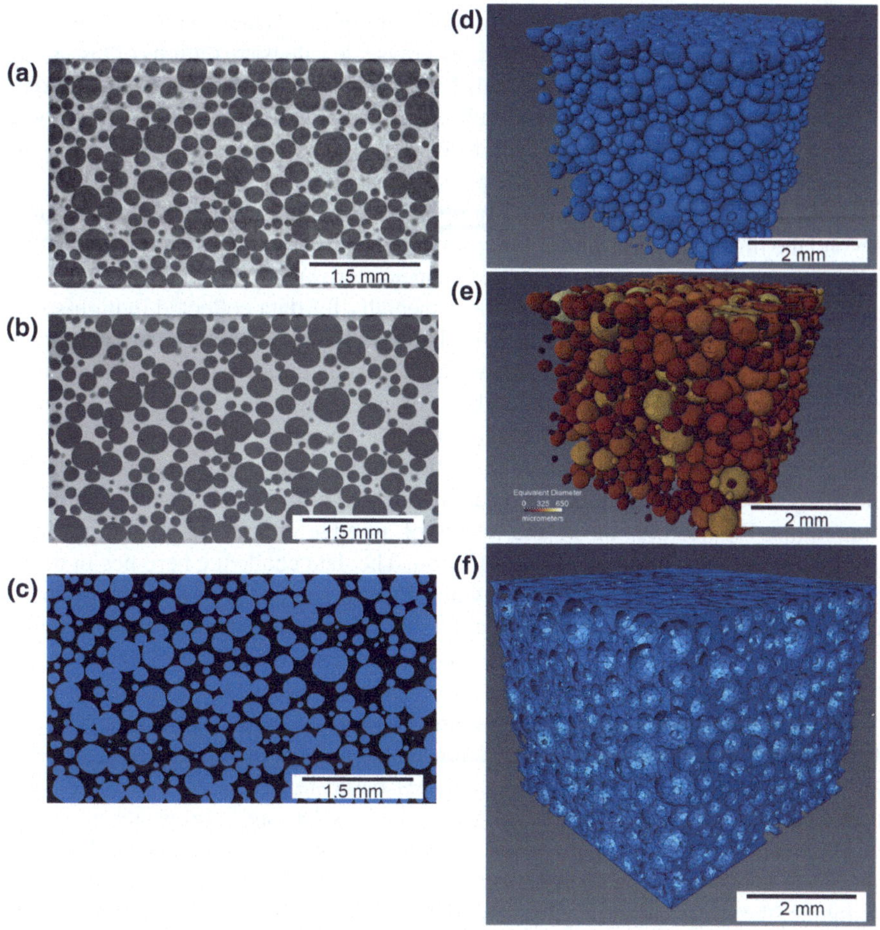

Fig. 6.9 The progression of a single image of an undeformed foam (silicone SX358) from an in situ data set from the reconstructed image (single slice shown, **a**), after filtering with an edge preserving smoothing (**b**), segmenting for the voids (**c**), rendering the voids (**d**), the voids rendered by each voids equivalent diameter (**e**), and finally converted into a mesh for finite element modeling (**f**)

the bulk polymeric material. For Al-Cu solidification experiments, there are often four phases: voids or cracks, aluminum, copper and liquid. For composite materials, there may be even more phases: voids, cracks, fibers, filler, and inclusions. There is a wide variety of techniques used to segment grayscale images. For the simplest segmentations, the grayscale should already consist of many separately grouped values. In practice, where the grayscale values may be convoluted, specialized techniques have been developed to obtain adequate segmentations.

Figure 6.9 shows the progression of a grayscale image through a simple grayscale value-based segmentation. Figure 6.9a shows one single reconstructed

16-bit grayscale slice from one in situ tomogram of a polymer foam used in a lab-based mechanical loading experiment. Low grayscale values are regions of low X-ray absorption (e.g., voids, cracks, air) while higher grayscale values represent materials of increasing X-ray absorption (e.g., bulk foam, metal inclusions). This section will describe many of the challenges in adequately segmenting the data, reducing the grayscale images, and identifying the phases present.

Often, images must be processed to optimize them before image segmentation. Beyond the image filters required for optimal reconstruction (e.g., ring removal, calculated center shifts, beam hardening), image noise reduction or image smoothing is often needed for adequate segmentation, especially for data collected in high-speed in situ X-ray CT imaging where the scintillator and detector are used at operational limits. A plethora of image filters are available that can improve the segmentation by improving the signal-to-noise ratio in the grayscale images as well as edge enhancement. Just as in 2D imaging, these filters include: mean, median, sharpening, edge preserving smoothing, Gaussian, interpolation (bilinear and bicubic), unsharpening, and many others. These filters can be applied to the full 3D data set for each of the tomograms. Figure 6.9b shows the results of an edge preserving smoothing filter [108], which mimics the process of diffusion. The data challenge here lies in determining which filter is appropriate, and which filter parameters produce the best image for segmentation. Because of the large number of options available, it is preferred that a raw reconstructed slice (before any filtering) be included in any X-ray CT manuscript in order to provide the reader an understanding of the data quality.

Once the data is appropriately smoothed, a variety of manual and automated segmentation techniques have been developed [109]. These include manual, adaptive thresholding, region growing, and techniques based upon machine learning. In a manual segmentation, the researcher may simply select an appropriate grayscale range that appears to capture the phase within the image. A simple manual threshold value was chosen for Fig. 6.9c and then rendered in 3D for Fig. 6.9d. With this technique, the distribution of grayscale values for the polymeric material and voids are separated sufficiently in grayscale and no overlap exists. For most materials, and depending upon the signal-to-noise ratio of the image, this may not be true. The segmentation conditions must be carefully chosen so that they are uniform for all of the in situ tomograms as the density of the material phases may change over the course of the experiment. Manual segmentation is only applicable for high contrast reconstructions. Automated segmentation techniques are being developed based upon the combination of several image processing steps as well as signal detection [110]. Recently, machine learning has been employed to segment X-ray tomograms [111, 112]. Training sets must be developed on separate phases within a slice or several slices and the remainder of the tomogram is used as the testing set. This technique has proven useful for both grayscale-based segmentation and texture-based (e.g., edge detection) segmentation. Most of the same software packages listed above for visualizing the data have some filtering and segmentation options available.

Upon successfully segmenting the data, it must also be prepared for modeling, quantification, and correlation to the in situ data. Of critical importance for modeling is preparing the data such that the number of facets adequately represent the 3D structure, while keeping the number of mesh faces as low as possible to reduce computation time. Quantifying the data requires separating segmented objects (e.g., splitting voids that may be connected due to resolution issues), sieving out objects that are a result of noise (e.g., single voxel objects), removing objects cut off by field of view limitations, and removing objects due to sampling errors (e.g., star features).

6.8 Modeling

The modeling and simulation of material behavior under an external stimulus is critical to understanding its properties. The solid description of its behavior is needed to make predictions, understand failure, develop improved synthesis and processing, and create better materials to meet society's needs. Modeling materials is a multiscale challenge, beginning at the atomic level, continuing through the microstructural scale, and including the bulk and system scale. The exploration of the elasticity, plasticity, fracture, thermal flow, and/or chemical changes within materials must be simulated to be understood. As seen in this chapter, nothing is more useful in verifying a model's robustness more than in the direct observation of the phenomena. Using the 3D microstructure of the material as the starting point of the modeling provides the opportunity for side-by-side, direct comparison [62] of the models'(Fig. 6.10) performance, validation, and robustness, to the experimentally-observed performance of the material. Directly visualizing the deformation in a foam, the solidification front in a metal eutectic, or the pull out of a fiber in a composite material [113] can aid in the refinement and the confirmation that the material scientist understands the physics of the material's behavior. Collecting tomographic in situ data adds a fourth dimension to the data interpretation and analysis. Having this fourth dimension of data allows the direct comparison between any processing based off the initial conditions and the true measured result. For example, using the initial structure of a polymer foam undergoing dynamic compression as a starting point for finite element analysis means that the structural changes in the material can be modeled and directly compared to its actual compression. The effects of heating and cooling upon materials can also be measured in situ. The in situ solidification of metals, metal alloys, and how the processing conditions (e.g., temperature gradient) affect the properties is critical to materials science. The challenge for the materials scientist is developing the experiments that can directly feed information (especially the physical microstructure) into the simulation code. This feed-forward process can then be used for code refinement.

3D image data are collected as isotropic voxels; each voxel has an x, y, and z coordinate and a grayscale value which is then segmented to label the phases. For this data to be used for modeling and simulation, the voxelized data must be converted into a data format that can be imported into a modelling program. This process is often referred to as 'meshing', in which the voxelized data are converted into tetrahedral elements that constitute the surface of material phases. Non-surface data are omitted from the mesh and the resulting volumes that the surfaces constitute are then considered as uniform bulk material. Once segmented and meshed, (depending upon

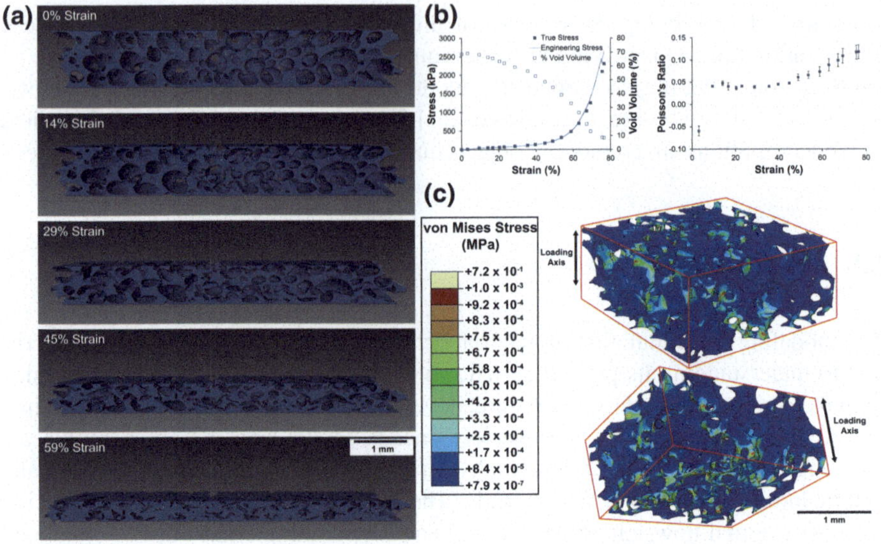

Fig. 6.10 A series of 3D reconstructed foams at increasing strains are shown (**a**). The stress-strain curve, change in percent void volume, and change in Poisson ratio are also shown (**b**). The undisturbed image was used for FEM modeling. The image had to be cropped and reduced in mesh faces to ease in 3D modeling computation (**c**)

the surface area of the interfaces within the material), tens of millions of tetrahedral elements can be created. To reduce the computational burden, the structure is often down-sampled (by many orders of magnitude), cropped (to reduce the volume of the sample), or reduced by removing small features that are less consequential to the overall performance of the modeled result. Each of these decisions can vary from researcher-to-researcher and can affect the quality of the model's robustness.

There are many software packages available for modeling, whether it is finite element modeling (FEM) (e.g., Aphelion [114], Abaqus [55, 115, 116], Python openCV [30]), microstructural modeling, particle-in-cell (e.g., CartaBlanca [117]), or others [118]. However, each program requires intensive computing resources and time (especially if the modeling is carried out in 3D). To the authors' knowledge, there is no metric for the direct comparison of a model's performance to the actual change in structure. Developing the ability to overlay the modeled FEM result to the experimental structure and obtaining a simple distance map could provide rigorous insight into the quality of the experiments and the modeling efforts.

6.9 In Situ Data

Collecting in situ data (e.g., force-displacement curves, thermal cycling profiles, or current-time curves) during the experiment and correlating the data to the images is critical for model development and making extrapolations to the causes of the changes within the material. For example, in a simple compression experiment, the compression motor can be calibrated and can compress to the sample at a time and rate of the experimenter's choosing. The true strain (in contrast to the engineering strain) can then be measured from the radiographs or tomograms. The force measured by the loading apparatus can be converted to true stress by taking the area of the sample from the reconstructed tomograms. These two simple conversions yield a stress-strain curve of the material deformation and are relatively easy to collect. For simple laboratory-based experiments, long signal cables are required; for dynamic experiments, slip-rings are required for the theta stage to rotate continuously. However, some in situ measurements are not so straightforward. In an in situ heating experiment, the true temperature of the sample may be difficult to measure. The heating of the sample is often conducted by a furnace [94], laser [49, 119], or high intensity lamps [76]. Calibrating and measuring this system can be a significant challenge as including thermocouples to the rotating sample is non-trivial. In operando experiments during thermal runaway of batteries using a thermal camera [120] eases the measurement of the temperature in that the stand-off camera can directly observe the rotating specimen, but certainly dealing with the decomposing battery creates its own unique challenge. Software is beginning to appear on the market for other in situ techniques, such as electron microscopy (e.g., Clarity Echo), but it is not currently automated in any tomography software package. Software will be needed to not only render and analyze the data but also correlate it back to the other measures.

6.10 Analyze and Advanced Processing

Taking the 3D image data beyond a qualitative understanding and turning it into a truly quantitative dataset requires the collection of measures and metrics of the material. For example:

- Polymer foams with large voids exhibit different compressive properties than foams with small voids [64]. Are voids that are $\pm 10\%$ in size enough to change the Poisson ratio of the material?
- How does the cooling rate of a metal affect the thickness of the eutectic structure [59]? How does this processing affect the mechanical, corrosive, and elastic properties?
- How far will a crack will travel through a metal during cyclical testing [46] and can it vary with its exposure to a corrosive environment?
- How much internal damage within a battery becomes catastrophic to cause thermal runaway [120]? What level of electrode breakdown is too much for the material to remain functional?

The ability to correlate quantitative numbers to morphological features within the 3D structure turns X-ray CT into a powerful analytical technique. As outlined by Liu et al. [121], many options are available after reconstruction for data analysis, including direct observation, morphological quantification, or network extraction. Other methods may include digital volume correlation, principal components analysis [122], or machine learning to extract quantification information from in situ X-ray CT data. Collecting multiple terabytes of X-ray radiographs, reconstructing them, processing them, segmenting them, rendering them, and converting the resulting data into movies can provide a great qualitative picture of what is occurring in a material while it is in operation.

After the reconstructed X-ray CT data is processed and segmented, 3D metrics of the phases of interest can be obtained. This is critical to obtaining quantitative information. Without solid quantitative information, it is not possible to visually compare samples and make determinations as to whether the material is different as a result of formulation or processing. The quantitative information may include thickness (e.g., thickness of the solidified eutectic), percent void volume, particle (or void) morphology (e.g., size, shape, equivalent diameter, Feret shape, orientation, center of mass, distance from other objects, and connectivity, just to list a few) can be obtained for each sample and at each step within the in situ experiment. It is possible to collect dozens of pieces of unique metrics on 10's of thousands of objects within the sample. Figure 6.11 shows the progression of some of the metrics from an in situ experiment of a polymer foam as it is being compressed. The initial results provide a tabulated list (Fig. 6.11b) of each object in the image and its metrics. Simple histogram plots (Fig. 6.11c) is used to give an idea of the distribution of each of the metrics (shown are the Feret shape, orientation theta, and equivalent diameter). These metrics show the increase in the Feret shape (aspect ratio), the increase in the randomness of the long axis of the void (orientation theta), and the decrease in the size of the voids (equivalent diameter). Additionally, each of the objects (voids) are individually labeled with each of the metrics and therefore a color scheme can be applied such that the objects can be colored by their metric values. Figure 6.11d shows the three images of the compressed foam with each of the voids colored by equivalent diameter. Lighter colors represent larger objects, darker represents smaller objects. Each of the metrics collected for each of the objects may be treated in this way. Correlating these changes to the in situ metrics can provide interesting insights into which metrics affect the changes in the material the most. For example, Fig. 6.11 in Patterson et al. [62], correlates the stress-strain curve with the change in percent void volume. Inflection points in each metric show how the void-collapse correlates to the bending, buckling, and densification of the ligaments within the structure to understand the changes in morphology upon the applied compressive strain and hyper elastic response.

The caveat in using these metrics is that the 3D objects must be imaged with sufficient resolution such that the voxel count within each object is high enough to remove the quantized nature of the measurement. This must be taken into account so that accurate metrics of the sample are obtained. For example, if an object is segmented that is only one voxel, the accuracy and precision of the measure of its surface area would not be believable. Filtering out objects below approximately 1000 voxels in size can reduce the absolute error of the measurement to below ~10% [123]. Sieving the objects can reduce the noise and improve the robustness of the

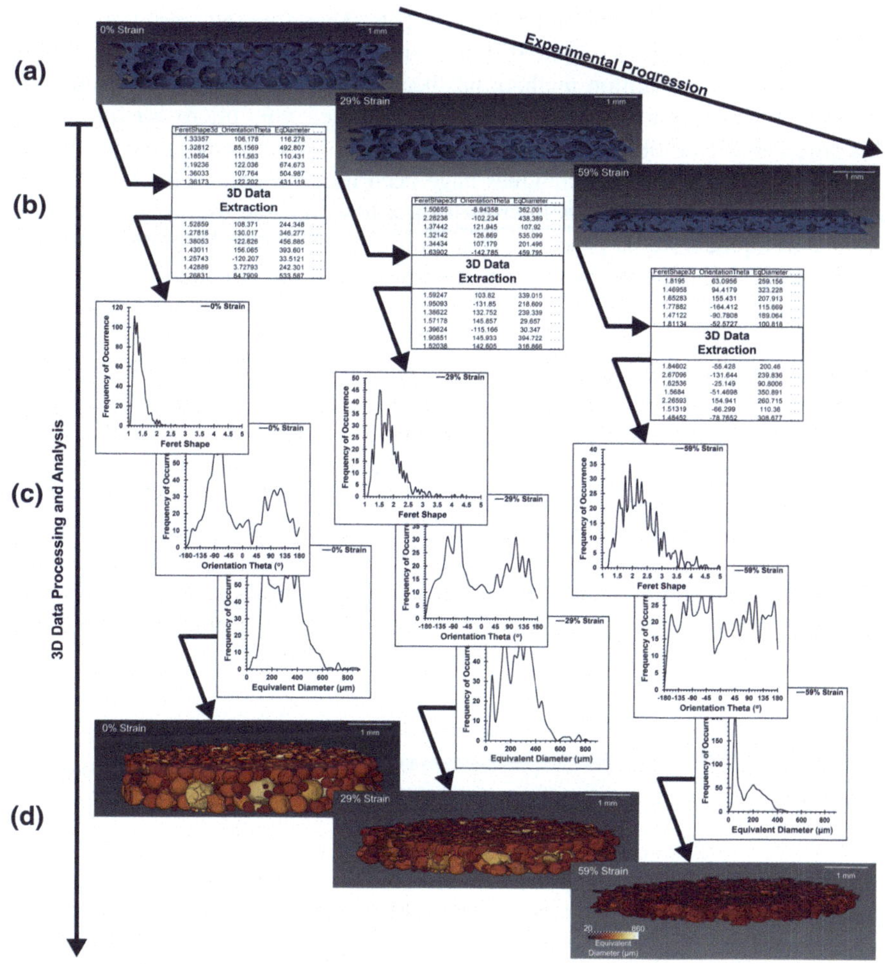

Fig. 6.11 Graphic showing the data challenge of in situ tomography. Dozens of 3D images may be collected (**a**), each one measured for a plethora of metrics (e.g., % volume of each phase, object/void size, shape, orientation, location, etc.) and put into tables (**b**), histogram graphics (**c**), and even color coding by one of these metrics. In this case, the voids are colored by equivalent diameter (**d**)

measurement. Proper sampling of the objects is critical to accurate measurements [124].

Some materials may contain 100's of thousands of objects [63]. In order to effectively collate and parse through this tremendous amount of information, higher order processing is needed. Simple histogram plots can illustrate shifts in these metrics, but discovering which metrics relate to material processing or which metrics adequately describe the experimental results can be difficult. Measuring a dozen different statistics will create too many values to provide a causal picture to relate the morphology to the results of the in situ experiments. Therefore, advanced processing and analysis steps may be required. For example, principal components analysis (PCA), a pattern recognition method, has been used to reduce the dimensionality and differentiate several polymer foams based on their void microstructure [125]. This differentiation is difficult to do with only one metric and impossible to conduct visually. Machine learning techniques have been used to relate a polymer foam's compressive performance to its microstructure; however, the use these techniques for the development of 3D property-structure-function relationships is an emerging sub-discipline and requires a significant research and development effort among 3D materials scientists.

The segmentation of phases also allows for other advanced analysis, including particle shape analysis [126, 127] and digital volume correlation (DVC). While statistics such as void axis ratios [128] in damaged materials can give great information regarding the growth mechanism, caution must be exhibited to assure that all of the voids are rendered with enough voxels to yield meaningful data, as mentioned previously. DVC is the 3D analogue to digital image correlation (DIC) and is used to track features between multiple images. This technique can track the evolution of strain, flow, crack propagation, or deformation in materials imaged while undergoing an external stimuli (Fig. 6.12). 3D studies of material damage by combined X-ray tomography and digital volume correlation [129–131], that correlate to the morphological statistics as well as to the modeled result can be used to measure the robustness of the model.

6.11 Conclusions

From its launch in 1990 through 2013, the Hubble Space Telescope collected approximately 45 terabytes of data on the universe [132], which is a rate of approximately two terabytes per year. Processing this data takes years before they are viewable to the public. At a synchrotron, it is possible to collect 3D X-ray CT data at a rate of greater than a terabyte per day. Add to it the challenge of reconstruction, rendering, segmenting, analyzing, modeling, and any advanced processing and correlation to the in situ data means that without automation, a very large percentage of the data may never even be examined. Additionally, the number of steps in each portion of the process means that dozens, if not hundreds, of decisions are made that can affect the quality and outcome of the analyzed data.

Ongoing work has focused on automating and batch processing many of the steps used in processing the data. Many of the commercial software packages are now including TCL and Python programming options for this batch processing. Once the appropriate processing conditions can be determined, applying them to the in situ data sets as well as multiple samples is possible. Future work needs to include removing the decisions for optimal data processing from the user and using machine learning to do this automatically.

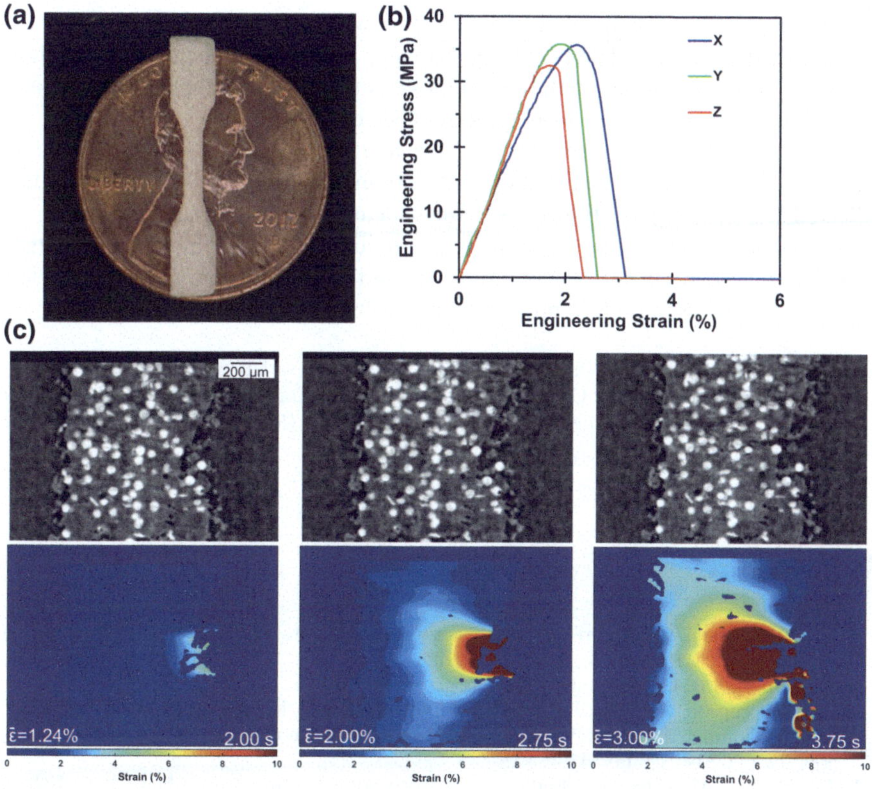

Fig. 6.12 Analysis of in situ tomographic images of a 3D printed tensile specimen using digital volume correlation. The specimen must be small to fit within the X-ray beam of the synchrotron (**a**). The stress-strain curves of three different formulations relating the elasticity of the material to its processing (**b**). Three reconstructed slices at increasing stress and the corresponding digital volume correlation maps (**c**) showing the propagation of the stress field from the notch. The glass bead inclusions provide a handy fiducial for the DVC. This data shows many interesting data points including the uniform distribution and size of the glass filler, the ultimate tensile strength of the material, the delamination of the filler from the nylon polymer, and the strain field progression during failure

Each of the steps in the process are often made in different software packages. Tracking which decisions are made, understanding how they affect the final outcome, saving the data at the appropriate processing steps, saving the software and conditions used to process the data, and doing so in a repeatable format is a daunting task. In addition to the challenge of data sharing, due to the multistep nature of this challenge, a knowledge of error propagation is critical [133]. In practice, manual segmentations may be practiced under various conditions to better understand how small value changes can affect the morphological statistics; but this is one decision out of a multitude of decisions. Some work has been published in which some of the processing steps may be skipped in order to reduce the processing time, but things may be missed. Knowledge as to whether this was successful may not occur for several months after the data is collected. Finally, linking the changes in morphology of the structure, observed during the in situ experiment, to the formulation and processing of the material is the holy grail of 3D materials science.

The combination of in situ experiments in real time with 3D imaging is an extremely powerful analytical technique. Processing the tremendous amount of data collected is a daunting and time consuming endeavor. With continued development, image analysis cycle time will continue to be reduced, allowing materials scientists to run multiple experiments for improved scientific integrity, and allowing a better understanding of the structure-property relationships within materials.

Funding Funding for the work shown in this chapter are from a variety of LANL sources including: the Enhanced Surveillance Campaign (Tom Zocco), the Engineering Campaign (Antranik Siranosian), DSW (Jennifer Young), and Technology Maturation (Ryan Maupin) in support of the Materials of the Future.

References

1. G.N. Hounsfield, Computerized transverse axial scanning (tomography): Part 1. Description of system. Br. J. Radiol. **46**(552), 1016–1022 (1973)
2. J.C. Elliott, S.D. Dover, X-ray microtomography. J. Microsc. **126**(2), 211–213 (1982)
3. A.C. Thompson, J. Llacer, L. Campbell Finman, E.B. Hughes, J.N. Otis, S. Wilson, H.D. Zeman, Computed tomography using synchrotron radiation. Nucl. Instrum. Methods Phys. Res. **222**(1), 319–323 (1984)
4. C. Bressler, M. Chergui, Ultrafast X-ray absorption spectroscopy. Chem. Rev. **104**(4), 1781–1812 (2004)
5. G. Renaud, R. Lazzari, F. Leroy, Probing surface and interface morphology with grazing incidence small angle X-ray scattering. Surf. Sci. Rep. **64**(8), 255–380 (2009)
6. F. Adams, K. Janssens, A. Snigirev, Microscopic X-ray fluorescence analysis and related methods with laboratory and synchrotron radiation sources. J. Anal. At. Spectrom. **13**(5), 319–331 (1998)
7. G.J. Havrilla, T. Miller, Micro X-ray fluorescence in materials characterization. Powder Diffr. **19**(2), 119–126 (2012)
8. A.M. Beale, S.D.M. Jacques, E.K. Gibson, M. Di Michiel, Progress towards five dimensional diffraction imaging of functional materials under process conditions. Coord. Chem. Rev. **277–278**, 208–223 (2014)
9. A. King, P. Reischig, J. Adrien, S. Peetermans, W. Ludwig, Polychromatic diffraction contrast tomography. Mater. Charact. **97**, 1–10 (2014)
10. D.J. Jensen, 4D characterization of metal microstructures, in *Microstructural Design of Advanced Engineering Materials* (Wiley-VCH Verlag GmbH & Co. KGaA, 2013), pp. 367–385
11. A.R. Woll, J. Mass, C. Bisulca, R. Huang, D.H. Bilderback, S. Gruner, N. Gao, Development of confocal X-ray fluorescence (XRF) microscopy at the Cornell high energy synchrotron source. Appl. Phys. A **83**(2), 235–238 (2006)
12. B. Kanngießer, W. Malzer, I. Reiche, A new 3D micro X-ray fluorescence analysis set-up—first archaeometric applications. Nucl. Instrum. Methods Phys. Res. Sect. B **211**(2), 259–264 (2003)
13. B. Laforce, B. Vermeulen, J. Garrevoet, B. Vekemans, L.V. Hoorebeke, C. Janssen, L. Vincze, Laboratory scale X-ray fluorescence tomography: instrument characterization and application in earth and environmental science. Anal. Chem. **88**(6), 3386–3391 (2016)

14. C. Yu-Tung, L. Tsung-Nan, S.C. Yong, Y. Jaemock, L. Chi-Jen, W. Jun-Yue, W. Cheng-Liang, C. Chen-Wei, H. Tzu-En, H. Yeukuang, S. Qun, Y. Gung-Chian, S.L. Keng, L. Hong-Ming, J. Jung Ho, M. Giorgio, Full-field hard X-ray microscopy below 30 nm: a challenging nanofabrication achievement. Nanotechnology **19**(39), 395302 (2008)
15. Y.S. Chu, J.M. Yi, F.D. Carlo, Q. Shen, W.-K. Lee, H.J. Wu, C.L. Wang, J.Y. Wang, C.J. Liu, C.H. Wang, S.R. Wu, C.C. Chien, Y. Hwu, A. Tkachuk, W. Yun, M. Feser, K.S. Liang, C.S. Yang, J.H. Je, G. Margaritondo, Hard-X-ray microscopy with Fresnel zone plates reaches 40 nm Rayleigh resolution. Appl. Phys. Lett. **92**(10), 103119 (2008)
16. G. Schneider, X-ray microscopy: methods and perspectives. Anal. Bioanal. Chem. **376**(5), 558–561 (2003)
17. A. Burteau, F. N'Guyen, J.D. Bartout, S. Forest, Y. Bienvenu, S. Saberi, D. Naumann, Impact of material processing and deformation on cell morphology and mechanical behavior of polyurethane and nickel foams. Int. J. Solids Struct. **49**(19–20), 2714–2732 (2012)
18. A. Elmoutaouakkil, L. Salvo, E. Maire, G. Peix, 2D and 3D characterization of metal foams using X-ray tomography. Adv. Eng. Mater. **4**(10), 803–807 (2002)
19. E. Maire, A. Elmoutaouakkil, A. Fazekas, L. Salvo, In situ X-ray tomography measurements of deformation in cellular solids. MRS Bull. **28**, 284–289 (2003)
20. K. Mader, R. Mokso, C. Raufaste, B. Dollet, S. Santucci, J. Lambert, M. Stampanoni, Quantitative 3D characterization of cellular materials: segmentation and morphology of foam. Colloids Surf. A **415**, 230–238 (2012)
21. K. Calvert, K. Trumble, T. Webster, L. Kirkpatrick, Characterization of commercial rigid polyurethane foams used as bone analogs for implant testing. J. Mater. Sci. Mater. Med. **21**(5), 1453–1461 (2010)
22. S.G. Bardenhagen, B.M. Patterson, C.M. Cady, W. Lewis Matthew, M. Dattelbaum Dana, The mechanics of LANL foam pads, in *ADTSC Nuclear Weapons Highlights 2007*, 07-041 (2007)
23. B.M. Patterson, G.J. Havrilla, J.R. Schoonover, Elemental and molecular characterization of aged polydimethylsiloxane foams. Appl. Spectrosc. **60**(10), 1103–1110 (2006)
24. M.P. Morigi, F. Casali, M. Bettuzzi, D. Bianconi, R. Brancaccio, S. Cornacchia, A. Pasini, A. Rossi, A. Aldrovandi, D. Cauzzi, CT investigation of two paintings on wood tables by Gentile da Fabriano. Nucl. Instrum. Methods Phys. Res. A **580**, 735–738 (2007)
25. G.R.S. Naveh, V. Brumfeld, R. Shahar, S. Weiner, Tooth periodontal ligament: direct 3D microCT visualization of the collagen network and how the network changes when the tooth is loaded. J. Struct. Biol. **181**(2), 108–115 (2013)
26. P. Schneider, M. Stauber, R. Voide, M. Stampanoni, L.R. Donahue, R. Müller, Ultrastructural properties in cortical bone vary greatly in two inbred strains of mice as assessed by synchrotron light based micro- and nano-CT. J. Bone Miner. Res. **22**(10), 1557–1570 (2007)
27. U. Bonse, F. Busch, O. Günnewig, F. Beckmann, R. Pahl, G. Delling, M. Hahn, W. Graeff, 3D computed X-ray tomography of human cancellous bone at 8 μm spatial and 10^{-4} energy resolution. Bone and Mineral **25**(1), 25–38 (1994)
28. K.G. McIntosh, N. Cordes, B. Patterson, G. Havrilla, Laboratory-based characterization of Pu in soil particles using micro-XRF and 3D confocal XRF. J. Anal. At. Spectrom. (2015)
29. P. Krüger, H. Markötter, J. Haußmann, M. Klages, T. Arlt, J. Banhart, C. Hartnig, I. Manke, J. Scholta, Synchrotron X-ray tomography for investigations of water distribution in polymer electrolyte membrane fuel cells. J. Power Sources **196**(12), 5250–5255 (2011)
30. V.W. Manner, J.D. Yeager, B.M. Patterson, D.J. Walters, J.A. Stull, N.L. Cordes, D.J. Luscher, K.C. Henderson, A.M. Schmalzer, B.C. Tappan, In situ imaging during compression of plastic bonded explosives for damage modeling. MDPI **10**(638) (2017)
31. C.A. Larabell, M.A. Le Gros, X-ray tomography generates 3D reconstructions of the yeast, Saccharomyces cerevisiae, at 60-nm resolution. Mol. Biol. Cell **15**, 957–962 (2004)
32. T.G. Holesinger, J.S. Carpenter, T.J. Lienert, B.M. Patterson, P.A. Papin, H. Swenson, N.L. Cordes, Characterization of an aluminum alloy hemispherical shell fabricated via direct metal laser melting. JOM **68**, 1–12 (2016)

33. J.C.E. Mertens, K. Henderson, N.L. Cordes, R. Pacheco, X. Xiao, J.J. Williams, N. Chawla, B.M. Patterson, Analysis of thermal history effects on mechanical anisotropy of 3D-printed polymer matrix composites via in situ X-ray tomography. J. Mater. Sci. **52**(20), 12185–12206 (2017)
34. P. Tafforeau, R. Boistel, E. Boller, A. Bravin, M. Brunet, Y. Chaimanee, P. Cloetens, M. Feist, J. Hoszowska, J.J. Jaeger, R.F. Kay, V. Lazzari, L. Marivaux, A. Nel, C. Nemoz, X. Thibault, P. Vignaud, S. Zabler, Applications of X-ray synchrotron microtomography for non-destructive 3D studies of paleontological specimens. Appl. Phys. A Mater. Sci. Process. **83**(2), 195–202 (2006)
35. N.L. Cordes, S. Seshadri, G. Havrilla, X. Yuan, M. Feser, B.M. Patterson, Three dimensional subsurface elemental identification of minerals using confocal micro X-ray fluorescence and micro X-ray computed tomography. Spectrochim. Acta Part B: At. Spectrosc. 103–104 (2015)
36. J. Nelson Weker, M.F. Toney, Emerging in situ and operando nanoscale X-ray imaging techniques for energy storage materials. Adv. Func. Mater. **25**(11), 1622–1637 (2015)
37. J. Wang, Y.-C.K. Chen-Wiegart, J. Wang, In situ three-dimensional synchrotron X-ray nanotomography of the (de)lithiation processes in tin anodes. Angew. Chem. Int. Ed. **53**(17), 4460–4464 (2014)
38. M. Ebner, F. Geldmacher, F. Marone, M. Stampanoni, V. Wood, X-ray tomography of porous, transition metal oxide based lithium ion battery electrodes. Adv. Energy Mater. **3**(7), 845–850 (2013)
39. I. Manke, J. Banhart, A. Haibel, A. Rack, S. Zabler, N. Kardjilov, A. Hilger, A. Melzer, H. Riesemeier, In situ investigation of the discharge of alkaline Zn–MnO_2 batteries with synchrotron X-ray and neutron tomographies. Appl. Phys. Lett. **90**(21), 214102 (2007)
40. E.S.B. Ferreira, J.J. Boon, N.C. Scherrer, F. Marone, M. Stampanoni, 3D synchrotron X-ray microtomography of paint samples. Proc. SPIE, 7391 (73910L) (2009)
41. C. Scheuerlein, M.D. Michiel, M. Scheel, J. Jiang, F. Kametani, A. Malagoli, E.E. Hellstrom, D.C. Larbalestier, Void and phase evolution during the processing of Bi-2212 superconducting wires monitored by combined fast synchrotron micro-tomography and X-ray diffraction. Supercond. Sci. Technol. **24**(11), 115004 (2011)
42. F. Meirer, D.T. Morris, S. Kalirai, Y. Liu, J.C. Andrews, B.M. Weckhuysen, Mapping metals incorporation of a whole single catalyst particle using element specific X-ray nanotomography. J. Am. Chem. Soc. **137**(1), 102–105 (2015)
43. J.-D. Grunwaldt, J.B. Wagner, R.E. Dunin-Borkowski, Imaging catalysts at work: a hierarchical approach from the macro- to the meso- and nano-scale. ChemCatChem **5**(1), 62–80 (2013)
44. S.S. Singh, J.J. Williams, X. Xiao, F. De Carlo, N. Chawla, In situ three dimensional (3D) X-ray synchrotron tomography of corrosion fatigue in Al7075 alloy, in *Fatigue of Materials II: Advances and Emergences in Understanding*, ed. by T.S. Srivatsan, M.A. Imam, R. Srinivasan (Springer International Publishing, Cham, 2016), pp. 17–25
45. H.X. Xie, D. Friedman, K. Mirpuri, N. Chawla, Electromigration damage characterization in Sn-3.9Ag-0.7Cu and Sn-3.9Ag-0.7Cu-0.5Ce solder joints by three-dimensional X-ray tomography and scanning electron microscopy. J. Electron. Mater. **43**(1), 33–42 (2014)
46. S.S. Singh, J.J. Williams, M.F. Lin, X. Xiao, F. De Carlo, N. Chawla, In situ investigation of high humidity stress corrosion cracking of 7075 aluminum alloy by three-dimensional (3D) X-ray synchrotron tomography. Mater. Res. Lett. **2**(4), 217–220 (2014)
47. J.C.E. Mertens, N. Chawla, A study of EM failure in a micro-scale Pb-free solder joint using a custom lab-scale X-ray computed tomography system (2014), pp. 92121E–92121E-9
48. J. Friedli, J.L. Fife, P. Di Napoli, M. Rappaz, X-ray tomographic microscopy analysis of the dendrite orientation transition in Al-Zn. IOP Conf. Ser.: Mater. Sci. Eng. **33**(1), 012034 (2012)
49. J.L. Fife, M. Rappaz, M. Pistone, T. Celcer, G. Mikuljan, M. Stampanoni, Development of a laser-based heating system for in situ synchrotron-based X-ray tomographic microscopy. J. Synchrotron Radiat. **19**(3), 352–358 (2012)
50. A. Clarke, S. Imhoff, J. Cooley, B. Patterson, W.-K. Lee, K. Fezzaa, A. Deriy, T. Tucker, M.R. Katz, P. Gibbs, K. Clarke, R.D. Field, D.J. Thoma, D.F. Teter, X-ray imaging of Al-7at.% Cu during melting and solidification. Emerg. Mater. Res. **2**(2), 90–98 (2013)

51. L. Jiang, N. Chawla, M. Pacheco, V. Noveski, Three-dimensional (3D) microstructural characterization and quantification of reflow porosity in Sn-rich alloy/copper joints by X-ray tomography. Mater. Charact. **62**(10), 970–975 (2011)
52. P. Hruby, S.S. Singh, J.J. Williams, X. Xiao, F. De Carlo, N. Chawla, Fatigue crack growth in SiC particle reinforced Al alloy matrix composites at high and low R-ratios by in situ X-ray synchrotron tomography. Int. J. Fatigue **68**, 136–143 (2014)
53. J.J. Williams, K.E. Yazzie, E. Padilla, N. Chawla, X. Xiao, F. De Carlo, Understanding fatigue crack growth in aluminum alloys by in situ X-ray synchrotron tomography. Int. J. Fatigue **57**, 79–85 (2013)
54. J. Williams, K. Yazzie, N. Connor Phillips, N. Chawla, X. Xiao, F. De Carlo, N. Iyyer, M. Kittur, On the correlation between fatigue striation spacing and crack growth rate: a three-dimensional (3-D) X-ray synchrotron tomography study. Metall. Mater. Trans. A **42**(13), 3845–3848 (2011)
55. E. Padilla, V. Jakkali, L. Jiang, N. Chawla, Quantifying the effect of porosity on the evolution of deformation and damage in Sn-based solder joints by X-ray microtomography and microstructure-based finite element modeling. Acta Mater. **60**(9), 4017–4026 (2012)
56. J.J. Williams, N.C. Chapman, V. Jakkali, V.A. Tanna, N. Chawla, X. Xiao, F. De Carlo, Characterization of damage evolution in SiC particle reinforced Al alloy matrix composites by in-situ X-ray synchrotron tomography. Metall. Mater. Trans. A. **42**(10), 2999–3005 (2011)
57. H. Bart-Smith, A.F. Bastawros, D.R. Mumm, A.G. Evans, D.J. Sypeck, H.N.G. Wadley, Compressive deformation and yielding mechanisms in cellular Al alloys determined using X-ray tomography and surface strain mapping. Acta Mater. **46**(10), 3583–3592 (1998)
58. A. Guvenilir, T.M. Breunig, J.H. Kinney, S.R. Stock, Direct observation of crack opening as a function of applied load in the interior of a notched tensile sample of Al-Li 2090. Acta Mater. **45**(5), 1977–1987 (1997)
59. B.M. Patterson, K.C. Henderson, P.J. Gibbs, S.D. Imhoff, A.J. Clarke, Laboratory micro- and nanoscale X-ray tomographic investigation of Al–7at.%Cu solidification structures. Mater. Charact. **95**, 18–26 (2014)
60. E. Maire, P.J. Withers, Quantitative X-ray tomography. Int. Mater. Rev. **59**(1), 1–43 (2014)
61. C. Gupta, H. Toda, P. Mayr, C. Sommitsch, 3D creep cavitation characteristics and residual life assessment in high temperature steels: a critical review. Mater. Sci. Technol. **31**(5), 603–626 (2015)
62. B.M. Patterson, N.L. Cordes, K. Henderson, J. Williams, T. Stannard, S.S. Singh, A.R. Ovejero, X. Xiao, M. Robinson, N. Chawla, In situ X-ray synchrotron tomographic imaging during the compression of hyper-elastic polymeric materials. J. Mater. Sci. **51**(1), 171–187 (2016)
63. B.M. Patterson, K. Henderson, R.D. Gilbertson, S. Tornga, N.L. Cordes, M.E. Chavez, Z. Smith, Morphological and performance measures of polyurethane foams using X-ray CT and mechanical testing. Microsc. Microanal. **95**, 18–26 (2014)
64. B.M. Patterson, K. Henderson, Z. Smith, Measure of morphological and performance properties in polymeric silicone foams by X-ray tomography. J. Mater. Sci. **48**(5), 1986–1996 (2013)
65. H. Bale, M. Blacklock, M.R. Begley, D.B. Marshall, B.N. Cox, R.O. Ritchie, Characterizing three-dimensional textile ceramic composites using synchrotron X-ray micro-computed-tomography. J. Am. Ceram. Soc. **95**(1), 392–402 (2012)
66. F. Awaja, M.-T. Nguyen, S. Zhang, B. Arhatari, The investigation of inner structural damage of UV and heat degraded polymer composites using X-ray micro CT. Compos. A Appl. Sci. Manuf. **42**(4), 408–418 (2011)
67. S.A. McDonald, M. Preuss, E. Maire, J.Y. Buffiere, P.M. Mummery, P.J. Withers, X-ray tomographic imaging of Ti/SiC composites. J. Microsc. **209**(2), 102–112 (2003)
68. J. Villanova, R. Daudin, P. Lhuissier, D. Jauffrès, S. Lou, C.L. Martin, S. Labouré, R. Tucoulou, G. Martínez-Criado, L. Salvo, Fast in situ 3D nanoimaging: a new tool for dynamic characterization in materials science. Mater. Today (2017)
69. B.M. Patterson, N.L. Cordes, K. Henderson, J.C.E. Mertens, A.J. Clarke, B. Hornberger, A. Merkle, S. Etchin, A. Tkachuk, M. Leibowitz, D. Trapp, W. Qiu, B. Zhang, H. Bale, X. Lu, R.

Hartwell, P.J. Withers, R.S. Bradley, In situ laboratory-based transmission X-ray microscopy and tomography of material deformation at the nanoscale. Exp. Mech. **56**(9), 1585–1597 (2016)

70. E. Maire, C. Le Bourlot, J. Adrien, A. Mortensen, R. Mokso, 20 Hz X-ray tomography during an in situ tensile test. Int. J. Fract. **200**(1), 3–12 (2016)
71. N.C. Chapman, J. Silva, J.J. Williams, N. Chawla, X. Xiao, Characterisation of thermal cycling induced cavitation in particle reinforced metal matrix composites by three-dimensional (3D) X-ray synchrotron tomography. Mater. Sci. Technol. **31**(5), 573–578 (2015)
72. P. Wright, X. Fu, I. Sinclair, S.M. Spearing, Ultra high resolution computed tomography of damage in notched carbon fiber—epoxy composites. J. Compos. Mater. **42**(19), 1993–2002 (2008)
73. A. Haboub, H.A. Bale, J.R. Nasiatka, B.N. Cox, D.B. Marshall, R.O. Ritchie, A.A. MacDowell, Tensile testing of materials at high temperatures above 1700 °C with in situ synchrotron X-ray micro-tomography. Rev. Sci. Instrum. **85**(8), 083702 (2014)
74. N. Limodin, L. Salvo, E. Boller, M. Suery, M. Felberbaum, S. Gailliegue, K. Madi, In situ and real-time 3D microtomography investigation of dendritic solidification in an Al-10wt.% Cu alloy. Acta Mater. **57**, 2300–2310 (2009)
75. S.D. Imhoff, P.J. Gibbs, M.R. Katz, T.J. Ott Jr., B.M. Patterson, W.K. Lee, K. Fezzaa, J.C. Cooley, A.J. Clarke, Dynamic evolution of liquid–liquid phase separation during continuous cooling. Mater. Chem. Phys. **153**, 93–102 (2015)
76. H.A. Bale, A. Haboub, A.A. MacDowell, J.R. Nasiatka, D.Y. Parkinson, B.N. Cox, D.B. Marshall, R.O. Ritchie, Real-time quantitative imaging of failure events in materials under load at temperatures above 1,600 °C. Nat. Mater. **12**(1), 40–46 (2013)
77. A. Bareggi, E. Maire, A. Lasalle, S. Deville, Dynamics of the freezing front during the solidification of a colloidal alumina aqueous suspension. In situ X-ray radiography, tomography, and modeling. J. Am. Ceram. Soc. **94**(10), 3570–3578 (2011)
78. A.J. Clarke, D. Tourret, S.D. Imhoff, P.J. Gibbs, K. Fezzaa, J.C. Cooley, W.-K. Lee, A. Deriy, B.M. Patterson, P.A. Papin, K.D. Clarke, R.D. Field, J.L. Smith, X-ray imaging and controlled solidification of Al-Cu alloys toward microstructures by design. Adv. Eng. Mater. **17**(4), 454–459 (2015)
79. B.J. Connolly, D.A. Horner, S.J. Fox, A.J. Davenport, C. Padovani, S. Zhou, A. Turnbull, M. Preuss, N.P. Stevens, T.J. Marrow, J.Y. Buffiere, E. Boller, A. Groso, M. Stampanoni, X-ray microtomography studies of localised corrosion and transitions to stress corrosion cracking. Mater. Sci. Technol. **22**(9), 1076–1085 (2006)
80. S.S. Singh, J.J. Williams, T.J. Stannard, X. Xiao, F.D. Carlo, N. Chawla, Measurement of localized corrosion rates at inclusion particles in AA7075 by in situ three dimensional (3D) X-ray synchrotron tomography. Corros. Sci. **104**, 330–335 (2016)
81. S.P. Knight, M. Salagaras, A.M. Wythe, F. De Carlo, A.J. Davenport, A.R. Trueman, In situ X-ray tomography of intergranular corrosion of 2024 and 7050 aluminium alloys. Corros. Sci. **52**(12), 3855–3860 (2010)
82. T.J. Marrow, J.Y. Buffiere, P.J. Withers, G. Johnson, D. Engelberg, High resolution X-ray tomography of short fatigue crack nucleation in austempered ductile cast iron. Int. J. Fatigue **26**(7), 717–725 (2004)
83. F. Eckermann, T. Suter, P.J. Uggowitzer, A. Afseth, A.J. Davenport, B.J. Connolly, M.H. Larsen, F.D. Carlo, P. Schmutz, In situ monitoring of corrosion processes within the bulk of AlMgSi alloys using X-ray microtomography. Corros. Sci. **50**(12), 3455–3466 (2008)
84. S.S. Singh, J.J. Williams, P. Hruby, X. Xiao, F. De Carlo, N. Chawla, In situ experimental techniques to study the mechanical behavior of materials using X-ray synchrotron tomography. Integr. Mater. Manuf. Innov. **3**(1), 9 (2014)
85. S.M. Ghahari, A.J. Davenport, T. Rayment, T. Suter, J.-P. Tinnes, C. Padovani, J.A. Hammons, M. Stampanoni, F. Marone, R. Mokso, In situ synchrotron X-ray micro-tomography study of pitting corrosion in stainless steel. Corros. Sci. **53**(9), 2684–2687 (2011)
86. J.C. Andrews, B.M. Weckhuysen, Hard X-ray spectroscopic nano-imaging of hierarchical functional materials at work. ChemPhysChem **14**(16), 3655–3666 (2013)

87. L. Salvo, M. Suéry, A. Marmottant, N. Limodin, D. Bernard, 3D imaging in material science: application of X-ray tomography. C R Phys. **11**(9–10), 641–649 (2010)
88. K.A. Mohan, S.V. Venkatakrishnan, J.W. Gibbs, E.B. Gulsoy, X. Xiao, M. De Graef, P.W. Voorhees, C.A. Bouman, TIMBIR: a method for time-space reconstruction from interlaced views. IEEE Trans. Comput. Imaging (99), 1–1 (2015)
89. P. Viot, D. Bernard, E. Plougonven, Polymeric foam deformation under dynamic loading by the use of the microtomographic technique. J. Mater. Sci. **42**(17), 7202–7213 (2007)
90. T.B. Sercombe, X. Xu, V.J. Challis, R. Green, S. Yue, Z. Zhang, P.D. Lee, Failure modes in high strength and stiffness to weight scaffolds produced by selective laser melting. Mater. Des. **67**, 501–508 (2015)
91. S.R. Stock, X-ray microtomography of materials. Int. Mater. Rev. **44**(4), 141–164 (1999)
92. A.C. Kak, M. Slaney, *Principles of Computerized Tomographic Imaging* (Society for Industrial and Applied Mathematics, 2001), p. 323
93. M.G.R. Sause, *Computed Tomography*. Springer Series in Materials Science (Springer, 2016), vol. 242
94. D. Bellet, B. Gorges, A. Dallery, P. Bernard, E. Pereiro, J. Baruchel, A 1300 K furnace for in situ X-ray microtomography. J. Appl. Crystallogr. **36**(2), 366–367 (2003)
95. J.Y. Buffiere, E. Maire, J. Adrien, J.P. Masse, E. Boller, In situ experiments with X-ray tomography: an attractive tool for experimental mechanics. Exp. Mech. **50**(3), 289–305 (2010)
96. F. De Carlo, X. Xiao, B. Tieman, X-ray tomography system, automation and remote access at beamline 2-BM of the Advanced Photon Source, in *Proceedings of SPIE* (2006), p. 63180K
97. R. Mokso, F. Marone, M. Stampanoni, Real time tomography at the swiss light source. AIP Conf. Proc. **1234**(1), 87–90 (2010)
98. M. Beister, D. Kolditz, W.A. Kalender, Iterative reconstruction methods in X-ray CT. Physica Med. **28**(2), 94–108 (2012)
99. D. Gursoy, F. De Carlo, X. Xiao, C. Jacobsen, TomoPy: a framework for the analysis of synchrotron tomographic data. J. Synchrotron Radiat. **21**(5), 1188–1193 (2014)
100. R.A. Brooks, G. Di Chiro, Beam hardening in X-ray reconstructive tomography. Phys. Med. Biol. **21**, 390–398 (1976)
101. R.A. Ketcham, W.D. Carlson, Acquisition, optimization and interpretation of X-ray computed tomographic imagery: applications to the geosciences. Comput. Geosci. **27**, 381–400 (2001)
102. W. Zbijewski, F. Beekman, Characterization and suppression of edge and aliasing artefacts in iterative X-ray CT reconstruction. Phys. Med. Biol. **49**, 145–157 (2004)
103. K.A. Mohan, S.V. Venkatakrishnan, L.F. Drummy, J. Simmons, D.Y. Parkinson, C.A. Bouman, Model-based iterative reconstruction for synchrotron X-ray tomography, in *2014 IEEE International Conference on Acoustics, Speech and Signal Processing (ICASSP)*, 4–9 May 2014 (2014), pp. 6909–6913
104. S. Soltani, M.S. Andersen, P.C. Hansen, Tomographic image reconstruction using training images. J. Comput. Appl. Math. **313**, 243–258 (2017)
105. A. Rosset, L. Spadola, O. Ratib, OsiriX: an open-source software for navigating in multidimensional DICOM images. J. Digit. Imaging **17**(3), 205–216 (2004)
106. E.R. Tufte, *Visual Explanations Images and Quantities, Evidence and Narrative*, 2nd edn. (Graphics Press, Chesire CT, 1997)
107. B.M. Patterson, C.E. Hamilton, Dimensional standard for micro X-ray computed tomography. Anal. Chem. **82**(20), 8537–8543 (2010)
108. J. Weickert, B.M.T.H. Romeny, M.A. Viergever, Efficient and reliable schemes for nonlinear diffusion filtering. IEEE Trans. Image Process. **7**(3), 398–410 (1998)
109. P. Iassonov, T. Gebrenegus, M. Tuller, Segmentation of X-ray computed tomography images of porous materials: a crucial step for characterization and quantitative analysis of pore structures. Water Resources Res. **45**(9), n/a–n/a (2009)
110. M. Freyer, A. Ale, R. Schulz, M. Zientkowska, V. Ntziachristos, K.H. Englmeier, Fast automatic segmentation of anatomical structures in X-ray computed tomography images to improve fluorescence molecular tomography reconstruction. J. Biomed. Opt. **15**(3), 036006 (2010)

111. M. Andrew, S. Bhattiprolu, D. Butnaru, J. Correa, The usage of modern data science in segmentation and classification: machine learning and microscopy. Microsc. Microanal. **23**(S1), 156–157 (2017)
112. N. Piche, I. Bouchard, M. Marsh, Dragonfly segmentation trainer—a general and user-friendly machine learning image segmentation solution. Microsc. Microanal. **23**(S1), 132–133 (2017)
113. A.E. Scott, I. Sinclair, S.M. Spearing, A. Thionnet, A.R. Bunsell, Damage accumulation in a carbon/epoxy composite: Comparison between a multiscale model and computed tomography experimental results. Compos. A Appl. Sci. Manuf. **43**(9), 1514–1522 (2012)
114. G. Geandier, A. Hazotte, S. Denis, A. Mocellin, E. Maire, Microstructural analysis of alumina chromium composites by X-ray tomography and 3-D finite element simulation of thermal stresses. Scripta Mater. **48**(8), 1219–1224 (2003)
115. C. Petit, E. Maire, S. Meille, J. Adrien, Two-scale study of the fracture of an aluminum foam by X-ray tomography and finite element modeling. Mater. Des. **120**, 117–127 (2017)
116. S. Gaitanaros, S. Kyriakides, A.M. Kraynik, On the crushing response of random open-cell foams. Int. J. Solids Struct. **49**(19–20), 2733–2743 (2012)
117. B.M. Patterson, K. Henderson, Z. Smith, D. Zhang, P. Giguere, Application of micro X-ray tomography to in-situ foam compression and numerical modeling. Microsc. Anal. **26**(2) (2012)
118. J.Y. Buffiere, P. Cloetens, W. Ludwig, E. Maire, L. Salvo, In situ X-ray tomography studies of microstructural evolution combined with 3D modeling. MRS Bull. **33**, 611–619 (2008)
119. M. Zimmermann, M. Carrard, W. Kurz, Rapid solidification of Al-Cu eutectic alloy by laser remelting. Acta Metall. **37**(12), 3305–3313 (1989)
120. D.P. Finegan, M. Scheel, J.B. Robinson, B. Tjaden, I. Hunt, T.J. Mason, J. Millichamp, M. Di Michiel, G.J. Offer, G. Hinds, D.J.L. Brett, P.R. Shearing, In-operando high-speed tomography of lithium-ion batteries during thermal runaway. Nat. Commun. **6**, 6924 (2015)
121. Y. Liu, A.M. Kiss, D.H. Larsson, F. Yang, P. Pianetta, To get the most out of high resolution X-ray tomography: a review of the post-reconstruction analysis. Spectrochim. Acta Part B **117**, 29–41 (2016)
122. N.L. Cordes, K. Henderson, B.M. Patterson, A route to integrating dynamic 4D X-ray computed tomography and machine learning to model material performance. Microsc. Microanal. **23**(S1), 144–145 (2017)
123. B.M. Patterson, J.P. Escobedo-Diaz, D. Dennis-Koller, E.K. Cerreta, Dimensional quantification of embedded voids or objects in three dimensions using X-ray tomography. Microsc. Microanal. **18**(2), 390–398 (2012)
124. G. Loughnane, M. Groeber, M. Uchic, M. Shah, R. Srinivasan, R. Grandhi, Modeling the effect of voxel resolution on the accuracy of phantom grain ensemble statistics. Mater. Charact. **90**, 136–150 (2014)
125. N.L. Cordes, Z.D. Smith, K. Henderson, J.C.E. Mertens, J.J. Williams, T. Stannard, X. Xiao, N. Chawla, B.M. Patterson, Applying pattern recognition to the analysis of X-ray computed tomography data of polymer foams. Microsc. Microanal. **22**(S3), 104–105 (2016)
126. E.J. Garboczi, Three-dimensional mathematical analysis of particle shape using X-ray tomography and spherical harmonics: application to aggregates used in concrete. Cem. Concr. Res. **32**(10), 1621–1638 (2002)
127. N. Limodin, L. Salvo, M. Suery, M. DiMichiel, In situ Investigation by X-ray tomography of the overall and local microstructural changes occuring during partial remelting of an Al-15.8wt.% Cu alloy. Acta Mater. **55**, 3177–3191 (2007)
128. A.D. Brown, Q. Pham, E.V. Fortin, P. Peralta, B.M. Patterson, J.P. Escobedo, E.K. Cerreta, S.N. Luo, D. Dennis-Koller, D. Byler, A. Koskelo, X. Xiao, Correlations among void shape distributions, dynamic damage mode, and loading kinetics. JOM **69**(2), 198–206 (2017)
129. J. Marrow, C. Reinhard, Y. Vertyagina, L. Saucedo-Mora, D. Collins, M. Mostafavi, 3D studies of damage by combined X-ray tomography and digital volume correlation. Procedia Mater. Sci. **3**, 1554–1559 (2014)
130. Z. Hu, H. Luo, S.G. Bardenhagen, C.R. Siviour, R.W. Armstrong, H. Lu, Internal deformation measurement of polymer bonded sugar in compression by digital volume correlation of in-situ tomography. Exp. Mech. **55**(1), 289–300 (2015)

131. R. Brault, A. Germaneau, J.C. Dupré, P. Doumalin, S. Mistou, M. Fazzini, In-situ analysis of laminated composite materials by X-ray micro-computed tomography and digital volume correlation. Exp. Mech. **53**(7), 1143–1151 (2013)
132. N.T. Redd, Hubble space telescope: pictures, facts and history. https://www.space.com/15892-hubble-space-telescope.html. Accessed 24 July 2017
133. L.T. Beringer, A. Levinsen, D. Rowenhorst, G. Spanos, Building the 3D materials science community. JOM **68**(5), 1274–1277 (2016)

Chapter 7
Overview of High-Energy X-Ray Diffraction Microscopy (HEDM) for Mesoscale Material Characterization in Three-Dimensions

Reeju Pokharel

Abstract Over the past two decades, several non-destructive techniques have been developed at various light sources for characterizing polycrystalline materials microstructure in three-dimensions (3D) and under various in-situ thermo-mechanical conditions. High-energy X-ray diffraction microscopy (HEDM) is one of the non-destructive techniques that facilitates 3D microstructure measurements at the mesoscale. Mainly, two variations of HEDM techniques are widely used: (1) Near-field (nf) and (2) far-field (ff) which are employed for non-destructive measurements of spatially resolved orientation ($\sim$1.5 μm and 0.01°), grain resolved orientation, and elastic strain tensor ($\sim 10^{-3}$–10^{-4}) from representative volume elements (RVE) with hundreds of bulk grains in the measured microstructure (mm^3). To date HEDM has been utilized to study variety of material systems under quasi-static conditions, while tracking microstructure evolution. This has revealed new physical mechanisms that were previously not observed through destructive testing and characterization. Furthermore, measured 3D microstructural evolution data obtained from HEDM are valuable for informing, developing, and validating microstructure aware models for accurate material property predictions. A path forward entails utilizing HEDM for initial material characterization for enabling microstructure evolution measurements under dynamic conditions.

7.1 Introduction

The understanding of materials at the mesoscale (1–100 μm) is of extreme importance to basic energy science because the properties of materials, critical to large-scale behavior, are impacted by local-scale heterogeneities such as grain boundaries, interfaces, and defects [1]. One challenge of mesoscale science is capturing a 3D view inside of bulk materials, at sub-grain resolution ($\sim$1 μm), while undergoing

R. Pokharel (✉)
Los Alamos National Laboratory, Los Alamos, NM 87544, USA
e-mail: reeju@lanl.gov

T. Lookman et al. (eds.), *Materials Discovery and Design*, Springer Series in Materials Science 280, https://doi.org/10.1007/978-3-319-99465-9_7

dynamic change. Techniques, such as electron microscopy, neutron scattering, or micro-computed tomography (μ-CT) are limited in being either destructive, providing only average data, or providing only density data, respectively. High energy X-ray diffraction microscopy (HEDM) is a novel, non-destructive method for capturing 3D mesoscale structure and evolution inside of material samples of $\sim$1 mm size, with $\sim$1 μm spatial and $\sim$0.1° grain orientation resolution. In this chapter, we give a brief overview of existing diffraction and imaging techniques for material characterization. In particular, HEDM datasets are discussed and a few examples of microstructure evolution under quasi-static conditions are presented to demonstrate the unique advantages provided by HEDM. However, this chapter will not attempt to summarize all the ongoing work on the subject. Additionally, future prospects of utilizing HEDM for enabling dynamic measurements are briefly discussed.

7.1.1 The Mesoscale

Multi-scale materials modeling is extremely challenging because it must cover many orders of magnitude of length scales ranging from the atomistic 10^{-10} m to the continuum $>10^{-3}$ m [1, 2]. Because of the difficulty of this task, there is a knowledge gap in terms of our ability to accurately pass insight from atomistic calculations/simulations to continuum scale predictions of engineering performance. At the lowest length scales, completely general and extremely accurate atomistic and molecular dynamics models exist that can simulate the behavior of many material systems based on fundamental physics simulations. Unfortunately, such models are extremely computationally intensive and limited to systems of hundreds to thousands of atoms even when using state of the art super computers. Therefore, while they can realistically predict the behavior of groups of atoms, it is impossible to scale them to sizes useful for manufacturing. At the other end of the spectrum, continuum mechanics models are mainly empirical and can reasonably predict bulk behaviors for a large family of materials. Extensive empirical tests have been carried out over many years to build databases on material properties such as ductility, elastic modulus, Poisson's ratio, shear modulus, yield strength for variety of material systems. The measured information are then incorporated into finite element models for predicting materials properties as well as engineering performance. However, such extensive experimental testing is inefficient, expensive, and time consuming.

Between these two extreme ends lies the mesoscale, a length scale at which current models are the least predictive and various model predictions exhibit extremely large variance. Structural materials are polycrystalline in nature with each individual grain experiencing constraints from its local neighborhood inducing heterogeneities and incompatibilities in adjacent grains. Complex properties and behaviors arise due to interaction between large population of heterogeneities such as defects, grain boundaries, phase boundaries, and dislocations. For instance, the relationship between "hot spots" in micro-mechanical fields and microstructural features such as grain boundaries and interfaces can be connected to material failure [3]. Therefore, the local variation in orientation and strain during plastic deformation are important in understanding damage nucleation in polycrystalline materials.

While constitutive relationships employed in most crystal plasticity simulations show some reasonable agreement with observation in terms of average properties, they are unable to reproduce local variations in orientation or strain [4]. This lack of agreement at the local scale is a direct evidence of our lack of physical understanding of the mesoscale regime. This missing link prevents material scientists from designing new, exotic materials with desired properties such as stronger, more durable, and lighter engineering components utilizing advanced manufacturing or accident tolerant nuclear fuels with higher thermal conductivity. As our understanding of a material's micro-mechanical properties relies heavily on the accurate knowledge of the underlying microstructure, spatially resolved information on evolution of microstructural parameters is imperative for understanding a material's internal response to accommodating imposed external loads. Therefore, a major goal of mesoscale science is capturing a 3D view inside of bulk materials, at sub-grain resolution ($\sim$1 μm), while undergoing dynamic change.

7.1.2 Imaging Techniques

Various material characterization techniques exist, of which one of the most popular is electron backscatter diffraction (EBSD), a standard technique for crystallographic orientation mapping and is heavily utilized by the materials community for surface characterization [5]. EBSD in concert with serial sectioning using focused ion beam (FIB) provides three-dimensional microstructure data; however, this route is destructive and mostly limited to post-mortem characterizations. Because this method is destructive, a single sample can only be fully characterized in 3D in one single state.

Non-destructive crystal structure determination techniques utilizing X-ray diffraction from a single crystal or powder diffraction for a large ensemble of crystals were first demonstrated over a century ago. However, most samples of interest are polycrystalline in nature, and therefore cannot be studied with a single crystal diffraction technique. In addition, powder diffraction is limited as it applies only to bulk samples with extremely large numbers of grains and provides only averaged measurements.

Nearly two decades ago, an alternate approach, multi-grain crystallography was successfully demonstrated [6], utilizing which 57 grains were mapped, for the first time, in an α-Al_2O_3 material [7, 8]. Since then, utilizing the third and fourth generation light sources, high-energy X-rays (in the energy range of 20–90 keV) based experimental techniques have enabled non-destructive measurements of a range of polycrystalline materials. These techniques have been transformational in advancing material microstructure characterization capability providing high-dimensional experimental data for microstructures in three-dimensions (3D) and their evolution under various in-situ conditions. Moreover, these datasets provide previously inaccessible information at the length scales (i.e. the mesoscale, 1–100 μm) relevant for informing and validating microstructure-aware models [2, 9–12] for linking materials processing-structure/property/performance (PSPP) relationships for advanced engineering applications [1, 13].

Since the first demonstration of 3D measurements, high-energy X-ray-based experimental techniques have advanced considerably, and 3D microstructure measurements are becoming routine. The multi-grain crystallography technique is now commonly referred to as high-energy X-ray diffraction microscopy (HEDM) or 3D X-ray diffraction (3DXRD) [14]. Various suites of HEDM techniques have been developed over the years for probing material microstructure and micro-mechanical field in polycrystals. Typically, the HEDM technique can probe 1 mm diameter samples and provide information on crystallographic orientation and elastic strain tensor averaged over a volume, commonly a grain. For example, near-field HEDM has been employed to study spatially resolved microstructures (orientation field, grains structure and morphology, sub-structure) and their evolution under thermo-mechanical conditions [15–22]. Far-field HEDM has been employed to study grain resolved micro-mechanics and variation in inter- and intra-granular stress states [6, 7, 23–32]. Utilizing both spatially resolved orientation and grain resolved elastic strains, stress evolution in Ti alloys have been studied [33–35]. Recently the HEDM technique has been extended to study deformation in shape memory alloys [36] and microstructure characterization of nuclear fuel materials [37, 38].

Apart from HEDM, other microstructure characterization techniques have been developed in parallel, utilizing either high-energy X-rays or neutrons for diffraction and imaging. Diffraction contrast tomography (DCT) is one such complementary non-destructive method that combines diffraction and tomographic techniques for mapping crystallographic orientation and grain morphology, in near pristine samples [39–41]. Micro-tomography provides additional density evolution information, ideal for imaging density contrast resulting from materials with high contrast in atomic number (Z) or contrast due to the presence of pores and cracks in the materials. Differential-aperture X-ray microscopy (DAXM) is another X-ray based technique for near-surface measurement, which enables in-situ material microstructure evolution measurements under various thermo-mechanical conditions [42]. Similarly, neutron diffraction and imaging based techniques can also provide non-destructive bulk measurements of structure and mechanical strains under in-situ conditions [43]. In this chapter, we will mainly focus on the HEDM technique and its application.

The remainder of the chapter is organized as follows. A brief background on physics of diffraction is presented in Sect. 7.2. In Sect. 7.3, the basic principle of HEDM technique and experimental geometry are presented. In addition, information on various tools that have been developed in the past decade for analyzing HEDM data are also provided. In Sect. 7.4, application of HEDM is presented where examples from literature on various experiments and material systems are presented and results are discussed. In Sect. 7.5, the current state is summarized and perspectives for future applications of HEDM and its relevance for future light sources are discussed.

7.2 Brief Background on Scattering Physics

In this section, basic concepts of the physics of elastic scattering are presented to establish a relationship between diffracted light and crystal structure, our approach is based on [44]. Diffraction is a result of constructive interference of the scattered wave after incident X-rays are scattered by electrons. Elastic scattering assumes that the incident and scattered X-ray photons have the same energy, that no energy is absorbed by the material during the scattering process. Consider an incident beam of X-rays as an electromagnetic plane wave:

$$E(t) = E_0 \cos(2\pi \nu t), \tag{7.1}$$

with amplitude E_0 and frequency ν. The interaction between the X-ray beam and an isolated electron can be approximated by forced simple harmonic motion of the form:

$$\ddot{x} = -\omega_0^2 x - b\dot{x} + \frac{q_e}{m_e} E(t), \tag{7.2}$$

where x is the displacement of the electron from equilibrium, ω_0 is the natural frequency of the system, q_e and m_e are the mass and charge of the electron, b is a damping term, and the third term on RHS is the force exerted on the electron by the electric field. According to the approximation (7.2), the electron oscillates according with the trajectory:

$$x(t) = A\cos(2\pi \nu t + \phi) + e^{-\frac{bt}{2}} f(t). \tag{7.3}$$

The term $e^{-\frac{bt}{2}} f(t)$ quickly decays and we are left with oscillations of the form

$$x(t) = A\cos(2\pi \nu t + \phi), \tag{7.4}$$

where both the amplitude $A = A(\nu)$ and the phase $\phi = \phi(\nu)$ depend on ν. The most important feature of (7.4) is that the electron oscillates at the same frequency as the driving force, and thereby emits light which has the same wavelength as the incident beam.

When a group of electrons $(e_1, \ldots, e_n)$ within an atom are illuminated by a plane wave of coherent light of the form (7.1), an observer at some location O will see, from each electron, a phase shifted electric field of the form:

$$\begin{aligned} \epsilon_j(t) &= A_j \cos\left(2\pi \nu t - \frac{2\pi l_j}{\lambda}\right) \\ &= A_j \cos(2\pi \nu t) \cos\left(\frac{2\pi l_j}{\lambda}\right) + A_j \sin(2\pi \nu t) \sin\left(\frac{2\pi l_j}{\lambda}\right), \end{aligned} \tag{7.5}$$

with λ being the light's wavelength and the amplitudes A_j and phase shifts $\frac{2\pi l_j}{\lambda}$ depend on the path lengths from the wave front to the observers, l_j. The total electric field observed at O is the sum of all of the individual electron contributions:

$$
\begin{aligned}
\epsilon(t) &= \sum_j \epsilon_j(t) \\
&= \sum_j A_j \cos(2\pi\nu t)\cos\left(\frac{2\pi l_j}{\lambda}\right) + \sum_j A_j \sin(2\pi\nu t)\sin\left(\frac{2\pi l_j}{\lambda}\right) \\
&= \cos(2\pi\nu t)\underbrace{\sum_j A_j \cos\left(\frac{2\pi l_j}{\lambda}\right)}_{A\cos(\phi)} + \sin(2\pi\nu t)\underbrace{\sum_j A_j \sin\left(\frac{2\pi l_j}{\lambda}\right)}_{A\sin(\phi)} \\
&= A\cos(2\pi\nu t)\cos(\phi) + A\sin(2\pi\nu t)\sin(\phi) \\
&= A\cos(2\pi\nu t - \phi).
\end{aligned}
\tag{7.6}
$$

The actual detected quantity is not the instantaneous diffracted electric field (7.6), but rather the intensity $I = \frac{cE^2}{8\pi}$, where

$$
E^2 = \left(\sum_n E_n \cos\left(\frac{2\pi l_n}{\lambda}\right)\right)^2 + \left(\sum_n E_n \sin\left(\frac{2\pi l_n}{\lambda}\right)\right)^2. \tag{7.7}
$$

Or, using complex notation,

$$
\epsilon_j = A_j e^{i(2\pi\nu t - 2\pi l_j/\lambda)} \tag{7.8}
$$

we can simply write $\epsilon\epsilon^\star = A^2$.

7.2.1 *Scattering by an Atom*

When considering scattering by a group of electrons in an atom the convention is to consider the center of the atom as the origin, O, the electrons located at positions $\mathbf{r}_n$, and an observer at position $\mathbf{P}$. We consider a plane wave of light incident on the plane passing through the origin, from which there is a distance l_1 to an electron at position $\mathbf{r}_n$, as shown on the left side of Fig. 7.1. Relative to the wavefront at O, the field acting on electron n is then given by

$$
\epsilon_n = A_n \cos\left(2\pi\nu t - \frac{2\pi l_1}{\lambda}\right) \tag{7.9}
$$

and for an observer at position P, the field is

$$
\epsilon_n = \frac{A_n e^2}{mc^2 l_2}\cos\left(2\pi\nu t - \frac{2\pi}{\lambda}(l_1 + l_2)\right). \tag{7.10}
$$

Assuming that both the source and observation distances are much larger than $|r_n|$, we make the simplifying assumptions

$$l_2 \to R, \quad l_1 + l_2 \to r_n \cdot s_0 + R - r_n \cdot s = R - (s - s_0) \cdot r_n. \tag{7.11}$$

Summing over all instantaneous fields at P we are left with

$$\epsilon = \frac{Ae^2}{mc^2 R} e^{2\pi i(\nu t - R/\lambda)} \sum_n e^{(2\pi i/\lambda)(s - s_0)\cdot r_n}. \tag{7.12}$$

Rather than considering each electron individually the quantum mechanics inspired approach is to consider a charge density ρ, such that ρdV is the ratio of charge in the volume dV relative to the charge of one electron. The sum (7.12) is then replaced with the integral

$$\epsilon_e = \frac{Ae^2}{mc^2 R} e^{2\pi i(\nu t - R/\lambda)} \int e^{(2\pi i/\lambda)(s - s_0)\cdot r_n} \rho dV, \tag{7.13}$$

where

$$f_e = \int e^{(2\pi i/\lambda)(s - s_0)\cdot r_n} \rho dV \tag{7.14}$$

is typically referred to as the scattering factor per electron. The equation for f_e is simplified by assuming spherical symmetry for the charge distribution $\rho = \rho(r)$. Then, considering right side of Fig. 7.1, $(s - s_0) \cdot r = 2 \sin\theta \cos\varphi$, and after performing integration with respect to φ we get

$$f_e = \int 4\pi r^2 \rho(r) \frac{\sin kr}{kr} dr, \tag{7.15}$$

where $k = \frac{4\pi \sin\theta}{\lambda}$.

For a collection of electrons in an atom we simply sum all of the contributions:

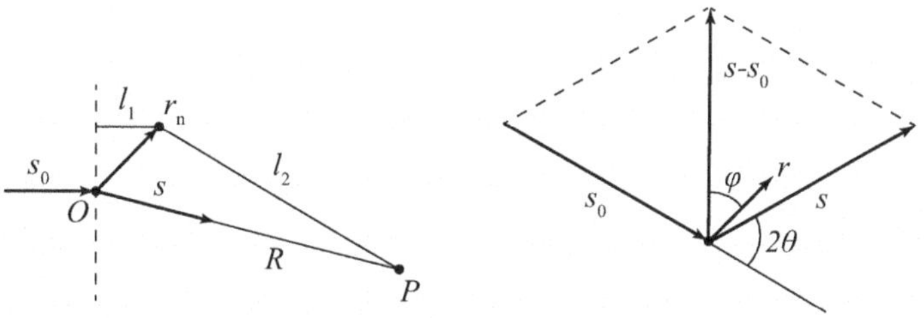

Fig. 7.1 Diffraction from the electrons in an atom with the approximation that $R \gg |\mathbf{r}_n|$

$$f = \sum_n f_{e,n} = \sum_n \int_0^\infty 4\pi r^2 \rho_n(r) \frac{\sin(kr)}{kr} dr, \tag{7.16}$$

this sum is known as the atomic scattering factor and gives the amplitude of scattered radiation per atom. The scattering factor given by (7.16) is only accurate when the X-ray wavelength is much smaller than any of the absorption edge wavelengths in the atom and when the electron distribution has spherical symmetry. For wavelengths comparable to absorption edge wavelengths, dispersion correction factors are necessary.

7.2.2 *Crystallographic Planes*

We consider a crystal with crystal axes $\{\mathbf{a}_1, \mathbf{a}_2, \mathbf{a}_3\}$, such that the position of an atom of type n in a unit cell $m_1m_2m_3$ is given by the vector $\mathbf{R}_m^n = m_1\mathbf{a}_1 + m_2\mathbf{a}_2 + m_3 + \mathbf{a}_3 + \mathbf{r}_n$. In order to derive the Bragg's law for such a crystal, we must consider the crystallographic planes hkl as shown in Fig. 7.2, where the first plane passes through the origin, O, and the next intercepts the crystal axes at locations a_1/h, a_2/k, a_3/l. The Bragg law depends on the orientation and spacing of these hkl, both properties are conveniently represented by the vector $\mathbf{H}_{hkl}$ which is normal to the planes and whose magnitude is reciprocal to the spacing, where the values (h, k, l) are commonly referred to as the Miller indices. In order to represent the $\mathbf{H}_{hkl}$ vectors for a given crystal, we introduce a reciprocal basis, $\{\mathbf{b}_1, \mathbf{b}_2, \mathbf{b}_3\}$, which is defined based on the crystal axes, given by:

$$\mathbf{b}_1 = \frac{\mathbf{a}_2 \times \mathbf{a}_3}{\mathbf{a}_1 \cdot \mathbf{a}_2 \times \mathbf{a}_3}, \qquad \mathbf{b}_2 = \frac{\mathbf{a}_3 \times \mathbf{a}_1}{\mathbf{a}_1 \cdot \mathbf{a}_2 \times \mathbf{a}_3}, \qquad \mathbf{b}_3 = \frac{\mathbf{a}_1 \times \mathbf{a}_2}{\mathbf{a}_1 \cdot \mathbf{a}_2 \times \mathbf{a}_3}. \tag{7.17}$$

These vectors are defined such that each reciprocal vector $\mathbf{b}_i$ is perpendicular to the plane defined by the two crystal axes of the other indices, $\mathbf{a}_{j \neq i}$. Furthermore, the $\mathbf{a}_i$ and $\mathbf{b}_j$ vectors satisfy the following scalar products:

$$\mathbf{a}_i \cdot \mathbf{b}_i = 1, \qquad \mathbf{a}_i \cdot \mathbf{b}_j = 0, \qquad \Longrightarrow \qquad \mathbf{a}_i \cdot \mathbf{b}_j = \begin{cases} 1 & i = j \\ 0 & i \neq j. \end{cases} \tag{7.18}$$

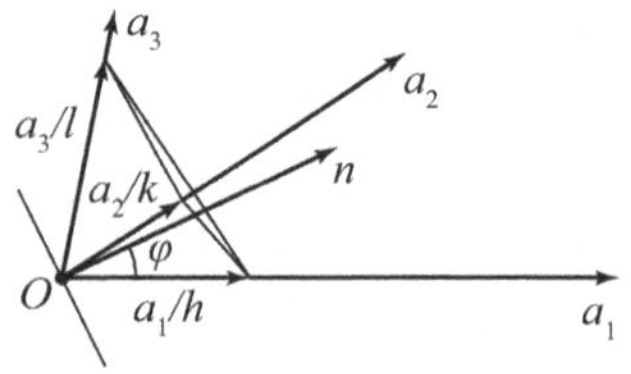

Fig. 7.2 Definition of the hkl planes relative to the crystal axes a_j

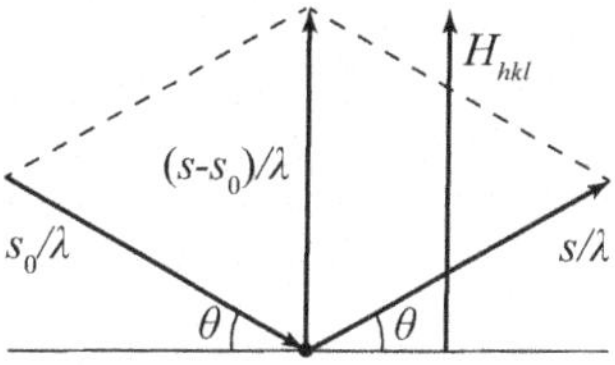

Fig. 7.3 Bragg law in terms of $\mathbf{H}_{hkl}$

Any $\mathbf{H}_{hkl}$ vector can then be written as the linear combination

$$\mathbf{H}_{hkl} = h\mathbf{b}_1 + k\mathbf{b}_2 + l\mathbf{b}_3, \tag{7.19}$$

and it can be easily calculated that if the perpendicular spacing between hkl planes is d_{hkl}, then

$$d_{hkl} = \frac{1}{|\mathbf{H}_{hkl}|}. \tag{7.20}$$

The usefulness of the $\mathbf{H}_{hkl}$ vector is that the Bragg condition can be concisely stated as:

$$\frac{\mathbf{s} - \mathbf{s}_0}{\lambda} = \mathbf{H}_{hkl}, \tag{7.21}$$

where $\mathbf{s}$ and $\mathbf{s}_0$ are unit vectors in the direction of the incident and diffracted light, as shown in Fig. 7.3. Equation 7.21 simultaneously guarantees that the incident and diffracted beam make equal angles with the diffracting planes and taking the magnitude of either side gives us:

$$\left|\frac{\mathbf{s} - \mathbf{s}_0}{\lambda}\right| = \frac{2\sin(\theta)}{\lambda} = |\mathbf{H}_{hkl}| = \frac{1}{d_{hkl}}, \tag{7.22}$$

which is equivalent to the usual form of the Bragg law $\lambda = 2d_{hkl}\sin(\theta)$.

7.2.3 *Diffraction by a Small Crystal*

Consider a monochromatic beam of wavelength λ with direction of propagation $\mathbf{s}_0$ incident on an atom at position $\mathbf{R}_m^n = m_1\mathbf{a}_1 + m_2\mathbf{a}_2 + m_3\mathbf{a}_3 + \mathbf{r}_n$. The diffracted light observed at point $\mathbf{P}$, as shown in Fig. 7.4, is given by

$$\epsilon_p = \frac{E_0 e^2}{mc^2 R} f_n \cos\left[2\pi\nu t - \frac{2\pi}{\lambda}(x_1 + x_2)\right], \tag{7.23}$$

where f_n is the atomic scattering factor. We assume the crystal to be so small relative to all distances involved that the scattered wave is also treated as a plane-wave and

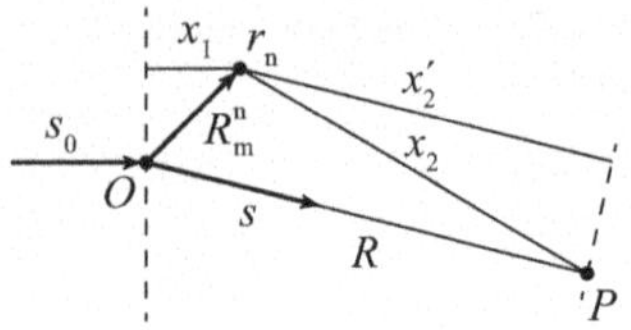

Fig. 7.4 Diffraction from the electrons in an atom with the approximation that $R \gg |\mathbf{r}_n|$

approximate $x_2 \sim x_2'$. The instantaneous field at position **P** due to atom (m, n) is then given by

$$\epsilon_p = \frac{E_0 e^2}{mc^2 R} f_n \exp\{i\,[2\pi\nu t - (2\pi/\lambda)\,(R - (\mathbf{s} - \mathbf{s}_0)\cdot(m_1\mathbf{a}_1 + m_2\mathbf{a}_2 + m_3\mathbf{a}_3 + \mathbf{r}_n))]\}\,. \tag{7.24}$$

If we sum (7.24) over all atoms in the crystal, we then get the total field at **P**. Assuming a crystal with edges $N_1\mathbf{a}_1$, $N_2\mathbf{a}_2$, $N_3\mathbf{a}_3$, and carrying out the sum, we can represent the observable quantity $\epsilon_p\epsilon_p^*$ which is proportional to the light intensity, as

$$\epsilon_p\epsilon_p^* = I_e F^2 \frac{\sin^2(\pi/\lambda)(\mathbf{s}-\mathbf{s}_0)\cdot N_1\mathbf{a}_1}{\sin^2(\pi/\lambda)(\mathbf{s}-\mathbf{s}_0)\cdot \mathbf{a}_1} \frac{\sin^2(\pi/\lambda)(\mathbf{s}-\mathbf{s}_0)\cdot N_2\mathbf{a}_2}{\sin^2(\pi/\lambda)(\mathbf{s}-\mathbf{s}_0)\cdot \mathbf{a}_2} \frac{\sin^2(\pi/\lambda)(\mathbf{s}-\mathbf{s}_0)\cdot N_3\mathbf{a}_3}{\sin^2(\pi/\lambda)(\mathbf{s}-\mathbf{s}_0)\cdot \mathbf{a}_3}, \tag{7.25}$$

where

$$F = \sum_n f_n e^{(2\pi i/\lambda)(\mathbf{s}-\mathbf{s}_0)\cdot\mathbf{r}_n} \tag{7.26}$$

is the structure factor which depends on the atomic positions $\mathbf{r}_n$ and

$$I_e = I_0 \frac{e^4}{m^2c^4R^2}\left(\frac{1+\cos^2(2\theta)}{2}\right), \tag{7.27}$$

where I_0 is the intensity of the primary beam and $(1+\cos^2(2\theta))/2$ is a polarization factor.

On the right hand side of (7.25) are terms of the form

$$\frac{\sin^2(Nx)}{\sin^2(x)}, \tag{7.28}$$

where $x_i = (\pi/\lambda)(\mathbf{s}-\mathbf{s}_0)\cdot\mathbf{a}_i$. Such functions have large peaks $\sim N^2$ at positions $x = n\pi$ and quickly fall to zero elsewhere. Therefore, the observed intensity will be zero almost everywhere except at those places satisfying the simultaneous Laue equations:

$$(\mathbf{s}-\mathbf{s}_0)\cdot\mathbf{a}_1 = h'\lambda, \qquad (\mathbf{s}-\mathbf{s}_0)\cdot\mathbf{a}_2 = k'\lambda, \qquad (\mathbf{s}-\mathbf{s}_0)\cdot\mathbf{a}_3 = l'\lambda, \tag{7.29}$$

where h', k', and l' are integers, a condition which is equivalent to the Bragg law.

Representing atomic positions relative to the crystal axes, we can write $\mathbf{r}_n = x_n\mathbf{a}_1 + y_n\mathbf{a}_2 + z_n\mathbf{a}_3$ and consider the value of the structure factor when the Bragg law is satisfied for a set of hkl planes, that is, when $\mathbf{s} - \mathbf{s}_0 = \lambda\mathbf{H}_{hkl}$. We then get

$$\begin{aligned} F_{hkl} &= \sum_n f_n \exp\left[2\pi i \left(h\mathbf{b}_1 + k\mathbf{b}_2 + l\mathbf{b}_3\right) \cdot \left(x_n\mathbf{a}_1 + y_n\mathbf{a}_2 + z_n\mathbf{a}_3\right)\right] \\ &= \sum_n f_n \exp\left[2\pi i \left(hx_n + ky_n + lz_n\right)\right], \end{aligned} \tag{7.30}$$

and if the structure factor for reflection hkl is zero, then so is the reflected intensity.

7.2.4 Electron Density

If we consider a small crystal relative with sides of length a, b, and c, we can represent the 3D electron density by its 3D Fourier transform

$$\rho(x, y, z) = \sum_p \sum_q \sum_r C_{pqr} \exp\left[-2\pi i \left(p\frac{x}{a} + q\frac{y}{b} + r\frac{z}{c}\right)\right], \tag{7.31}$$

where the Fourier coefficients C_{pqr} can be found by integrating

$$\int_0^a \int_0^b \int_0^c \rho(x, y, z) \exp\left[2\pi i \left(h\frac{x}{a} + k\frac{y}{b} + l\frac{z}{c}\right)\right] = abcC_{hkl}. \tag{7.32}$$

If we now replace the coordinates x_n, y_n, and z_n in (7.30) by x_n/a, y_n/b, and z_n/c, we can rewrite the discrete structure factor as

$$F_{hkl} = \sum_n f_n \exp\left[2\pi i \left(h\frac{x_n}{a} + k\frac{y_n}{b} + l\frac{z_n}{c}\right)\right], \tag{7.33}$$

which we can then rewrite in terms of a continuous electron density:

$$F_{hkl} = \int_0^a \int_0^b \int_0^c \rho(x, y, z) \exp\left[2\pi i \left(h\frac{x}{a} + k\frac{y}{b} + l\frac{z}{c}\right)\right] dV, \tag{7.34}$$

and so the electron density in electrons per unit volume is given by the Fourier coefficients of the structure factors F_{hkl} according to

$$\rho(x, y, z) = \frac{1}{V} \sum_h \sum_k \sum_l F_{hkl} \exp\left[-2\pi i \left(h\frac{x}{a} + k\frac{y}{b} + l\frac{z}{c}\right)\right]. \tag{7.35}$$

Therefore, according to (7.35), the observed hkl reflections from a crystal correspond to the Fourier series of the crystal's electron density and therefore X-ray diffraction of a crystal can be thought of as a Fourier transform of the crystal's electron density. Each coefficient in the series for $\rho(x, y, z)$ corresponds to a point hkl in the reciprocal lattice. Unfortunately, rather than observing the F_{hkl} values directly, which would allow for the direct 3D calculation of the electron density according to (7.35), the quantities that are actually observed are 2D projections of the intensities $F_{hkl}F^*_{hkl} = |F_{hkl}|^2$, in which all phase information is lost and must be recovered via iterative phase retrieval techniques provided that additional boundary condition and support information about the crystal structure are available.

7.3 High-Energy X-Ray Diffraction Microscopy (HEDM)

7.3.1 Experimental Setup

There are mainly two experimental setups utilized for performing HEDM measurements: (1) Near-field (nf-) HEDM and (2) far-field (ff-) HEDM, where the main difference between the two setups is the sample to detector distance. In the case of nf-HEDM, the sample to detector distance range from 3 to 10 mm while the ff-HEDM setup can range anywhere from 500 to 2500 mm. The schematic of the experimental setup is shown in Fig. 7.5. A planar focused monochromatic beam of X-rays is incident on a sample mounted on the rotation stage, where crystallites that satisfy the Bragg condition give rise to diffracted beams that are imaged on a charge coupled detector (CCD).

HEDM employs a scanning geometry, where the sample is rotated about the axis perpendicular to the planar X-ray beam and diffraction images are acquired over integration intervals $\delta\omega = 1°$ and 180 diffraction images are collected. Note that the integration interval can be decreased if the sample consist of small grains and large orientation mosaicity. During sample rotation, it is important to ensure that the sample is not precessing in and out of the beam, as some fraction of the Bragg scattering would be lost from that portion that passes out of the beam. Mapping the full sample requires rotating the sample about the vertical axis (ω-axis) aligned perpendicular to the incident beam. Depending on the dimensions of the parallel beam, translation of the sample along the z-direction might be required to map the full 3D volume.

The near-field detector at APS 1ID-E and CHESS comprise of an interline CCD camera, which is optically coupled through 5$\times$ (or 10$\times$) magnifying optics to image fluorescent light from a 10 μm thick, single crystal, Lutecium aluminum garnet scintillator. This results in a final pixel size that is approximately 1.5 μm ($\sim 3 \times 3\,\text{mm}^2$ field of view). The far-field data is also recorded on an area detector with an active area of $\sim 410 \times 410\,\text{mm}^2$ (2K $\times$ 2K pixel array). The flat panel detector has a layer of cesium iodide and a-silicon scintillator materials for converting X-ray photons to visible light. The final pixel pitch of the detector is 200 μm. Research is

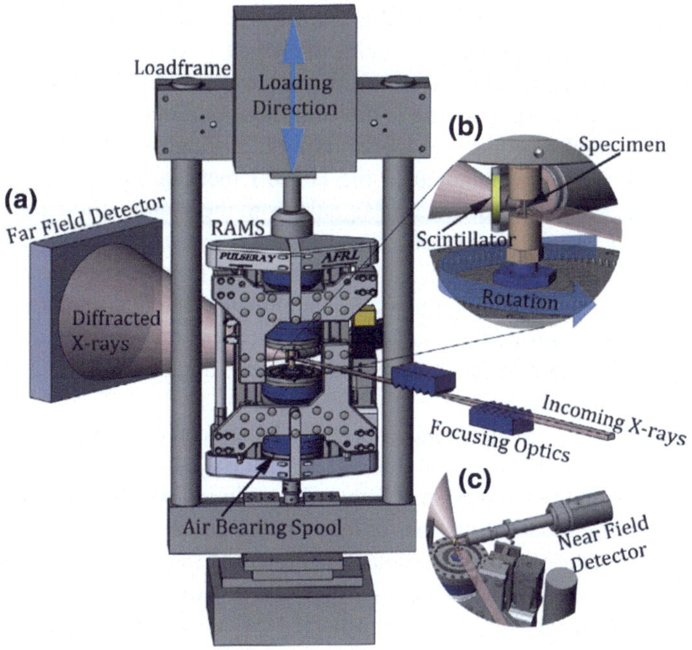

Fig. 7.5 HEDM setup at APS beamline 1-ID E. **a** Far-field detector setup, **b** specimen mounted on a rotation stage, and **c** near-field detector setup [33]

underway for developing in-situ and ex-situ environments as well as area detectors with improved efficiency and data collection rates [45].

To obtain spatially resolved information on local orientation field, near-field geometry is utilized where the diffraction image is collected at more than one sample to detector distances per rotation angle, to aid in high-fidelity orientation reconstructions. Ff-HEDM provides center of mass position of individual grains, average orientations, relative grain sizes, and grain resolved elastic strain tensors. The ff- detector can be translated farther back along the beam path (i.e. very far-field geometry), if higher strain resolution is desirable and if permitted by the beam/beamline specifications. Therefore, HEDM measurements can be tailored to fit individual experimental needs and as necessitated by the science case by tuning parameters such as beam dimensions, setups, and data collection rates.

7.3.2 Data Analysis

In the case of nf-HEDM, the diffraction spots seen on the detector are randomly positioned and the spot size and shape correlates directly with the grain size and morphology. Since the grain shape is projected on the detector, spatially resolved orientation field reconstruction is possible using the near-field geometry. In contrast, the diffraction spots in the far-field geometry sit on the Debye-Scherrer ring, similar to what is observed during the powder diffraction measurements. The difference is that in ff-HEDM measurements, the ring is discontinuous and individual spots are more or less isolated, which is important for obtaining high-fidelity data reconstructions.

The diffraction images obtained in the HEDM measurements need to be pre-processed in order to extract diffraction signals from the sample. First, as a clean-up step, background and stray scattering are removed from the raw detector images and hot pixels can be removed using median filtering, if required. One of the most critical steps in the reconstruction process is identifying the instrument parameters. A calibration sample is used for this purpose. Critical parameters include calibrated beam energy, sample to detector distance, rotation axis and detector tilts with respect to the incident beam plane.

Several orientation and strain indexing tools have been developed for analyzing HEDM data. Fully automated beam line experiment (FABLE) software was initially developed for analyzing far-field data. Recently, grain indexing tool has been added that enables near-field data reduction for a box-beam geometry, where the incoming beam is incident on the middle of the detector, allowing Friedel pairs detection. Hexrd software [46] was developed in parallel at Cornell University for reconstructing grain orientations and strain tensors from ff-HEDM data. This software is currently maintained and updated by Lawrence Livermore National Laboratory. Integrating nf-HEDM data reconstruction capability in Hexrd in collaboration with CHESS is underway.

IceNine software developed at Carnegie Mellon University [47] operates mainly on the nf-HEDM data collected using a planar focused beam. Therefore, both the data collection and reconstruction take longer compared to the other two methods. However, forward model method utilized by the IceNine software enables high resolution spatially resolved orientation field reconstructions and provides unique capability to characterize heavily deformed materials.

Figure 7.6 shows a schematic demonstrating the nf-HEDM measurements using planar beam and 3D orientation field reconstruction. The raw diffraction data is background subtracted and the peaks are segmented. The image is then utilized by the reconstruction software for 2D microstructure reconstructions. These steps are repeated for all the 2D layers measured by translating the sample along the z-direction. Finally, 3D microstructure map is obtained by stacking the 2D layers on top of each other. Since the sample is not touched during the full volume mapping, the stacking procedure does not require registration, which are otherwise needed in EBSD+FIB type measurements.

Recently, Midas software [49] was developed at APS for simultaneous reconstruction of nf-HEDM and ff-HEDM data. In this case, the average grain orientation information from the ff-HEDM reconstruction is given as guess orientations for spatially resolved orientation reconstructions in nf-HEDM. Such seeding significantly reduces the search space for both spatial and orientation reconstruction and significantly speeds up the reconstruction process. However, the seeding results in overestimating some grain sizes, while missing grains that were not indexed in

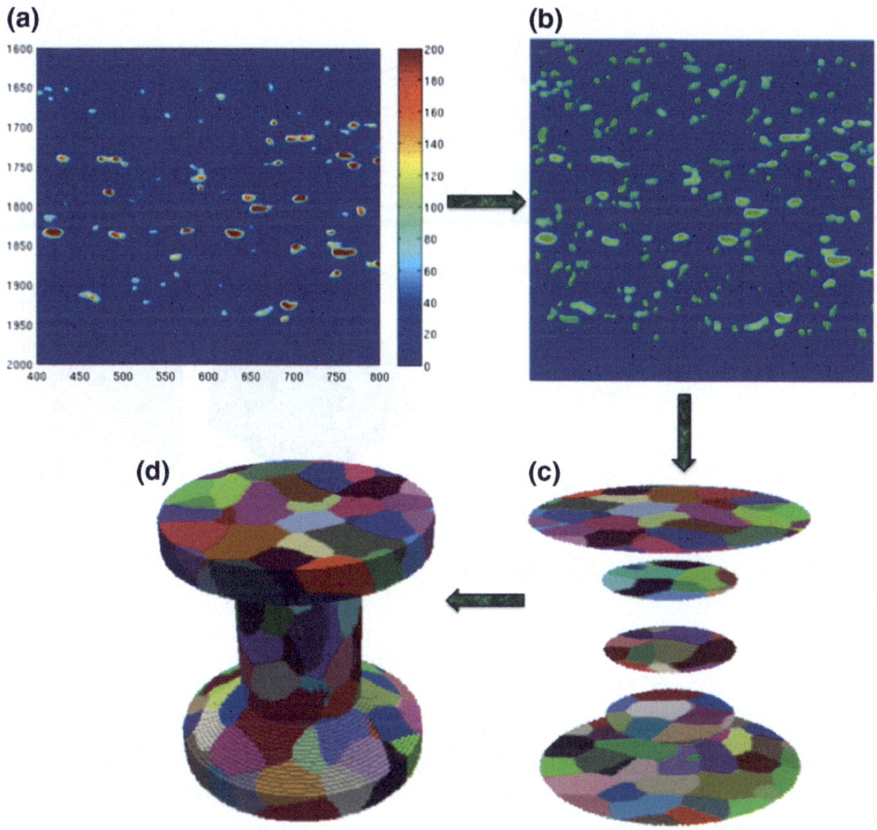

Fig. 7.6 Reconstruction yields 2D orientation maps, which are stacked to obtain a 3D volume [48]

the far-field. Another drawback is that the technique does not work well for highly deformed materials, as far-field accuracy drops with increasing peaks smearing and overlap that occurs with increasing deformation level.

Continued improvement and development are underway.

7.4 Microstructure Representation

Figure 7.7 schematically demonstrate the orientation and misorientation representations used in crystallography.

Crystallographic orientation or rotation required to bring a crystal in coincidence with another (termed as misorientation) can be represented as a proper rotation matrix R in basis $B_{x,y,z}$, which can be written in terms of the basic rotations matrices as:

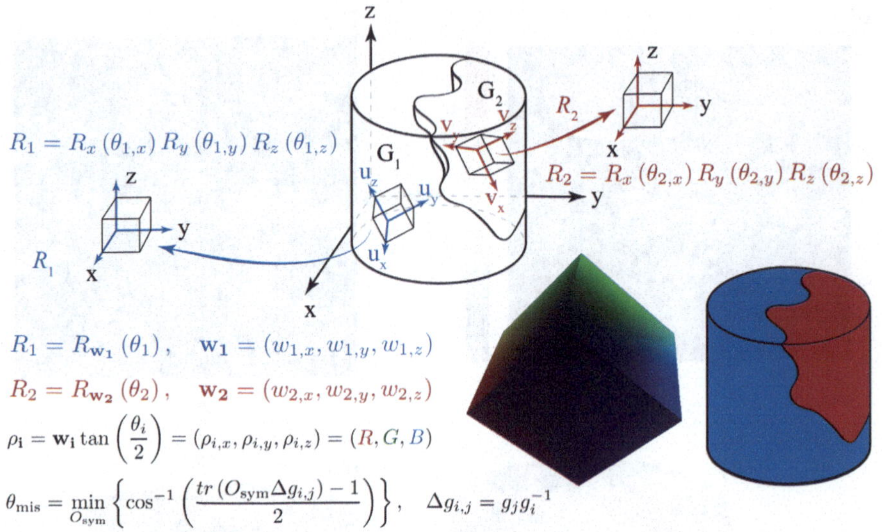

Fig. 7.7 Conventions used for microstructure representation

$$R = R_x(\alpha) R_y(\beta) R_z(\gamma), \tag{7.36}$$

where R_x, R_y, and R_z are 3D rotations about x, y, and z axes, respectively. It is convenient to represent the final rotation matrix R as an axis/angle pair, where axis is a rotation axis in some other basis $B_{u,v,w}$ at some angle θ. Additionally, for any proper rotation matrix, there exists Eigen value $\lambda = 1$, such that

$$Ru = \lambda u = u. \tag{7.37}$$

The vector u is the rotation axis of the rotation matrix R. We also want to find θ, the rotation angle. We know that if we start with u and choose two other orthonormal vectors v and w, then the rotation matrix can be written in the u, v, w basis, $B_{u,v,w}$, as

$$M_u(\theta) = \begin{pmatrix} 1 & 0 & 0 \\ 0 & \cos(\theta) & -\sin(\theta) \\ 0 & \sin(\theta) & \cos(\theta) \end{pmatrix} \tag{7.38}$$

Since the trace of a matrix is invariant to change of basis, we know

$$tr(R) = tr(M) = 1 + 2\cos(\theta). \tag{7.39}$$

This allows us to calculate the rotation angle, θ, without ever expressing M_u in the form (7.38), we simply use the given form, R, and calculate

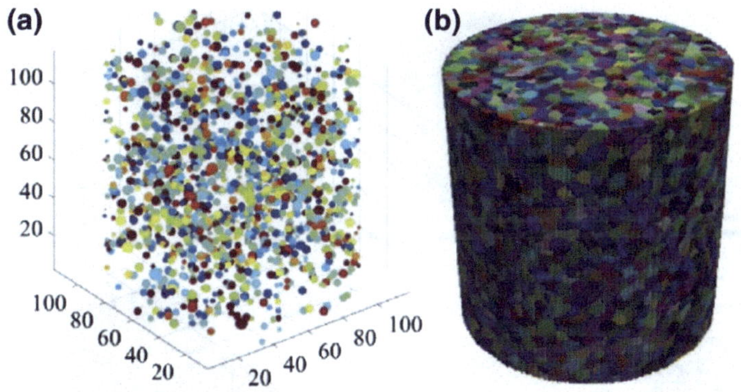

Fig. 7.8 Synthetic microstructure resembling microstructure maps obtained from HEDM data. **a** Ff-HEDM and **b** nf-HEDM [10]

$$\theta = \arccos\left(\frac{tr(R) - 1}{2}\right), \tag{7.40}$$

which is known as misorientation angle in crystallography.

Figure 7.8 illustrates the type of microstructure data that HEDM technique provides. Figure 7.8a represents the grain average information that ff-HEDM provides, where the colors correspond to either orientation or components of elastic strain tensor. Figure 7.8b shows a spatially resolved 3D orientation field that could be obtained from nf-HEDM measurements. From the spatially resolved microstructure map, individual 3D grains are segmented as a post-processing step by clustering points belonging to similar orientations within some specified threshold misorientation angle.

7.5 Example Applications

7.5.1 Tracking Plastic Deformation in Polycrystalline Copper Using Nf-HEDM

Nf-HEDM is mainly suitable for structure determination of individual crystallites as well as their local neighborhood in a polycrystalline material. Utilizing nf-HEDM, Pokharel et al. [4, 17] demonstrated characterization of 3D microstructure evolution due to plastic deformation in a single specimen of polycrystalline material. A 99.995% pure oxygen free electrical (OFE) Cu was used for this study, where a tensile specimen with a gage length of 1 mm and a cylindrical cross section of 1 mm diameter was prepared. The tensile axis was parallel to the cylindrical axis of the sample. The Cu specimen was deformed in-situ under tensile loading and nf-HEDM

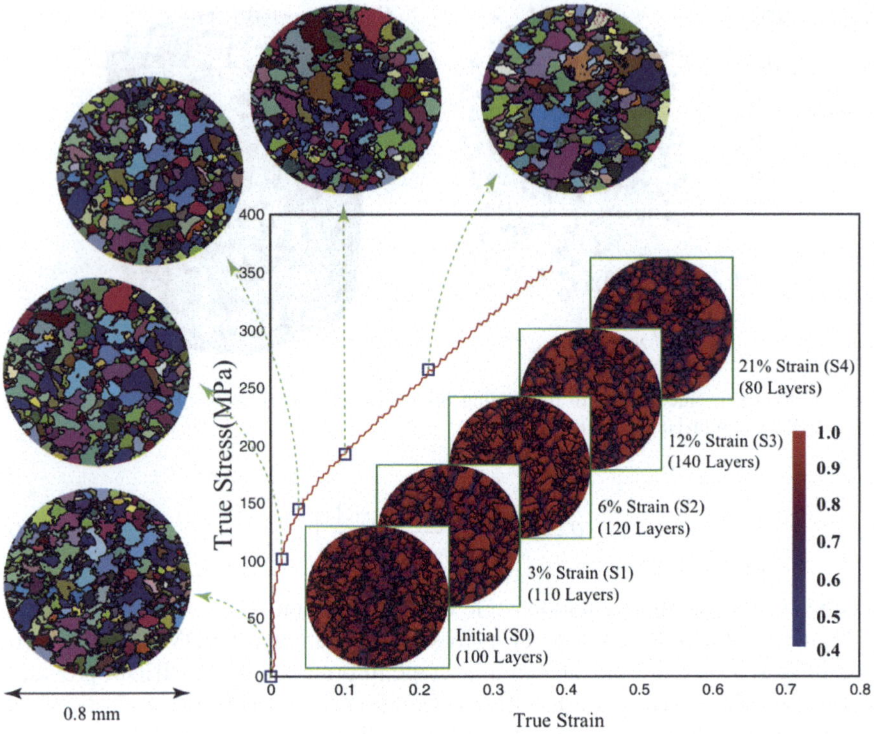

Fig. 7.9 Experimental stress-strain curve along with one of the 2D slices of orientation and confidence maps from each of the five measured strain states. Nf-HEDM measurements were taken at various strain levels ranging from 0 to 21% tensile strain. IceNine software was used for data reconstruction. The 2D maps plotted outside the stress-strain curve represent the orientation fields from each of the corresponding strain levels obtained using forward modeling method analysis software. The 2D maps plotted inside the stress-strain curve are the confidence, C, maps for the reconstructed orientation fields at different strain levels. Confidence values of the five plots range from 0.4 to 1, where C = 1 means all the simulated scatterings coincide with the experimental diffraction data and C = 0.4 corresponds to 40% overlap with the experimental diffraction peaks. For each strain level a 3D volume was measured, where each strain state consists on average 100 layers [17]

data were collected at various strain levels. Figure 7.9 [17] shows the stress-strain curve along with the example 2D orientation field maps and corresponding confidence maps for strain levels up to 21% tensile strain. Figure 7.10 [17] shows the corresponding 3D volumetric microstructure maps for 3 out of 5 measured strain states, where ~5000 3D grains were tracked through initial, 6, and 12% tensile deformation. The measured microstructure evolution information was used to study spatially resolved orientation change and grain fragmentation due to intra-granular misorientation development during tensile deformation.

Figure 7.11 [4, 17] shows the ability to track individual 3D grains at different strain levels. Figure 7.11a shows the kernel average misorientation (KAM) map indicating local orientation change development due to plastic deformation. The higher KAM

Fig. 7.10 Three 3D volumes of the measured microstructures **a** initial, **b** 6% strain, and **c** 12% strain. Colors correspond to an RGB mapping of Rodrigues vector components specifying the local crystal orientation [17]

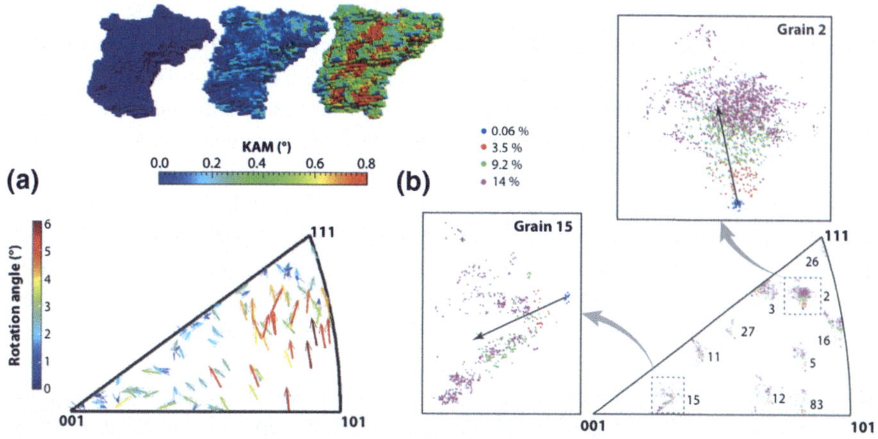

Fig. 7.11 Tracking deformation in individual grains through deformation [4, 17]

value indicates that the intra-granular misorientation between adjacent crystallite orientation is high. Figure 7.11b shows the inverse pole figure (left) where average grain reorientation of 100 largest grains in the material was tracked. The tail of the arrow corresponds to the average orientation in the initial state and the arrow head represents average orientation after subjecting the sample to 14% tensile strain. Inverse pole figure (right) shows the trajectory of individual voxels in a grain tracked at 4 different strain levels. The insets show the grain rotations for two grains near <111>- and <001>-corners of the stereographic triangle. The black arrow shows the grain averaged rotation from the initial to the final strain. It is observed that the two grains, #2 and #15, show very different intra-granular orientation change, where grain fragmentation is observed for grain #15. It is evident that spatially resolved information is needed to capture the local details within a grain. These plots further indicate that only grain averaged orientation information was insufficient to capture the local heterogeneities that develop in individual grains due to plastic deformation. Variation in the combinations of slip systems activated during plastic deformation can lead to such heterogeneous internal structure development in a polycrystalline material. In addition, a strong dependence was observed between orientation change and grain size, where larger grains developed higher average local orientation change in comparison to smaller grains. This suggests that the type of deformation structures formation is also dependent on the initial orientation and grain size. Moreover, decrease in average grain size was observed with deformation due to grain fragmentation and sub-grain formation.

7.5.2 *Combined nf- and ff-HEDM for Tracking Inter-granular Stress in Titanium Alloy*

A proof-of-principle combined nf- and ff-HEDM measurements were reported by Schuren et al. [33], where microstructure and micro-mechanical field evolution were measured in a single sample undergoing creep deformation. In-situ measurements of titanium alloy (Ti-7Al) were performed, where HEDM data were collected during quasi-static loading. Experimental setup employed for this multi-modal diffraction and tomography measurements is shown in Fig. 7.5. Nf- and ff- data were reconstructed using IceNine and Hexrd software, respectively. Spatially resolved grain maps and corresponding grain cross-section averaged stress field were used for studying local neighborhood effect on observed anisotropic elastic and plastic properties.

Figure 7.12 shows the microstructure and micro-mechanical properties obtained from the HEDM measurements. Figure 7.12a shows the spatially resolved orientation field map from nf-HEDM with corresponding COM positions for individual grains obtained from ff-HEDM. Figure 7.12b shows the spatial maps colored by hydrostatic and the effective stresses. Figure 7.12c plots the hydrostatic and effective stresses versus the coaxiality angle defined as the angle between the grain scale stress vector and applied macroscopic stress direction. In pre-creep state clear evidence of higher hydrostatic stress was observed for grains with stress states aligned with the applied

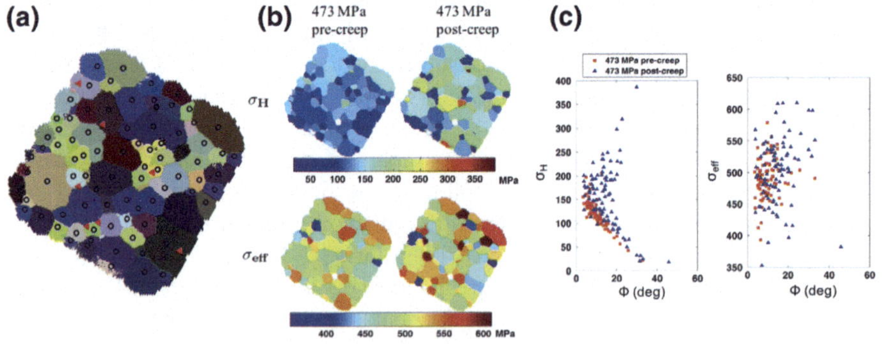

Fig. 7.12 Combined nf- and ff-HEDM measurements of Ti microstructure. **a** ff-COM overlaid on nf-orientation map. **b** Hydrostatic and deviatoric stress evolution pre- and post-creep. **c** Hydrostatic and deviatoric stresses versus coaxiality angle [33]

macroscopic stress. In post-creep state, bifurcation of hydrostatic stress was observed where grain scale stress deviated away from the applied macroscopic stress.

The same experimental setup was utilized by Turner et al. [50] to perform in-situ ff-HDEM measurements during tensile deformation of the Ti-7Al sample, previously measured during the creep deformation [33]. 69 bulk grains in the initial state of the nf-HEDM volume (200 μm $\times$ 1 mm $\times$ 1 mm) were matched with the ff-HEDM data at various stages of tensile loading. Nf-HEDM data were not collected at the loaded states as the measurements are highly time intensive (24 h/volume). Due to the complexity of the experimental setup, the tensile specimen was subjected to axial load of 23 MPa during mounting the sample in the load frame. Therefore, the initial state of the material was not in a fully unloaded state. The grain averaged elastic strain tensor were tracked through deformation and distinct inter-granular heterogeneity was observed, which seems to have resulted directly from the strain heterogeneity in the unloaded state (23 MPa). This indicated that the initial residual stresses present in the material influenced the strain and corresponding stress evolution during deformation.

Combined nf- and ff-HEDM in-situ data enabled polycrystal model instantiation and validation, where crystal plasticity simulation of tensile deformation of Ti-7Al was performed using the Ti-7Al data [34]. Predicted strain and stress evolution showed good qualitative agreement with measurements; however, grain scale stress heterogeneity was not well captured by the crystal plasticity simulations. The comparison could be improved by incorporating initial residual stresses present in the material along with measured 3D microstructure as input to simulation.

7.5.3 *Tracking Lattice Rotation Change in Interstitial-Free (IF) Steel Using HEDM*

Lattice rotation in polycrystalline material is a complex phenomenon influenced by factors such as microstructure, grain orientation, interaction between neighboring grains, which result in grain level heterogeneity. 3D X-ray diffraction microscopy (3DXRD) was employed by Oddershede et al. and Winther et al. [29, 30] to study lattice rotation evolution in 3D bulk grains of IF steel. Monochromatic X-ray beam energy, E = 69.51 keV, and beam height of 10 μm were used for microstructure measurements. Initial microstructure of a tensile specimen with dimensions $0.7 \times 0.7 \times 30$ mm^3 were mapped via HEDM, then the sample was re-measured after subjecting it to 9% tensile deformation. FABLE software was used for 3D microstructure reconstruction.

Three bulk grains with similar initial orientations, close to <522> orientation located between [001]–[-111] line in a stereographic triangle, were identified for detailed study of intra-granular variation in lattice rotation. It was observed that the tensile axes of all three deformed grains rotated towards the [001] direction, which was also the macroscopic loading direction. To investigate the intra-granular variation in rotation, raw diffraction spots were tracked before and after deformation. Three

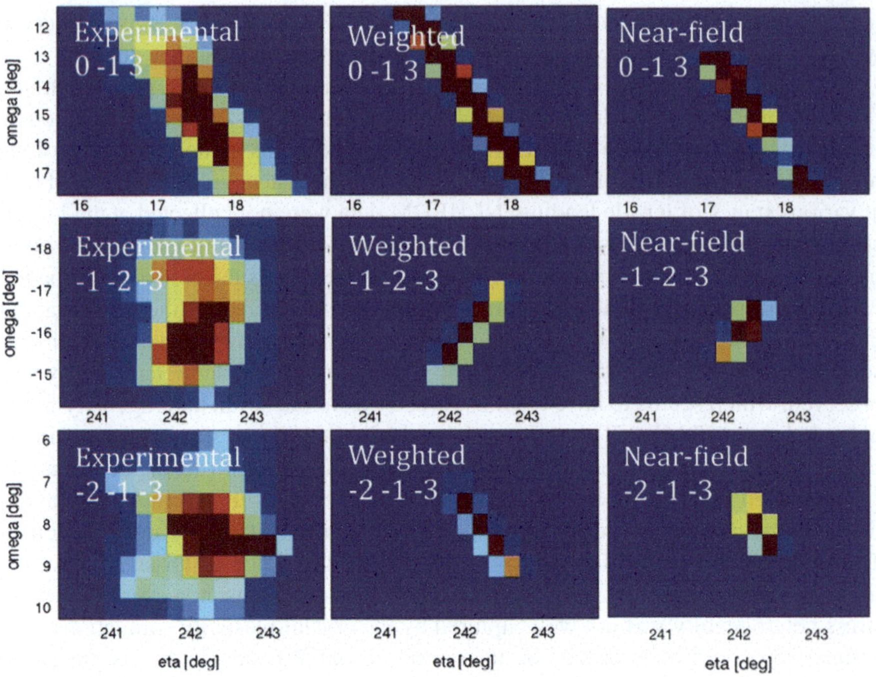

Fig. 7.13 Tracking deformation in individual grains through deformation [29, 30]

different reflections for each grain orientation were considered, where the observed change in location and morphology of the diffraction spots were linked to intra-granular orientation change in individual grains.

Crystal plasticity simulations were performed to identify slip systems activity that led to orientation spread. The peak broadening effect was quantified by integrating the diffraction spots along the ω (rotation about the tensile loading direction) and η (along the Debye-Scherrer ring) directions. Figure 7.13 shows the measured and predicted reflections for one of the three grains after deformation. Predicted orientation spread were in good agreement with measurements, where large spread in diffraction spots were observed along both ω and η directions. Four slip systems were predicted to be active based on both Schmid and Taylor models. However, large intra-granular variation in the spread was attributed mostly to the activity of (0-1-1)[11-1] and' (-101)[11-1] slip systems, which also had the highest Schmid factors. Moreover, the results indicated that the initial grain orientation played a key role in the development of intra-granular orientation variation in individual grains.

7.5.4 *Grain-Scale Residual Strain (Stress) Determination in Ti-7Al Using HEDM*

HEDM technique was employed by Chatterjee et al. [35] to study deformation induced inter-granular variation in orientation and micro-mechanical field in Ti-7Al material. Tensile specimen of Ti-7Al material consisting of fully recrystallized grains with 100 μm average grain size were prepared for grain scale orientation and residual stress characterization. Planar-focused high-energy X-ray beam of 1.7 μm height and 65.351 keV energy were used for probing 2D cross-section (layer) of the 3D sample. 3D data were collected by translating the material along the vertical axis of the tensile specimen, mapping volumetric region of $1.5 \times 1.5 \times 0.54\,\text{mm}^3$. Total of 15 layers around the gage volume were measured with 40 μm vertical spacing between layers. Diffraction data were collected at various load steps as well as during loading and unloading of the material. FABLE software was used for diffraction data analysis for grain center of mass and grain cross-section averaged strain determination.

Figure 7.14 shows the grain scale stress states developed in three neighboring grains in the sample. Upon unloading, grain scale residual stresses were observed in the material. Although uniaxial load was applied to the tensile specimen, complex multi-axial stress-states resembling combined ‘bending’ and tension were observed

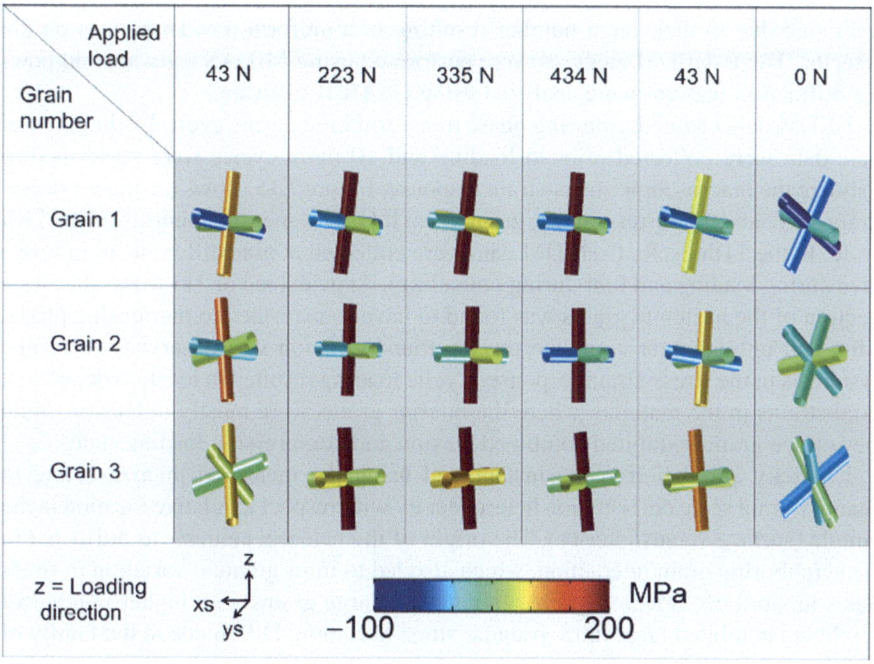

Fig. 7.14 Stress jacks to demonstrate complex grain scale stress states development for three neighboring grains in a sample subjected to uniaxial macroscopic load [35]

in individual grains. Co-axiality angle was calculated as an angle between the macroscopic loading direction and the grain scale loading state. Spatial variation in the co-axiality angle indicated variation in inter-granular stress states, which was mainly attributed to the local interactions between neighboring grains irrespective of the macroscopic loading conditions. Such local heterogeneity that develop at the grain scale influences macroscopic behavior and failure mechanisms in polycrystalline material.

7.5.5 In-Situ ff-HEDM Characterization of Stress-Induced Phase Transformation in Nickel-Titanium Shape Memory Alloys (SMA)

Paranjape et al. [36] studied the variation in super-elastic transformation strain in shape memory alloys (SMA) materials utilizing ff-HEDM technique with 2 mm wide by 0.15 mm tall beam of energy 71.676 keV. In-situ diffraction data were collected during cyclic loading in tension (11 cycles of loading and unloading) of Ti-50.9at.%Ni samples, exhibiting super-elasticity property at room temperature. Two phases were present in the material: austenite and martensite, upon loading and unloading, where 3D microstructure and micro-mechanical field were analyzed only for the austenite phase. Martensitic grains were not resolved by the ff-HEDM technique due to their large number resulting in a uniform powder pattern on the detector. The ff-HEDM analyses were performed using MIDAS software and powder diffraction patterns were analyzed using GSAS-II software.

HEDM data enabled capturing phase transformation during cyclic loading. Initial state data were collected prior to loading and 10 more cycles were performed to stabilize the macroscopic stress-strain response. Figure 7.15 shows the macro stress-strain curve and the corresponding grains from ff HEDM measurements from the 11th cycle. In the 11th cycle, ff-HEDM data were collected at nine different strain levels (five during loading and four during unloading). At peak load of 311 MPa (state 4), a fraction of the austenite grains was found to have transformed to martensitic phase. After full unload, near complete reverse transformation was observed with some hysteresis in the stress-strain response. Cyclic loading resulted in location dependent axial strains in the material, where the interior grains were mostly in tension while the surface grains exhibited combined tension and compression loading states.

Elasticity simulations were instantiated using the measured microstructure to quantify grain scale deformation heterogeneity with respect to relative location in the sample (surface versus interior). The origin of the heterogeneities was attributed to the neighboring grain interaction, which also led to intra-granular variation in stress states in similarly oriented grains. In addition, large grains with higher number of neighbors exhibited large intra-granular stress variation. Difference in the family of slip systems activated in the interior versus the surface grains were also suspected to play a role in variation in intra-granular stress states. Resulting stress heterogeneity influenced the strain induced phase transformation in SMA materials.

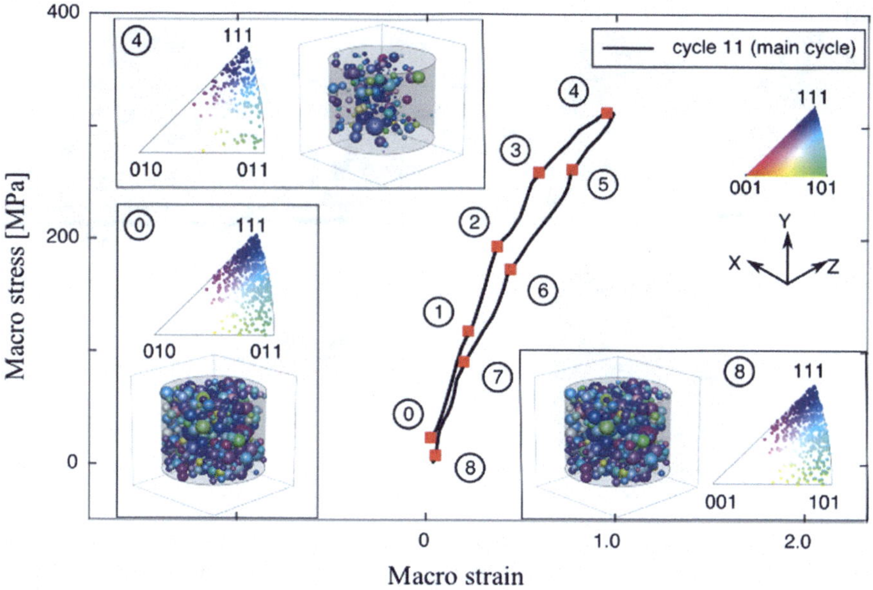

Fig. 7.15 Inverse pole figure and a 3D view of the grain center of mass is shown at three key stages: 0 load (0), peak load (4) showing fewer B2 grains remaining due to phase transformation, and full unload (8) showing near-complete reverse transformation to B2. The grains are colored according to an inverse pole figure colormap [36]

7.5.6 HEDM Application to Nuclear Fuels

Properties of nuclear fuels strongly depend on microstructural parameters, where residual porosity reduces thermal conductivity of the fuel and grain size and morphology dictates fission gas release rates as well as dimensional change during operation. Both of these factors can greatly limit the performance and life-time of nuclear fuel. HEDM technique is well-suited for characterizing microstructures of ceramics and metallic nuclear fuels due to minimal amount of plastic deformation exhibited by these materials. In addition, nuclear fuel sample preparation for conventional metallography and microstructure characterization are both costly and hazardous. In contrast, HEDM requires little to no sample preparation where a small parallelepiped can be cut from the fuel pellet for 3D characterization. Recently, conventional, UO_2, and candidate accident tolerant fuels (ATF), UN-U_3Si_5, materials have been characterized utilizing nf-HEDM technique. Brown et al. [37] employed the nf-HEDM technique, for the first time, to non-destructively probe 3D microstructure of nuclear fuel materials. High-energy X-ray beam of 1.3 mm wide by 3 μm tall and 85.53 keV energy were used to measure 3D microstructure of ceramic UO_2. Similarly, the nf-HEDM technique was also utilized to characterized 3D microstructure of ATF fuels [43]. The 3D microstructures were reconstructed using IceNine software.

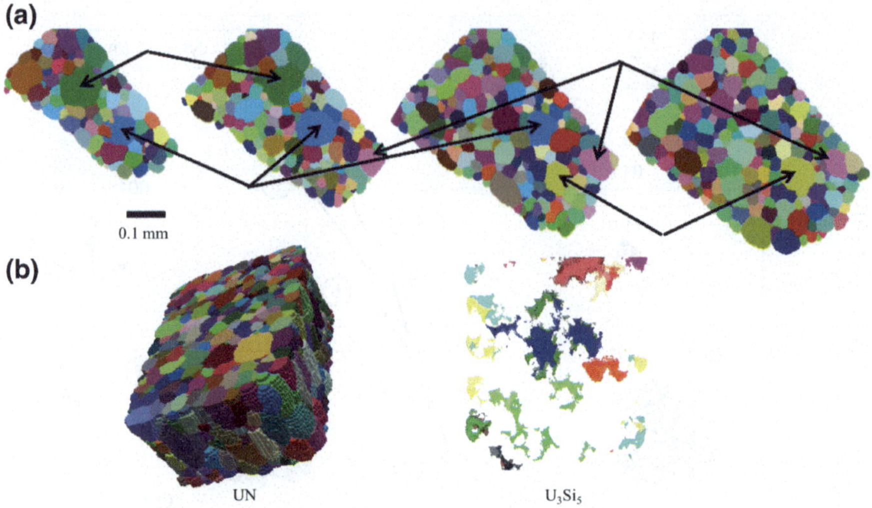

Fig. 7.16 Grain orientation maps for **a** UO_2 at 25 μm intervals from near the top (left) of the sample. Arrows indicate grains that span several layers are indicated [37], and **b** UN-USi ATF fuel, where 3D microstructure is shown for the major phase (UN) and 2D projection of 10 layers are shown for the minor phase (USi) [43]

Figure 7.16a shows of 3D characterization of a UO_2 materials, where 2D maps from different region on the samples are plotted. Similarly, Fig. 7.16b shows the orientation field maps of the two-phase ATF fuel, where the 3D microstructure of the major phase, UN, is shown on the left and the 2D projection of 10 layers of the U_3Si_5 phase is shown on the right. Note that in both case no intra-granular orientation gradient were present in the grains, which suggest that minimal dislocation density or plastic deformation is present in these materials.

Figure 7.17 shows the orientation maps for UO_2 material before and after heat treatment. Visual inspection indicates significant grain growth after heat treatment, where the initial residual porosity disappeared resulting in a near fully dense material. The measured microstructures were utilized for instantiating grain growth models for nuclear fuels [51].

7.5.7 *Utilizing HEDM to Characterize Additively Manufactured 316L Stainless Steel*

Additive manufacturing (AM) is a process of building 3D materials in a layer by layer manner. Variety of AM processing techniques and AM process parameters are employed for material fabrication, which lead to AM materials with large variation in materials properties and performance. AM materials could greatly benefit

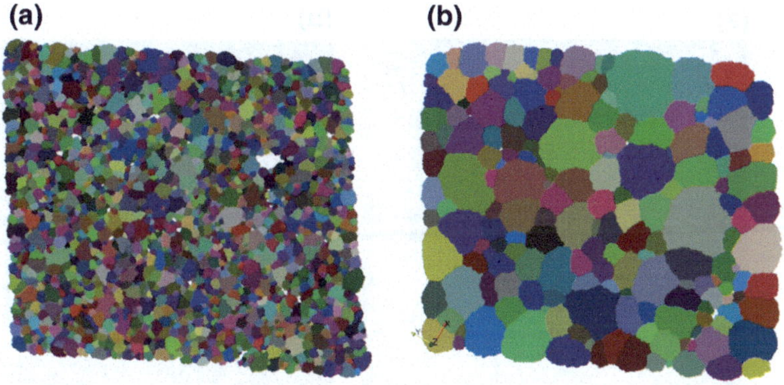

Fig. 7.17 Microstructure evolution in UO_2. **a** As-sintered and **b** after heat-treatment to 2200 °C for 2.5 h [51]

from HEDM technique, where in-situ or ex-situ measurements of microstructure and residual stress could improve the current understanding of SPP relationships in AM materials.

As a feasibility test, nf-HEDM measurements were performed on AM 316L stainless steel (SS) materials before and after heat treatment. Figure 7.18a, b show the detector images for as-built and annealed 316L SS sample, respectively. In the case of as-built material, the detector image is complex, resembling either diffraction from a powder sample with large number of small grains or diffraction from highly deformed material. On the other hand, annealed material exhibited sharp isolated peaks common of recrystallized materials. IceNine software was employed for microstructure reconstruction. Complimentary powder diffraction measurements revealed the presence of austenite and ferrite phase in the initial microstructure.

In the as-built state, the secondary ferrite phase with fine grain size was not resolved by nf-HEDM measurements; therefore, only austenite phase was reconstructed. Figure 7.18c shows the first attempt at reconstructing the austenite phase in the as-built microstructure. As the sample was >99.5% dense, the white spaces shown in the microstructure map corresponds to either small/deformed austenite grains or small ferrite grains. Upon annealing, complete ferrite to austenite phase transformation was observed along with recovery of austenite grains. Figure 7.18d shows the austenite phase orientation maps resulting in equiaxed and recrystallized austenite grains.

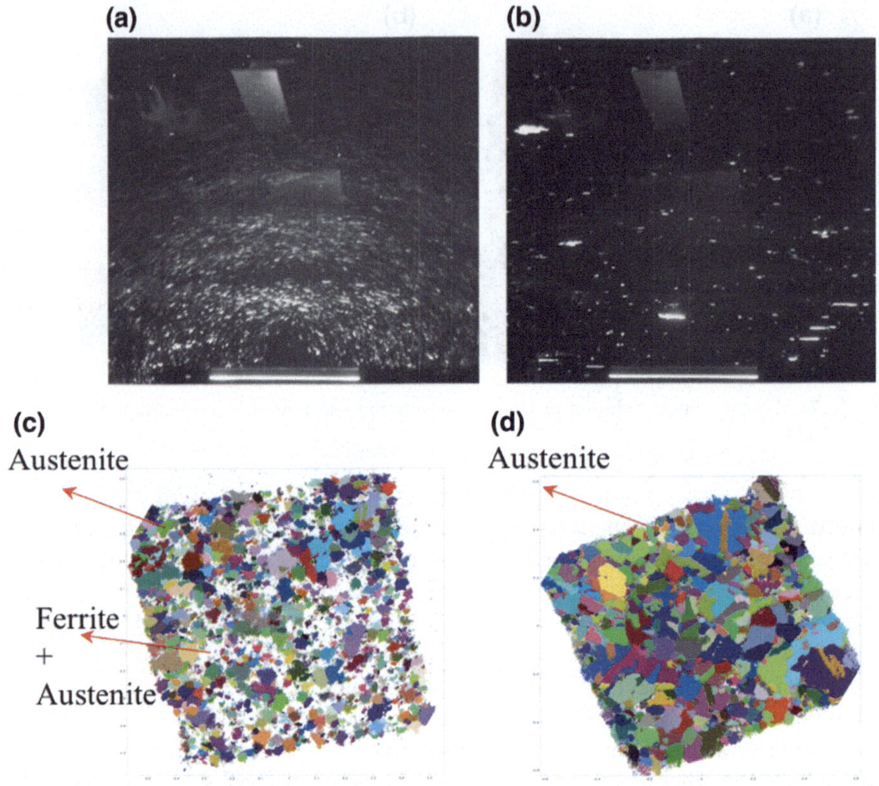

Fig. 7.18 Near-field detector image for AM 304L SS steel for **a** as-built and **b** after heat treatment to 1060 °C for 1 h. Before and after detector images show sharpening of diffraction signals after annealing. Nf-HEDM orientation maps are shown for **c** as-built and **d** annealed material. Small austenite and ferrite grains were not resolved in the reconstruction. After annealing the residual ferrite phase in the initial state completely phase transformed to austenite phase, resulting in a fully dense material

7.6 Conclusions and Perspectives

The following are some of the conclusions that can be drawn from the literature employing the HEDM technique for microstructure and micro-mechanical field measurements:

- HEDM provides previously inaccessible mesoscale data on a microstructure and its evolution under operating conditions. Such data is unprecedented and provides valuable insight for microstructure sensitive model development for predicting material properties and performance.
- HEDM provides the flexibility to probe a range of material systems, from low-Z to high-Z. One of the major limitations of HEDM technique in terms of probing high-Z material is that the signal to noise ratio drastically decreases due to high

absorption cross-section of high-Z materials. In addition, the quantum efficiency of the scintillator deteriorates with increasing energy (>80 keV used for nuclear materials). As a result, longer integration times per detector image are required for high-quality data acquisition. This means that the data collection time can easily increase by factor of 3–4 for uranium in comparison to low-Z materials such as titanium and copper.

- Nf-HEDM is ideal for probing spatially resolved 3D microstructure and provides information on sub-structure formation within a grain as well as evolution of its local neighborhood under in-situ conditions. As defect accumulation and damage nucleation are local phenomena, spatially resolved microstructure and internal structure evolution information from experiments are valuable for providing insight into physical phenomena that affect materials properties and behavior.
- Ff-HEDM data provides center of mass of a grain and grain resolved elastic strain for thousands of grains in a polycrystalline material. Employing a box beam geometry, statistically significant numbers of grains can be mapped in a limited beam time. In addition, due to faster data collection rates of ff-HEDM in comparison to nf-HEDM, a large number of material states can be measured while the sample is subjected to external loading conditions. This enables detailed view of microstructure and micro-mechanical field evolution in a single sample.
- Various HEDM studies elucidated development of mesoscale heterogeneities in polycrystalline materials subjected to macroscopic loading conditions. Variations in intra-granular stress states were observed in various material systems. This variation was mainly attributed to the local interaction between neighboring grains with minor effects from initial grain orientations and loading conditions.
- In the case of deformed materials or AM materials with large deformation and complex grain size and morphology, the diffracted peaks smeared out and the high order diffraction intensities dropped. Development of a robust method for background subtraction and diffraction peak segmentation will be crucial for high-fidelity microstructure reconstructions for highly deformed samples..

The main insight from various applications utilizing the HEDM techniques was that macroscopic responses of polycrystalline materials were affected by the heterogeneities in microstructure and micro-mechanical fields at the local scale. All the examples presented here demonstrated in-situ uniaxial loading of the polycrystalline materials; however, experimental setups for more complex loading conditions such as bi-axial loading are currently being explored [52]. Furthermore, major challenges remain in terms of characterization of more complex materials such as additively manufactured materials where large variations in initial grain orientation as well as grain morphology are observed. The technique is still highly limited to polycrystals with relatively large grains (>10 μm as pixel pitch of the near-field detector is ~ 1.5 μm) and with low deformation level (e.g. $<20\%$ tensile strain). Advancement in detector technology as well as data reduction tools will be required to apply HEDM techniques for materials with small grains and large deformation. Note that in the current data analysis framework uncertainty quantification is one of the areas that is not fully explored. Therefore, how errors propagate from measurements to data reconstruction and when the experimental data is used for model instantiation how that affects predicted material properties and behaviors are not yet understood.

7.6.1 Establishing Processing-Structure-Property-Performance Relationships

The advent of 3rd and 4th generation light sources has enabled the development of advanced non-destructive microstructure characterization techniques such as HEDM. As a result, high-resolution and high-dimensional data acquisition have been made possible. The goal is to utilize these techniques to obtain high-fidelity information for establishing processing-structure-property-performance (PSPP) relationships in materials. However, the ability to design material microstructures with desired properties and performance is still limited. Any advancement in material design requires the development of multi-mechanism and multi-physics predictive models, which can rely heavily on experimental testing and measurements. Because HEDM type data collection is expensive, only a limited number of sample states can be experimentally tested and the resulting data sets are extremely sparse in the vast PSPP material space.

Currently, lengthy measurements (hours) severely limit the time scales at which mesoscale 3D microstructure evolution data can be collected with high spatial ($\sim$1 μm) and orientation, ($\sim$0.01°) resolution. Furthermore, extremely long reconstruction times (days) prevent sample evolution-based feedback during an experiment. The reconstruction techniques currently used are brute force and the turnaround time from data collection to reconstruction is very long. For example, the 2D spatially resolved orientation field reconstruction shown in Fig. 7.9 took $\sim$20 mins per sample cross-section on 512 processors, requiring $\sim$650 K core-seconds/layer on a system rated at 9.2 Gflops/core. Typically, there are 50–100 such cross-sections in a full 3D volume, which would require several minutes of reconstruction on a $\sim$6.5 Pflop machine. In addition, the first step of reconstruction requires lengthy, manual calibration to find appropriate instrument parameters for a given experiment.

The Advanced Photon Source (APS), Linac Coherent Light Source (LCLS) and Cornell High Energy Synchrotron Source (CHESS) are upgrading their X-ray sources and detector technologies over the next few years to obtain better temporal resolution in imaging and diffraction, which means faster data collection rates. Therefore, it is important that more focus is placed on meeting the data reduction and reconstruction demands created by these increasing data collection rates. This will not only improve the information extraction capability from high dimensional data but also provide faster feedback to drive experiments.

For instance, the current Edisonian approach in materials measurement and testing needs to be replaced with more strategic approach to maximize information extraction capability from sparse datasets. Furthermore, the current norm in HEDM type measurements is to collect large amounts of data (several hundred GBs to TBs) for a

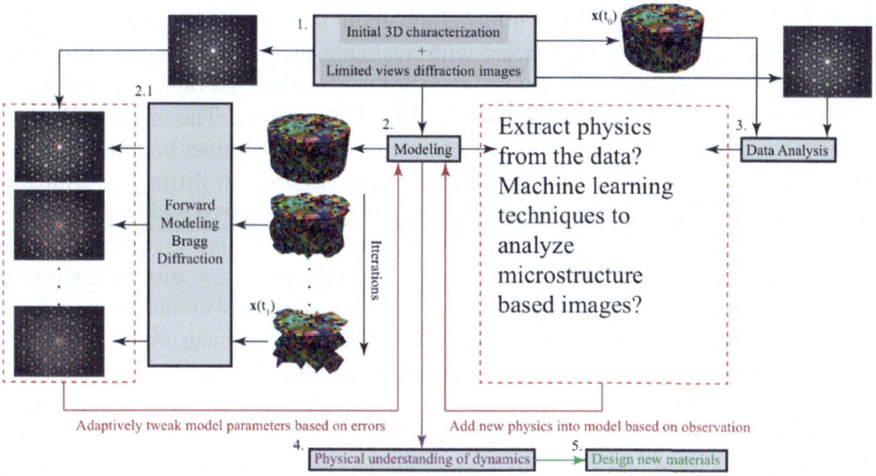

Fig. 7.19 Schematic to illustrate data analysis pipeline for physics based model development

given sample state, only to later discard or never analyze most of the acquired data. The main reason for such an inefficient measurement protocol is due to the lack of real-time analyses tools that can guide measurements during the limited available beam time. Investment in the development of efficient, fast, and user-friendly data reduction and reconstruction software has the potential to change how experiments (analysis, throughput) are performed. This could improve the data quality as well as enable multiple sample state measurements, lending to high-temporal resolution.

For dynamic conditions, both spatially and temporally resolved (~1 μm and ~1 ps) microstructure information is desired to understand materials properties and performance for engineering applications. Current literature demonstrates that non-destructive techniques can be successfully utilized for studying 3D microstructure evolution under quasi-static conditions. However, extending such studies to dynamic loading conditions is still a challenge. Currently, the most common approach to mapping 3D microstructures is by rotating the sample and collecting multiple views. However, to capture the dynamics during shock loading or during high strain-rate loading conditions, data needs to be acquired at the same temporal scale as the dynamic process. Waiting for a sample to be rotated and imaged from multiple angles during such in-situ measurements is just too slow. In order to speed up the HEDM measurement process, we must consider more sophisticated approaches to measurement and reconstruction, including iterative techniques which utilize past sample state data and dynamic models for subsequent reconstructions.

Figure 7.19 demonstrates a possible work flow for enabling dynamic measurements. HEDM and various other 3D characterization techniques discussed earlier can be utilized to fully characterize the initial state of the material before dynamic loading. This would provide information such as chemistry, composition, microstructure, phase, and defect structures of the material of interest. Note that Fig. 7.19 is a vastly simplified vision, where this prior information about the sample would help develop a data analysis framework in concert with available microstructure based models and the forward modeling method for direct simulation of diffraction. The measured initial 3D structure will be used as input to an existing model. The model will then evolve the structure based on some governing equations and proper boundary conditions. The forward modeling method can then be used to simulate diffraction from the evolved structure. The simulated diffraction can then be compared with experiments. Given the physics in the model is adequate, iteratively changing associated model parameters could give us a reasonable match between the observation and simulation, at least for the initial time steps. A feedback loop would be created for iterating and updating each step. Note that there are uncertainties in the measured data (detector images in our case) that will propagate in the predicted features (reconstructed microstructural properties), which are then used for predicting the corresponding material properties. Therefore, measurement uncertainty needs to be accounted for when adaptively tweaking the model parameters. Such an approach could provide new insight into and understanding of the mechanisms driving dynamic processes in polycrystalline materials. However, it is highly unlikely that the existing models have adequate physics to accurately capture the complex micro-mechanical field development throughout the whole dynamic process. Given the possibility of acquiring data with high temporal resolution, albeit sparse spatial views, an assumption can be made that the material change from one state to the next is relatively small. Therefore, utilizing the initial characterization and linking the dynamic measurements, diffraction simulations, data mining tools, and existing models could enable extraction of 3D information from limited views and highly incomplete datasets.

Acknowledgements The author gratefully acknowledges the Los Alamos National Laboratory for supporting mesoscale science technology awareness and this work. Experimental support on the measurements of ATF fuel and AM samples from the staff of the APS-1-ID-E beamline is also acknowledged. The author is also thankful to Alexander Scheinker and Turab Lookman for their valuable inputs during the course of writing this chapter.

References

1. G. Crabtree, J. Sarrao, P. Alivisatos, W. Barletta, F. Bates, G. Brown, R. French, L. Greene, J. Hemminger, M. Kastner et al., From quanta to the continuum: opportunities for mesoscale science. Technical report, USDOE Office of Science (SC) (United States) (2012)
2. D.L. McDowell, A perspective on trends in multiscale plasticity. Int. J. Plast. **26**(9), 1280–1309 (2010)
3. D. Krajcinovic, Damage mechanics: accomplishments, trends and needs. Int. J. Solids Struct. **37**(1), 267–277 (2000)
4. R. Pokharel, J. Lind, A.K. Kanjarla, R.A. Lebensohn, S.F. Li, P. Kenesei, R.M. Suter, A.D. Rollett, Polycrystal plasticity: comparison between grain-scale observations of deformation and simulations. Annu. Rev. Condens. Matter Phys. **5**(1), 317–346 (2014)
5. R. A. Schwarzer, D. P. Field, B. L. Adams, M. Kumar, A. J. Schwartz, Present state of electron backscatter diffraction and prospective developments in *Electron backscatter diffraction in materials science*. (Springer, 2009), pp. 1–20

6. H.F. Poulsen, S.F. Nielsen, E.M. Lauridsen, S. Schmidt, R.M. Suter, U. Lienert, L. Margulies, T. Lorentzen, D.J. Jensen, Three-dimensional maps of grain boundaries and the stress state of individual grains in polycrystals and powders. J. Appl. Crystallogr. **34**(6), 751–756 (2001)
7. S. Schmidt, H.F. Poulsen, G.B.M. Vaughan, Structural refinements of the individual grains within polycrystals and powders. J. Appl. Crystallogr. **36**(2), 326–332 (2003)
8. H.F. Poulsen, *Three-Dimensional X-Ray Diffraction Microscopy: Mapping Polycrystals and Their Dynamics*, vol. 205 (Springer Science & Business Media, 2004)
9. R.A. Lebensohn, R. Pokharel, Interpretation of microstructural effects on porosity evolution using a combined dilatational/crystal plasticity computational approach. JOM **66**(3), 437–443 (2014)
10. R. Pokharel, R.A. Lebensohn, Instantiation of crystal plasticity simulations for micromechanical modelling with direct input from microstructural data collected at light sources. Scr. Mater. **132**, 73–77 (2017)
11. K. Chatterjee, J.Y.P. Ko, J.T. Weiss, H.T. Philipp, J. Becker, P. Purohit, S.M. Gruner, A.J. Beaudoin, Study of residual stresses in Ti-7Al using theory and experiments. J. Mech. Phys. Solids (2017)
12. D.C. Pagan, P.A. Shade, N.R. Barton, J.-S. Park, P. Kenesei, D.B. Menasche, J.V. Bernier, Modeling slip system strength evolution in Ti-7Al informed by in-situ grain stress measurements. Acta Mater. **128**, 406–417 (2017)
13. D.L. McDowell, Multiscale crystalline plasticity for materials design, in *Computational Materials System Design* (Springer, 2018), pp. 105–146
14. U. Lienert, S.F. Li, C.M. Hefferan, J. Lind, R.M. Suter, J.V. Bernier, N.R. Barton, M.C. Brandes, M.J. Mills, M.P. Miller, High-energy diffraction microscopy at the advanced photon source. JOM J. Miner. Metals Mater. Soc. **63**(7), 70–77 (2011)
15. C.M. Hefferan, J. Lind, S.F. Li, U. Lienert, A.D. Rollett, R.M. Suter, Observation of recovery and recrystallization in high-purity aluminum measured with forward modeling analysis of high-energy diffraction microscopy. Acta Mater. **60**(10), 4311–4318 (2012)
16. S.F. Li, J. Lind, C.M. Hefferan, R. Pokharel, U. Lienert, A.D. Rollett, R.M. Suter, Three-dimensional plastic response in polycrystalline copper via near-field high-energy X-ray diffraction microscopy. J. Appl. Crystallogr. **45**(6), 1098–1108 (2012)
17. R. Pokharel, J. Lind, S.F. Li, P. Kenesei, R.A. Lebensohn, R.M. Suter, A.D. Rollett, In-situ observation of bulk 3D grain evolution during plastic deformation in polycrystalline Cu. Int. J. Plast. **67**, 217–234 (2015)
18. J. Lind, S.F. Li, R. Pokharel, U. Lienert, A.D. Rollett, R.M. Suter, Tensile twin nucleation events coupled to neighboring slip observed in three dimensions. Acta Mater. **76**, 213–220 (2014)
19. C.A. Stein, A. Cerrone, T. Ozturk, S. Lee, P. Kenesei, H. Tucker, R. Pokharel, J. Lind, C. Hefferan, R.M. Suter, Fatigue crack initiation, slip localization and twin boundaries in a nickel-based superalloy. Curr. Opin. Solid State Mater. Sci. **18**(4), 244–252 (2014)
20. J.F. Bingert, R.M. Suter, J. Lind, S.F. Li, R. Pokharel, C.P. Trujillo, High-energy diffraction microscopy characterization of spall damage, in *Dynamic Behavior of Materials*, vol. 1 (Springer, 2014), pp. 397–403
21. B. Lin, Y. Jin, C.M. Hefferan, S.F. Li, J. Lind, R.M. Suter, M. Bernacki, N. Bozzolo, A.D. Rollett, G.S. Rohrer, Observation of annealing twin nucleation at triple lines in nickel during grain growth. Acta Mater. **99**, 63–68 (2015)
22. A. D. Spear, S. F. Li, J. F. Lind, R. M. Suter, A. R. Ingraffea, Three-dimensional characterization of microstructurally small fatigue-crack evolution using quantitative fractography combined with post-mortem X-ray tomography and high-energy X-ray diffraction microscopy. Acta. Materialia. **76**, 413–424 (2014)
23. J. Oddershede, S. Schmidt, H.F. Poulsen, H.O. Sorensen, J. Wright, W. Reimers, Determining grain resolved stresses in polycrystalline materials using three-dimensional X-ray diffraction. J. Appl. Crystallogr. **43**(3), 539–549 (2010)
24. J.V. Bernier, N.R. Barton, U. Lienert, M.P. Miller, Far-field high-energy diffraction microscopy: a tool for intergranular orientation and strain analysis. J. Strain Anal. Eng. Des. **46**(7), 527–547 (2011)

25. J. Oddershede, S. Schmidt, H.F. Poulsen, L. Margulies, J. Wright, M. Moscicki, W. Reimers, G. Winther, Grain-resolved elastic strains in deformed copper measured by three-dimensional X-ray diffraction. Mater. Charact. **62**(7), 651–660 (2011)
26. N.R. Barton, J.V. Bernier, A method for intragranular orientation and lattice strain distribution determination. J. Appl. Crystallogr. **45**(6), 1145–1155 (2012)
27. D.C. Pagan, M.P. Miller, Connecting heterogeneous single slip to diffraction peak evolution in high-energy monochromatic X-ray experiments. J. Appl. Crystallogr. **47**(3), 887–898 (2014)
28. M. Obstalecki, S.L. Wong, P.R. Dawson, M.P. Miller, Quantitative analysis of crystal scale deformation heterogeneity during cyclic plasticity using high-energy X-ray diffraction and finite-element simulation. Acta Mater. **75**, 259–272 (2014)
29. J. Oddershede, J.P. Wright, A. Beaudoin, G. Winther, Deformation-induced orientation spread in individual bulk grains of an interstitial-free steel. Acta Mater. **85**, 301–313 (2015)
30. G. Winther, J.P. Wright, S. Schmidt, J. Oddershede, Grain interaction mechanisms leading to intragranular orientation spread in tensile deformed bulk grains of interstitial-free steel. Int. J. Plast. **88**, 108–125 (2017)
31. D.C. Pagan, M. Obstalecki, J.-S. Park, M.P. Miller, Analyzing shear band formation with high resolution X-ray diraction. Acta Mater. (2018)
32. D. Naragani, M. D. Sangid, P. A. Shade, J. C. Schuren, H. Sharma, J. S. Park, ..., I. Parr, Investigation of fatigue crack initiation from a non-metallic inclusion via high energy x-ray diffraction microscopy. Acta. Materialia. **137**, 71–84 (2017)
33. J.C. Schuren, P.A. Shade, J.V. Bernier, S.F. Li, B. Blank, J. Lind, P. Kenesei, U. Lienert, R.M. Suter, T.J. Turner, New opportunities for quantitative tracking of polycrystal responses in three dimensions. Curr. Opin. Solid State Mater. Sci. **19**(4), 235–244 (2015)
34. T.J. Turner, P.A. Shade, J.V. Bernier, S.F. Li, J.C. Schuren, P. Kenesei, R.M. Suter, J. Almer, Crystal plasticity model validation using combined high-energy diffraction microscopy data for a Ti-7Al specimen. Metall. Mater. Trans. A **48**(2), 627–647 (2017)
35. K. Chatterjee, A. Venkataraman, T. Garbaciak, J. Rotella, M.D. Sangid, A.J. Beaudoin, P. Kenesei, J.-S. Park, A.L. Pilchak, Study of grain-level deformation and residual stresses in Ti-7Al under combined bending and tension using high energy diffraction microscopy (HEDM). Int. J. Solids Struct. **94**, 35–49 (2016)
36. H.M. Paranjape, P.P. Paul, H. Sharma, P. Kenesei, J.-S. Park, T.W. Duerig, L.C. Brinson, A.P. Stebner, Influences of granular constraints and surface effects on the heterogeneity of elastic, superelastic, and plastic responses of polycrystalline shape memory alloys. J. Mech. Phys. Solids **102**, 46–66 (2017)
37. D.W. Brown, L. Balogh, D. Byler, C.M. Hefferan, J.F. Hunter, P. Kenesei, S.F. Li, J. Lind, S.R. Niezgoda, R.M. Suter, Demonstration of near field high energy x-ray diffraction microscopy on high-z ceramic nuclear fuel material, in *Materials Science Forum*, vol. 777 (Trans Tech Publications, 2014), pp. 112–117
38. R. Pokharel, D. W. Brown, B. Clausen, D. D. Byler, T. L. Ickes, K. J. McClellan, ..., P. Kenesei, Non-destructive characterization of UO2 + x nuclear fuels. Microsc. Today **25**(6), 42–47 (2017)
39. W. Ludwig, S. Schmidt, E.M. Lauridsen, H.F. Poulsen, X-ray diffraction contrast tomography: a novel technique for three-dimensional grain mapping of polycrystals. I. Direct beam case. J. Appl. Crystallogr. **41**(2), 302–309 (2008)
40. W. Ludwig, P. Reischig, A. King, M. Herbig, E.M. Lauridsen, G. Johnson, T.J. Marrow, J.-Y. Buffiere, Three-dimensional grain mapping by X-ray diffraction contrast tomography and the use of friedel pairs in diffraction data analysis. Rev. Sci. Instrum. **80**(3), 033905 (2009)
41. L. Renversade, R. Quey, W. Ludwig, D. Menasche, S. Maddali, R.M. Suter, A. Borbély, Comparison between diffraction contrast tomography and high-energy diffraction microscopy on a slightly deformed aluminium alloy. IUCrJ **3**(1), 32–42 (2016)
42. B.C. Larson, W. Yang, G.E. Ice, J.D. Budai, J.Z. Tischler, Three-dimensional X-ray structural microscopy with submicrometre resolution. Nature **415**(6874), 887–890 (2002)
43. S.C. Vogel, M.A. Bourke, A.S. Losko, R. Pokharel, T.L. Ickes, J.F. Hunter, D.W. Brown, S.L. Voit, K.J. Mcclellan, A. Tremsin, Non-destructive pre-irradiation assessment of un/u-si lanl1 atf formulation. Technical report, Los Alamos National Laboratory (LANL) (2016)

44. B.E. Warren, *X-ray Diffraction* (Courier Corporation, 1969)
45. J.-S. Park, J. Okasinski, K. Chatterjee, Y. Chen, J. Almer, Non-destructive characterization of engineering materials using high-energy X-rays at the advanced photon source. Synchrotron Radiat. News **30**(3), 9–16 (2017)
46. D. E. Boyce, J. V. Bernier, heXRD: Modular, open source software for the analysis of high energy x-ray diffraction data (No. LLNL-SR-609815) (Lawrence Livermore National Laboratory (LLNL), Livermore, CA, 2013)
47. S. F. Li, R. M. Suter, Adaptive reconstruction method for three-dimensional orientation imaging. J Appl. Crystallogr. **46**(2), 512–524 (2013)
48. R. Pokharel, Spatially resolved in-situ study of plastic deformation in polycrystalline copper using high-energy X-rays and full-field simulations. Ph.D. thesis (Carnegie Mellon University, 2013)
49. MIDAS, Microstructural Identification using Diffraction Analysis Software. https://github.com/marinerhemant
50. T.J. Turner, P.A. Shade, J.V. Bernier, S.F. Li, J.C. Schuren, J. Lind, U. Lienert, P. Kenesei, R.M. Suter, B. Blank, Combined near-and far-field high-energy diffraction microscopy dataset for Ti-7Al tensile specimen elastically loaded in situ. Integr. Mater. Manuf. Innov. **5**(1), 5 (2016)
51. B. Fromm, Y. Zhang, D. Schwen, D. Brown, R. Pokharel, Assessment of marmot grain growth model. Technical report, Idaho National Lab. (INL), Idaho Falls, ID (United States) (2015)
52. G.M. Hommer, J.S. Park, P.C. Collins, A.L. Pilchak, A.P. Stebner, A new in situ planar biaxial far-field high energy diffraction microscopy experiment, in *Advancement of Optical Methods in Experimental Mechanics*, vol. 3 (Springer, 2017), pp. 61–70

Chapter 8
Bragg Coherent Diffraction Imaging Techniques at 3rd and 4th Generation Light Sources

Edwin Fohtung, Dmitry Karpov and Tilo Baumbach

Abstract Although X-ray crystallography is established as a state of the art imaging technique that has been revolutionary across materials sciences, physics, chemistry, biology and medicine, the imaging of non-crystalline objects is inaccessible by this method. A promising approach that can overcome this challenge is coherent diffractive imaging (CDI). CDI is a lensless microscopy technique that can provide nanoscale images of both non-crystalline and crystalline objects. The morphology, structure and evolution of an object of interest is probed using a coherent source of photons (often X-rays, visible light) or electrons. Coherency is needed for the interference to produce a usable diffraction pattern. While the diffraction pattern contains the magnitude information of the object in reciprocal space, the phase information can be recovered using iterative feedback algorithms, allowing the reconstruction of the image of an object. As no lenses are used, the image is free of aberrations and hence the resolution is limited only by the wavelength of the probe, exposure, and the robustness of the reconstruction algorithm. This technique has proven crucial for imaging of variety of samples, from nanostructures to bio-tissues and individual cells. The aim of this chapter is to provide a clear picture of recent state-of-the-art developments in CDI techniques, and particularly in Bragg Coherent Diffraction Imaging, applied to oxide nanostructures.

E. Fohtung (✉) · D. Karpov
Department of Physics, New Mexico State University, Las Cruces, NM 88003, USA
e-mail: efohtung@nmsu.edu

E. Fohtung
Los Alamos National Laboratory, Los Alamos, NM 87545, USA

D. Karpov
Physical-Technical Institute, National Research Tomsk Polytechnic University, Tomsk 634050, Russia
e-mail: dkarpov@nmsu.edu

T. Baumbach
Institute for Photon Science and Synchrotron Radiation, Kasrlsruhe Institute of Technology, 76344 Eggenstein-Leopoldshafen, Germany
e-mail: tilo.baumbach@kit.edu

T. Lookman et al. (eds.), *Materials Discovery and Design*, Springer Series in Materials Science 280, https://doi.org/10.1007/978-3-319-99465-9_8

8.1 Introduction

X-ray Bragg Coherent Diffractive Imaging (BCDI) method has been developed for nondestructive imaging of three-dimensional (3D) displacement fields and strain evolution within microscale and nanoscale crystals [1–3]. BCDI has been widely utilized by researchers and scientists in academy, industry and governmental laboratories resulting in a vast user community at photon factories or light sources worldwide. BCDI relies on the fact that given a spatially coherent beam of X-rays illuminating a specimen, so that scattering from all crystal extremities interfere, the diffraction patterns contain enough information to be inverted to real space images. The technique is based on general principles of coherent diffractive imaging and iterative phase retrieval methodology which were first suggested by Sayre in 1953 [4] and first demonstrated by Miao in 1999 [5].

The diffraction pattern is measured in reciprocal space. The reciprocal space in the BCDI experimental geometry is largely empty, allowing the investigation of individual nanoparticles and grains. A polycrystalline sample will have closely-packed grains with numerous different orientations. The Bragg diffraction from a polycrystalline sample will resemble that of a powder but, given a small enough beam and typical grain sizes close to a micron, individual grain diffraction can be isolated by an area detector [6]. Even highly textured samples can still have enough distribution of orientations so that the grains are usually distinguishable. Once a Bragg peak is isolated and aligned, its 3D intensity distribution can be recorded by means of an area detector placed on a long motorized arm. A rocking series of images passing through the Bragg peak center provides 3D data, as shown in Fig. 8.1c, consisting of characteristic rings resembling the Airy pattern of a compact solid object and streaks attributed to its prominent facets.

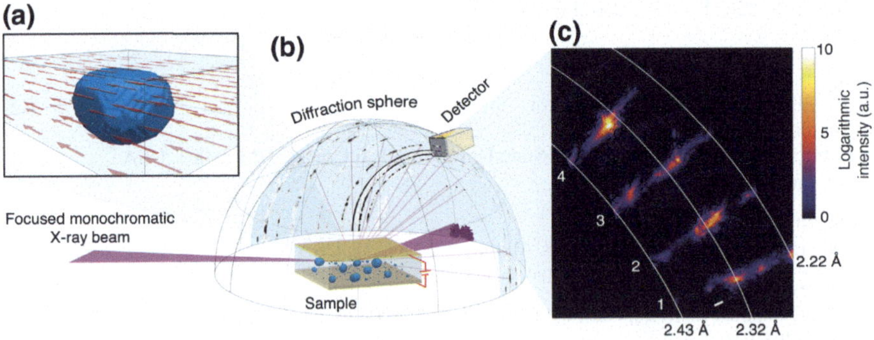

Fig. 8.1 Bragg X-ray coherent diffraction experiment. **a** Isosurface of the magnitude of the Bragg electron density of a $BaTiO_3$ nanocrystal is reconstructed from the diffraction patterns measured in reciprocal space. **b** A focused monochromatic beam illuminates the sample and a single peak reflected from a nanocrystal is isolated in reciprocal space. **c** Evolution of the diffraction intensities from the nanoparticle undergoing phase transitions induced by external electric field. White scale bar corresponds to 0.1 $Å^{-1}$. Illustration is taken from [9]

The reconstruction (also known as "inversion") of the data into the real space images of an object is a critical step that uses a computer algorithm that takes advantage of internal redundancies in the data, when the measurement points are spaced close enough together and satisfy the *oversampling* requirement. The first step in the reconstruction procedure is to postulate a 3D *support* volume in which all the sample density will be constrained to physically exist. Arguably, the best method so far for phase retrieval that allows to avoid *stagnation* of the reconstruction is the Fienup's Hybrid Input-Output (HIO) algorithm [7]. Thanks to continues improvements in algorithms developments [8], we now consider the communitys phase retrieval inversion routines to be a trustworthy *black box* tools for data evaluation, which soon will be made available for real time reconstructions at the light sources and dedicated BCDI beamlines.

We illustrate the capabilities of the BCDI method in Fig. 8.2 with the example of a $BaTiO_3$ nanocrystal, which is taken from the recent publication on topological vortex dynamics [9]. The physical density of the crystal was almost uniform except at the regions where topological defect was predicted by phase field simulation (see Fig. 8.2b). However, there was a prominent imaginary part, the origin of which is attributed to an internal ferroelectric displacement field. The figure shows the reconstructed ferroelectric displacement fields and a comparison with a theoretically simulated model based on Landau theory [9]. The maximum displacement

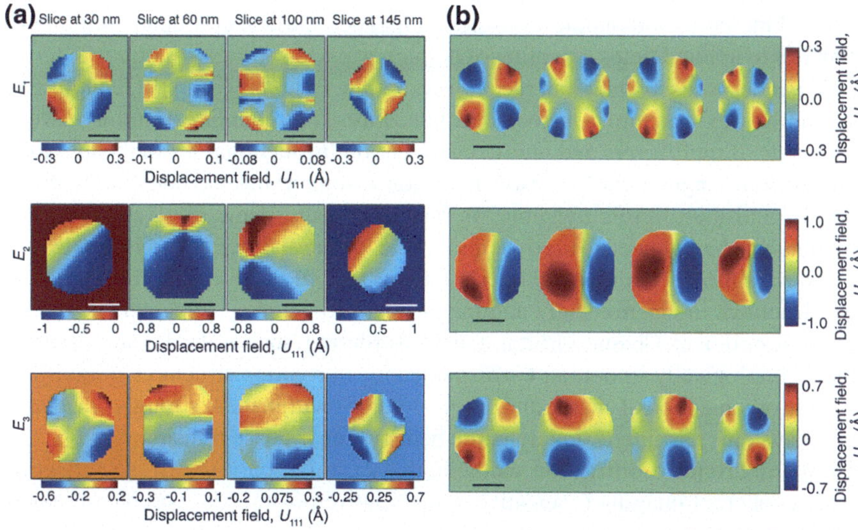

Fig. 8.2 X-ray BCDI experiments and theoretical results. **a** Reconstructions of a single $BaTiO_3$ nanocrystal showing the ferroelectric displacement field distributions at various 2D cut planes in the nanocrystal. **b** Simulated model of $BaTiO_3$ nanocrystal. The reconstructions show the evolution of the displacement field under the influence of an external electric field E_1 depicting the virgin state of the nanocrystal at 0V, E_2 maximum field state at 10V, and E_3 the remnant field state at 0V. Scale bars correspond to 60 nm. Illustration is taken from [9]

component of the tetragonal structural phase as seen in the reconstructions of the complex density of 0.03 nm, corresponding to a total displacement (relative to the ideal crystal lattice) of about a half of a $BaTiO_3$111 atomic spacing, or 0.06 nm. BCDI has matured over the last decade and is now capable of playing a crucial role in solving important problems in materials research and condensed matter physics.

Complex Ferroic Oxides can be driven far from equilibrium by external perturbations such as heat, light, electric and magnetic fields. Four primary ferroic orders namely toroidal, dipole, elastic and magnetic moments can be tuned at the microscopic level and used in the design of novel devices and functional properties leading to high-TC superconducting cuprates and colossal magnetoresistance.

Multiferroics have attracted enormous attention in the past decade, due to their fascinating physics and applicable magnetoelectric functionalities. A variety of promising technological applications include energy transformation, signal generation and processing, information storage, etc. X-ray scattering techniques couple directly to these order parameters relevant to ferroelectricity and magnetism, and are fully quantitative [10]. BCDI can resolve small strains with spatial resolution far better than area-selective diffraction approaches. In addition, coherent scattering experiments can have time resolutions limited only by the bunch length of the pulses from the storage ring, which will be on the order of 30ps for NSLS-II and other 3rd and 4th generation light sources such as the LCLS, APS, XFEL, ESRF, SLAC, ANKA, BESSY, PETRA-IV, ALS.

Phenomena accessible with the dramatic advance in spatial and temporal resolution include domain dynamics, physics of magnetoelectric coupling, coupling of soft modes to applied fields, coupling of domain walls and vortex structures in ferroelectric nanocrystals as shown in recent work by the Fohtung group [9], coupling of vortex-antivortex structure in ferroelectric nanowires [11], coupling of strain between components of multilayers and multicomponent multifunctional ferroic materials.

BCDI from magnetic order (resonant or non-resonant) has the potential to probe magnetic, and multifferoic dynamics in buried systems. Time-resolved coherence techniques can be extended to the dynamics of electronic and magnetic systems driven far from equilibrium with external perturbations. BCDI methods will be applied to phase structures in electronically and magnetically ordered systems, ferroic oxides showing Topoligical, Charge, Orbital, and Spin ordering, such as the domain structure of the technologically important CMR manganites or vortex structures in multiferroics. The organization of charge and orbital domains within these strongly correlated electron systems has interesting dynamic behavior near phase transitions [11].

Unlike micro/nano diffraction methods that examine only the Bragg reflection intensity, coherent imaging is sensitive to the relative positions of ordered domains. BCDI beamlines at 3rd and 4th generation sources could provide answers to questions such as: are the walls of a vortex-core in multiferroic nanostructure ferroelectric, paraelectric or even semi-metallic, why is there substantial inhomogeneously distributed distortion of the Charge Density Wave (CDW)/Spin Density Wave (SDW) wavevector in Cr and other incommensurate magnetic materials, for example.

Here we demonstrate that ferroic materials can be imaged under operando conditions in a functional capacitor (as shown in Fig. 8.4) to study the formation and

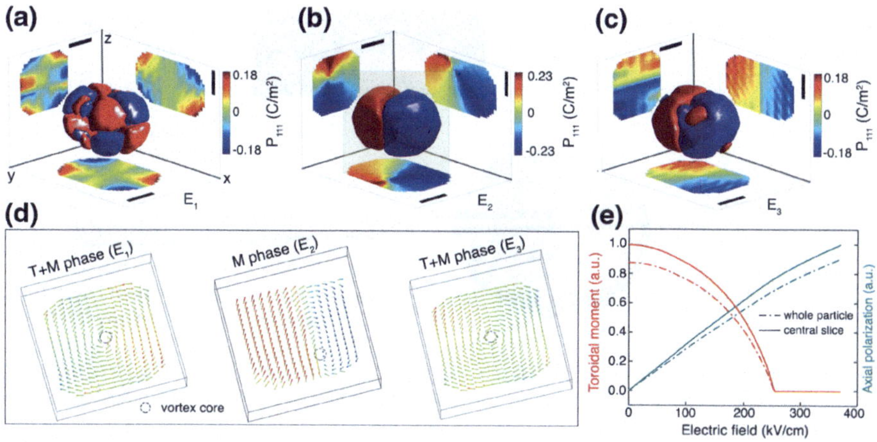

Fig. 8.3 3D reconstructions of ferroelectric polarization. Isosurfaces of the domain distribution and the ferroelectric polarization within an individual nanocrystal under an applied electric field of **a** $E_1 = 0$ kV/cm, **b** $E_2 = 223$ kV/cm, and **c** $E_3 = 0$ kV/cm remnant state. **d** Phase field simulation of the polarization maps depicting the structural phases within the nanocrystal. With M denoting the monoclinic phase and T the tetragonal phase. **e** Simulated behavior of the toroidal moment and axial polarization at different external electric fields. Note that experimental maximum of 223 kV/cm is confirmed by the phase field model. Scale bars correspond to 60 nm. Illustration is taken from [9]

evolution of complex configurations of electric polarizations such as a vortex structure other structures shown in Fig. 8.3 associated with the electric filed structural phase transitions.

Topological defects of spontaneous polarization are extensively studied as templates for unique physical phenomena and in the design of reconfigurable electronic devices. Experimental investigations of the complex topologies of polarization have been limited to surface phenomena, which has restricted the probing of the dynamic volumetric domain morphology in operando. Here, we utilize X-ray BCDI of individual $BaTiO_3$ nanoparticle in a composite polymer/ferroelectric capacitor shown in Fig. 8.4 to study the behavior of a three-dimensional vortex formed due to competing interactions involving ferroelectric domains. Our investigation of the structural phase transitions under the influence of an external electric field shows a mobile vortex core exhibiting a reversible hysteretic transformation path. We also study toroidal moment of the vortex under the action of the field. Our results open avenues for the study of the structure and evolution of polar vortices and other topological structures in operando in functional materials under cross field configurations.

In a typical BCDI experiment performed ar 3rd and/or 4th generation light source, a monochromator tunes into the required for such ferroic nanocrystals [9] X-ray energy of 9 keV with 1 eV bandwidth. The beam is then focused by Fresnel Zone plates or KB-mirrors as shown in Fig. 8.5. The focused monochromatic beam is then scattered from the sample extremities (Fig. 8.1) and recorded by the detector mounted

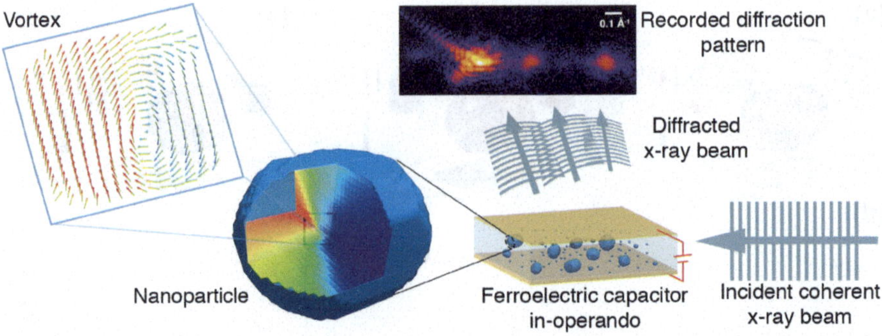

Fig. 8.4 Experimental scheme of BCDI and inoperando functional capacitor. Incident coherent X-ray beam is scattered by a nanoparticle embedded in conducting non-polarizing polymer with attached electrodes. Constructive interference patterns are recorded during application of an external electric field on the particle. Recorded high-resolution Bragg-peak diffraction carries information on the electron density and atomic displacement variations, allowing to reconstruct the complex process of defect evolution and monitoring of vortex. Illustration is taken from [9]

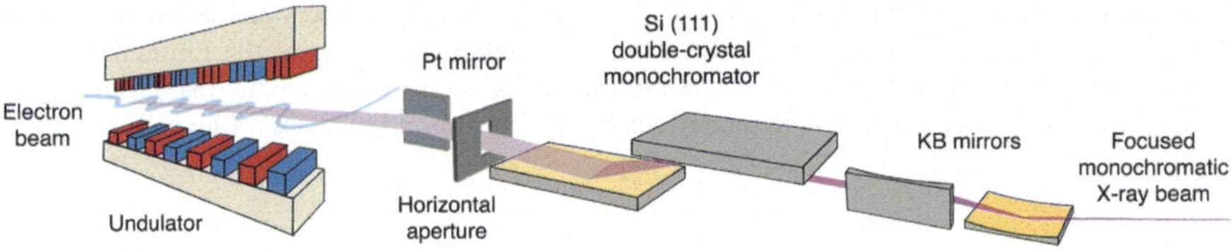

Fig. 8.5 Principle schemes of the beamline at the station 34-ID-C up to the experimental table

on the motorized arm that can be positioned in a spherical coordinates around the sample. The size of focused beam is usually chosen to fully illuminate the individual nanocrystal.

To record the diffraction patterns we used a Medipix2 detector composed of 256 by 256 picture elements, with individual pixel size of 55 μm by 55 μm. Experimental positioning system allows to translate the sample on 3 rectangular coordinate axes as well as to adjust roll and pitch. When the desired Bragg peak is found, the nanocrystal is *rocked* about the Bragg reflection by subtle rotation with respect to the X-ray beam. The rocking curve in the cited work [9] was collected in the vicinity of the (111) Bragg peak with the scanning range of $\Delta\theta = \pm 0.3^\circ$.

Using iterative phase retrieval algorithms, we can reconstruct 3D distribution of the displacement fields as shown in (Fig. 8.2a).Theoretical simulations based on Landau Phase-field model was used to interpret the reconstructions as shown in (Fig. 8.2b). The following relationship can be used to extract the stain tensor if multiple Bragg peaks can be measured from the same nanocrystal:

$$\epsilon_{ij} = 1/2(\partial \mathbf{u}^i_{111}/\partial x_j + \partial \mathbf{u}^j_{111}/\partial x_i). \tag{8.1}$$

From the reconstructed strain field, ferroelectric polarization maps P can be evaluated using the relationship:

$$\epsilon^{o}_{ij} = Q_{ijkl} P_k P_l, \tag{8.2}$$

where Q_{ijkl} is the electrostrictive tensor and ϵ^{o}_{ij} is the spontaneous strain. This approach allows to visualize three-dimensional shape, morphology and the evolution of the observed ferroelectric vortex phase as predicted by phase field model (Fig. 8.3d, e).

We observed that complex topologies of polarizations can be engineered and controlled by external perturbations such as an applied electric, magnetic fields, stresses, as well as by built-in interface effects at the mesoscale. Since the vortex core can be displaced and erased through a reversible hysteretic transformation path, it can be thought of as a conductive channel inside a monolith of nominally insulating ferroelectric material. This could be exploited in the design of integrated electronic devices based on polar vortices and the possibility of creating artificial states of matter through the control of related phase transitions. Advances in the development of bright coherent light sources allow greater temporal resolution in dynamical studies of these phenomena. Our findings can help to pave the way for future studies of shape and size dependence along with hysteretic behavior under temperature and other external perturbations to unveil what is yet hidden in vortex dynamics.

Energy Storage Materials and Nanocatalysts can be imaged under operando conditions to study dislocations and other phase defects associated with their function. These deeply buried systems present an opportunity where electron microscopy can only be used under unrealistic conditions, while BCDI is amenable to a realistic battery environment. Recent work by the Shpyrko group [12–15] has shown dislocation motions in heavily-cycled cathode nanoparticles driven by charge/discharge cycles in operando batteries. Their group showed BCDI visualization of defects in the nanostructured disordered spinel material $LiNi_{0.5}Mn_{1.5}O_4$ (LNMO), which is a promising high-voltage cathode material.

Lithium diffusion, responsible for the charge storage, was found to drive dislocation motions. Nanoparticles were also observed to exhibit phase separation during initial charge and discharge cycles. Other critical imaging efforts include the transport of oxygen vacancies in electrolytes for solid-oxide fuel cells, imaged by probing the vacancy-induced lattice distortion. In addition, BCDI will enable imaging of catalyst nanoparticles during reactions and the strain fields associated with intrinsic and extrinsic defects in solar absorbers. It is envisaged there will be a large user community interested in battery materials, for whom we will provide standardized sample environments for in-operando studies, similar to *coin-cells*, in which structural changes in anode and cathode materials can be probed in the presence of an electrolyte under moderate pressure. By using Bragg diffraction we can obtain steep angles of the X-ray beam through the window materials and avoid most of the background signal. Individual grains in a polycrystalline material can be selected from the population of grain orientations and imaged by BCDI to learn how the local crystal distortions change during charge/discharge.

Zeolite and perovskite oxide materials have both been proposed as alternatives to Platinum-Group Metal (PGM) automotive emission catalysts with the advantage of being substantially cheaper [16]. Perovskites are classical mixed-ion oxides with composition ABO_3.The oxygen coordination is cuboctahedral (12-fold) around the A site and octahedral around B. The discovery in 2010 that $La_{1-x}Sr_xMnO_3$ (LSMO) and $La_{1-x}Sr_xCoO_3$ (LSCO) perovskites were as effective as PGMs for removing NO_x from diesel exhaust was an important breakthrough [17].

This class of oxides had been found ineffective before and thought to become active catalysts by virtue of having Sr^{++} active sites [17]. Zeolites are crystalline aluminosilicate composites with a regular nano-porous lattice structure which allows gases to reach active sites within their framework. The zeolites most relevant to automotive catalysts are ZSM-5 and SAPO-34. We will provide gas handling sample environments for in-operando experiments to observe the strains within these microcrystalline oxide catalysts while they perform their reactions. BCDI can obtain 3D images of distortions of the internal "plumbing" of the nano-porous network which provides their large surface area and selectivity during reactions. The expected resolution, in the 20–50 nm range, is insufficient to see the 0.5 nm pores directly, but the strain sensitivity is much better than a lattice constant; crystal distortions in the picometer range can be detected when they extend over the resolution range. The 20–50 nm length scale is well-matched to the expected scale of distortions due to "coking" which results in catalyst lifetimes too short for commercial exploitation. A potentially new area of laser-promoted ultrafast catalysis can be explored using time domain BCDI experiments.

Laser-excited materials will be investigated by using stroboscopic pump-probe techniques with a picosecond timing laser to overlap with the 30 ps X-ray pulse structure of NSLS-II. This can be used to measure optically driven phase transitions and to explore coherent vibrational properties of nanocrystals, as achieved recently at Stanfords Linac Coherent Light Source (LCLS) [18].

Robinson's group showed that 3D cross sectional images of a snapshot of a shear-wave vibration in a Au nanocrystal can be observed [19]. The vibration period in this example is 100 ps, which will be accessible at NSLS-II. Transient melting phenomena [19] and metastable "hidden" phases of matter can be systematically explored in this way. New laser-driven phenomena are starting to be observed, like the hidden magnetic state seen in $Nd_{0.5}Sr_{0.5}MnO_3$ (NSMO) thin films [20] and "enhanced" superconductivity in $YBa_2Cu_3O_{6.5}$ [21]. While the fastest, femtosecond phenomena tend to be purely electronic in nature and would only be accessible by XFEL methods, time-resolved BCDI addresses structural changes involving atomic motions, on the timescale of phonons, which are possible with 30 ps time resolution. The lifetimes of these new transient states [20] were around 1 ns. We note that timing options are less interesting at new or upgraded Multibend-Acromat (MBA) sources because of their longer pulse lengths.There is a large, unexplored opportunity to observe materials in the ultrafast time domain close to phase transitions. Such experiments are only possible using Bragg diffraction because only this offers sensitivity to the sub-Angstrom distortions associated with displacive transitions. If these novel excited states.can be observed and found to be useful or interesting, they can be stabilized by traditional doping methods. One example is the reported core-shell structure of nanocrystalline Barium Titanate (BTO) [22], which shows a threefold enhancement of its dielectric constant, important in the development of supercapacitors.

Earth Materials behavior at pressure and temperature governs a considerable number of large scale processes in the Earth, such as plate tectonics, volcanism, and mantle convection. Despite considerable experimental and theoretical progress in the past decade, many of these processes are not fully understood. In Earth Sciences, BCDI is a new experimental tool to study the stress and strain relationships and strain fields in Earth Materials at pressure and temperature on the level of individual grains [23]. For example, BCDI will be used to determine the strain field in individual grains of porous rocks at relevant pressure and temperature conditions of water, oil, gas and CO_2 storage reservoirs. The detailed knowledge of the stress and strain field in reservoir rocks will dramatically increase our abilities to model the complex hydromechanical processes in the reservoirs, leading to more efficient gas and oil extraction, and safer storage capabilities for CO_2.

Semiconductor strains are important for mobility and band gap engineering techniques used in the present generation of CMOS devices and in the emerging world of silicon quantum electronics. BCDI has been used by Thomas et al. to examine individual strained Silicon-on-Insulator (SOI) structures and by the Baumbach group in GaMnAs nanowires [24–26]. Strain patterns can be created in model devices with sizes more relevant to current technology (22 nm), that penetrate partially into the thickness of the SOI layer, as is relevant. As Thomas et al. found, BCDI is particularly valuable in devices fabricated using SOI because the active layer of Si has a different orientation from the much thicker handle [27]. The challenge of manipulating the structure of silicon includes creating interfaces with graphene and other emerging two-dimensional electronic materials and integrating other functionalities into silicon electronics. BCDI studies of nanowire structures have come out of ID01 at ESRF, which has become a major center for this activity [28]. The laser-excitation of nanowire structures is a large untapped area for pump-probe BCDI.

8.2 BCDI Methods at Light Sources

User demand is best assessed by the citation rates of the key papers defining the BCDI methods. References [1–3] have been cited 303, 171 and 145 times respectively. These citations represent the level of interest in the method. Informal inquiries suggest that major obstacles to more widespread uptake of the method are (i) the small number of suitable facilities and (ii) the lack of easy-to-use data analysis software packages for generating and viewing the 3D image information that results. Both of these issues will be solved by the dedicated BCDI beamlines at 3rd and 4th generation light sources. Once the BCDI diffraction pattern is inverted (see below) the strain, displacement field is mapped in the resulting complex real-space 3D image as the phase of the complex number at each location. The phase sensitivity is good enough that distortions can be mapped down to a level below 10 pm, over regions as small as 20 nm, limited by the spatial resolution. If needed, all components of **u** can be measured by triangulation from three or more Bragg peaks [29]. The following enlargements of the basic BCDI scopes are available at light sources.

Fresnel Coherent Diffraction Imaging (FCDI) is one example of modulation-based imaging methods that can be accommodated, although not strictly as part of the Bragg setup because they do not use the diffractometer. FCDI methods use a coherently modulated incident beam, often with a spherical wavefront, and capture part of the sample information holographically and part by diffraction [30]. FCDI experiments requires additional detector system in the forward direction, extending to about 8 m from the sample, which can be accommodated with a hutch extension. A worthwhile new direction would be modulation based detector/analyzer systems that can be combined with Bragg to aid phasing. These are currently long-term future directions that will be developed at light sources.

Bragg Ptychography using measurement of coherent diffraction in the forward direction, is a strong interest of several BCDI team members. In ptychography, the phasing of the coherent diffraction is achieved by an overlap constraint between patterns recorded from overlapping adjacent regions of the sample, and is also capable of recovering the full phase structure of the beam illuminating the sample [31].Bragg ptychography is also currently available at some BCDI beamlines for extended samples such as thin films, which do not fit within the 1–7 μm crystal size range. Bragg ptychography uses the same precise piezo scanning stage for the sample on the diffractometer in addition to other components of the BCDI beamline to hold the detector at the Bragg angle [32]. Bragg ptychography was recently demonstrated [33] and even though it has still not reached its full potential, possibly because of its extreme sensitivity to sample-optics vibrations, progress can be made on this method.

Grazing incidence BCDI. By upgrading the detector system in the forward direction, it should also be possible to apply coherent imaging techniques in the grazing incidence geometry, enhancing the science case by increasing the sensitivity to interface phenomena. There is a specific interest in the grazing geometry for time-resolved BCDI studies of thin film growth, where the grazing angle is varied to adjust the probing depth within the film [22].

Time-resolved BCDI. Pump-probe timing experiments are already a key part of the science case using the dedicated laser hutch. Time-resolved BCDI is a major interest of some scientists and researchers. These experiments would use the so-called "Co-GISAXS" formalism [34] and require the ability to mount UHV chambers to the beamline to study the films during growth.

8.3 Big Data Challenges in BCDI

We already know that Big Data is a big deal, and its here to stay. In fact, 65% of companies fear that they risk becoming irrelevant or uncompetitive if they don't embrace it. But despite the hype surrounding Big Data, companies struggle to make use of the data they collate. With an increase in coherence at 3rd and 4th generation light sources, it is expected that more data will to be accumulated leading to similar Big data challenges as that faced by social media and the tech industries. This challenge provides new opportunities as about 61% of companies state that Big Data is driving revenue because it is able to deliver deep insights into customer behavior. For most businesses, this means gaining a 360° of their customers, by analyzing and integrating existing data. In BCDI experiments there is a huge gap between the theoretical knowledge of material, big data and actually putting this theory into practice. So what is the problem? (i) Finding the Signal in the Noise, (ii) Inaccurate Data and (iii) Lack of Skilled Work force.

While to derive opportunities it is enough to look on common trends where the availability of information (even unstructured) brings potential for clarification of concepts and ideas. Indeed, it is hard to argue that when the phenomenon is captured in the data it will soon be uncovered, explained and merged with the scientific framework in the particular field. At the same time the challenges are not so straightforward to foresee since their nature is different in that it arises from extreme quantity of the data and the diversity of the data sources.

Inaccurate Data is simply the large amount of data. Assuming a state-of-the-art Dectris PILATUS3 S 2M X-ray detector we can calculate its data throughput in idealized experimental situation. With the dynamic range of 20 bits and 1475 by 1679 picture elements, single file 32-bit unsigned tiff file "weights" 9 MB. Assuming one week of the beamtime which can result in around 500 datasets measured, with each datasets containing 140 frames, we can estimate the outcome of the experiment to be around 600 GB of data. For the high-end PILATUS3 S 6M this value will double. This brings estimated output of the CDI station to the approximate level of 20–50 TB of data per year. With increase in brightness in coherence these values can be expected to grow by at least one magnitude due to decreases in acquisition time. Analyzing amount of data like this is a challenging task on its own, but taking the nature of the data the complexity is only growing.

Another complication arises from the additional datasets being accumulated by other methods. Often the sample is studied by such means as High Resolution Transmission Electron Microscopy, Scanning Electron Microscopy, X-ray Laboratory Diffraction, various spectroscopic techniques etc. Working with datasets obtained with multiple methods not only increases the amount of data, but also the complexity of the analysis. Taking into consideration the fabrication techniques only adds up to the challenge of analysis and interpretation of the results since the end goal of scientific research is not merely a production of a single graph but a new knowledge acquired in deep systematic studies confirmed by multiple experimental and theoretical groups.

8.4 Conclusions

It is often claimed that once the source is coherent, any beamline can do coherent scattering experiments. While this is partly true, the specialization of most beamline for BCDI can allow to focus on key performance issues, such as vibrational stability of the beam on the sample. It is expected that the BCDI performance of the beamline will be significantly better than multipurpose beamlines and that the throughput of user experiments will be better because reconfigurations will not be needed.

It is important to note that being inherently optimization based, CDI techniques are suitable for integration with different information-theoretic tools covered in this book. Availability of open-source packages for data analysis and packages that allow cross-language use of different libraries makes the integration even more tempting. Generation of structured and deep probed knowledge performed with Data Science and Optimal Learning approaches will benefit every area of modern materials science.

Acknowledgements This work was supported by the Air Force Office of Scientific Research (AFOSR) under Award No. FA9550-14-1-0363 (Program Manager: Dr. Ali Sayir) and by LDRD program at LANL. We also acknowledge support, in part from the LANSCE Professorship sponsored by the National Security Education Center at Los Alamos National Laboratory under subcontract No. 257827.

References

1. G. Williams, M. Pfeifer, I. Vartanyants, I. Robinson, Phys. Rev. Lett. **90**, 175501 (2003)
2. I. Robinson, R. Harder, Nat. Mater. **8**, 291 (2009)
3. M.A. Pfeifer, G.J. Williams, I.A. Vartanyants, R. Harder, I.K. Robinson, Nature **442**, 63 (2006)
4. D. Sayre, Acta Crystallogr. **5**, 843 (1952)
5. J. Miao, P. Charalambous, J. Kirz, D. Sayre, Nature **400**, 342 (1999)
6. A. Yau, W. Cha, M.W. Kanan, G.B. Stephenson, A. Ulvestad, Science **356**, 739 (2017)
7. J.R. Fienup, Appl. Opt. **21**, 2758 (1982)
8. M. Köhl, A. Minkevich, T. Baumbach, Opt. Exp. **20**, 17093 (2012)
9. D. Karpov, Nat. Commun. **8**, 1 (2017)
10. A. Grigoriev, Phys. Rev. Lett. **100**, 027604 (2008)
11. Z. Liu, B. Yang, W. Cao, E. Fohtung, T. Lookman, Phys. Rev. Appl. **8**, 034014 (2017)
12. A. Ulvestad et al., Science **348**, 1344 (2015)
13. A. Ulvestad, Nano Lett. **14**, 5123 (2014)
14. A. Ulvestad, Appl. Phys. Lett. **104**, 073108 (2014)
15. A. Singer, Nano Lett. **14**, 5295 (2014)
16. J.E. Parks, Science **327**, 1584 (2010)
17. C.H. Kim, G. Qi, K. Dahlberg, W. Li, Science **327**, 1624 (2010)
18. J.N. Clark et al., Science **341**, 1 (2013)
19. J.N. Clark, Proc. Natl. Acad. Sci. **112**, 7444 (2015)
20. H. Ichikawa, Nat. Mater. **10**, 101 (2011)
21. R. Mankowsky et al., Nature **516**, 71 (2014)
22. T. Hoshina, S. Wada, Y. Kuroiwa, T. Tsurumi, Appl. Phys. Lett. **93**, 192914 (2008)
23. W. Yang et al., Nat. Commun. **4**, 1680 (2013)
24. A. Minkevich, EPL (Europhys. Lett.) **94**, 66001 (2011)
25. A. Minkevich, Phys. Rev. B **84**, 054113 (2011)

26. A. Minkevich, M. Köhl, S. Escoubas, O. Thomas, T. Baumbach, J. Synchrotron Radiat. **21**, 774 (2014)
27. M. Gailhanou, Appl. Phys. Lett. **90**, 111914 (2007)
28. M. Heurlin, Nano Lett. **15**, 2462 (2015)
29. M.C. Newton, S.J. Leake, R. Harder, I.K. Robinson, Nat. Mater. **9**, 120 (2010)
30. G. Williams, H. Quiney, A. Peele, K. Nugent, New J. Phys. **12**, 035020 (2010)
31. P. Thibault et al., Science **321**, 379 (2008)
32. S. Hruszkewycz, Nano Lett. **12**, 5148 (2012)
33. S. Hruszkewycz, Nat. Mater. **16**, 244 (2017)
34. M.G. Rainville, Phys. Rev. B **92**, 214102 (2015)

Chapter 9
Automatic Tuning and Control for Advanced Light Sources

Alexander Scheinker

Abstract The next generation of X-ray Free Electron Laser (FEL) advanced light sources allow users to drastically change beam properties for various experiments. The main advantage of FELs over synchrotron light sources is their ability to provide more coherent, brighter flashes of light by tens of orders of magnitude with custom bunch lengths down to tens of femtoseconds. The wavelength of the brighter, more coherent light produced by an FEL is extremely dependent on both the electron beam energy, which must be adjusted between different experiments, and maintaining minimal electron bunch emittance. A large change in beam energy and bunch length usually requires a lengthy manual re-tuning of almost the entire accelerator. Therefore, unlike traditional machines which can operate for months or years at fixed energies, RF, and magnet settings FELs must have the ability to be completely re-tuned very quickly. For example, the Linac Coherent Light Source (LCLS) FEL can provide electrons at an energy range of 4–14 GeV and 1 nC pulses with 300 fs pulse width down to 20 pC pulses with 2 fs pulse width. The next generation of X-ray FELs will provide even bright, shorter wave-length (0.05 nm at EuXFEL, 0.01 nm at MaRIE), more coherent light, and at higher repetition rates (2 MHz at LCLS-II and 30,000 lasing bunches/second at EuXFEL, 2.3 ns bunch separation at MaRIE) than currently possible, requiring smaller electron bunch emittances than achievable today. Therefore, the next generation of light sources face two problems in terms of tuning and control. In parallel with the difficulties of improving performance to match tighter constraints on energy spreads and beam quality, existing and especially future accelerators face challenges in maintaining beam quality and quickly tuning between various experiments. We begin this chapter with a brief overview of some accelerator beam dynamics and a list of control problems important to particle accelerators. In the second half of this chapter we introduce some recently developed model-independent techniques for the control and tuning of accelerators with a focus on a feedback based extremum seeking method for automatic tuning and optimization which can tune multiple coupled parameters simultaneously and is incredibly robust to time-variation of system components and noise.

A. Scheinker (✉)
Los Alamos National Laboratory, Los Alamos, NM, USA
e-mail: ascheink@lanl.gov

T. Lookman et al. (eds.), *Materials Discovery and Design*, Springer Series in Materials Science 280, https://doi.org/10.1007/978-3-319-99465-9_9

9.1 Introduction

Particle accelerators are large complex systems composed of many thousands of coupled components which include radio frequency (RF) electromagnetic accelerating cavities, magnets, cooling systems, and detectors. For many decades accelerators have been designed with specific, static, operating conditions in mind, such as specific beam energies, currents, repetition rates, and bunch separations. For example, the Los Alamos Neutron Science Center accelerator is a $\sim$1 km long linear accelerator that has two fixed design energies of 100 and 800 MeV, and 8 fixed beam types which vary in terms of bunch length, charge/bunch, and repetition rate. Once the accelerator is tuned up following a maintenance outage, it is mostly continuously run with the various beam types accommodated by a fixed magnet/RF system setup, with intermediate tuning by operators to make up for small disturbances and fluctuations. The advanced photon source (APS) is a $\sim$1 km circumference synchrotron with magnet and RF systems tuned for a fixed 7 GeV electron beam which can be sent to various user stations with unique magnet and optic systems including monochromators for the production of specific light energy ranges from 3.5 to 100 keV. The Large Hadron Collider at CERN is the world's most powerful accelerator with a circumference of 27 km and beam energy of 6.5 TeV per beam for two counter circulating proton beams. The machine is run at a fixed energy for years at a time while massive detectors at four collision points collect data for fundamental particle physics research. The three machines described above encompass a majority of existing accelerators, which are designed for and operated at fixed settings, providing very specific beam types and energies.

Unlike the static machines described above, the next generation of X-ray Free Electron Laser (FEL) advanced light sources are being designed and operated with the fundamentally different approach of allowing users to drastically change beam properties for various experiments. The main advantage of FELs over synchrotron light sources such as the APS is their ability to provide more coherent, brighter flashes of light by tens of orders of magnitude with custom bunch lengths down to tens of femtoseconds. The wavelength of the brighter, more coherent light produced by an FEL is extremely dependent on the electron beam energy, which must be adjusted between different experiments. A large change in beam energy and bunch length requires the re-tuning of almost the entire accelerator. For example, the shortest, few femtosecond electron bunches require an adjustment of the source in lowering the total electron bunch charge so that the space charge forces of such short pulses are manageable. The bunch compressor system and RF energy settings and offsets must then also be adjusted to provide the new, shorter bunch length. Finally, depending on the required light and therefore electron beam energy of the given experiment, the magnet focusing systems throughout the accelerator and the undulator must be retuned. Therefore, unlike traditional machines which can operate for months or years at fixed energies, RF, and magnet settings FELs must have the flexibility to be completely re-tuned. For example, the Linac Coherent Light Source (LCLS) FEL can provide electrons at an energy range of 4–14 GeV and 1 nC pulses with 300 fs pulse width down to 20 pC pulses with 2 fs pulse width.

The next generation of X-ray FELs will provide even bright, shorter wave-length (0.05 nm at EuXFEL, 0.01 nm at MaRIE), more coherent light, and at higher repetition rates (2 MHz at LCLS-II and 30000 lasing bunches/second at EuXFEL, 2.3 ns bunch separation at MaRIE) than currently possible, requiring smaller electron bunch emittances than achievable today. Existing light sources are also exploring new and exotic schemes such as two-color operation (LCLS, FLASH, SwissFEL). To achieve their performance goals, the machines face extreme constraints on their electron beams. The LCLS-II requires $<0.01\%$ rms energy stability, a factor of $>10\times$ more than the existing LCLS [1], while the EuXFEL requires <0.001 deg rms RF amplitude and phase errors, respectively (current state of the art is ~ 0.01) [2].

Therefore, the next generation of light sources face two problems in terms of tuning and control. In parallel with the difficulties of improving performance to match tighter constraints on energy spreads and beam quality, existing and especially future accelerators face challenges in maintaining beam quality and quickly tuning between various experiments. It can take up to 10 h to retune the low energy beam sections (<500 MeV) and they still achieve sub-optimal results, wasting valuable beam time. Future accelerators require an ability to quickly tune between experiments and to compensate for extremely closely spaced electron bunches, such as might be required for MaRIE, requiring advanced controls and approaches such as droop correctors [3, 4].

While existing and planned FELs have automatic digital control systems, they are not controlled precisely enough to quickly switch between different operating conditions [5]. Existing controls maintain components at fixed set points, which are set based on desired beam and light properties, such as, for example, the current settings in a bunch compressor's magnets. Analytic studies and simulations initially provide these set points. However, models are not perfect and component characteristics drift in noisy and time-varying environments; setting a magnet power supply to a certain current today does not necessarily result in the same magnetic field as it would have 3 weeks ago. Also, the sensors are themselves noisy, limited in resolution, and introduce delays. Therefore, even when local controllers maintain desired set points exactly, performance drifts. The result is that operators continuously tweak parameters to maintain steady state operation and spend hours tuning when large changes are required, such as switching between experiments with significantly different current, beam profile (2 color, double bunch setups), or wavelength requirements. Similarly, traditional feed-forward RF beam loading compensation control systems are limited by model-based beam-RF interactions, which work extremely well for perfectly known RF and beam properties, but in practice are limited by effects which include un-modeled drifts and fluctuations and higher order modes excited by extremely short pulses. These limitations have created an interest in model-independent beam-based feedback techniques that can handle time-varying uncertain nonlinear systems [6–13], as well as machine learning, and other optimization techniques [14–18].

We begin this chapter with a list of control problems important to particle accelerators and a brief overview of simple beam dynamics, including longitudinal and transverse effects and the coupling between them and an overview of RF systems. The second half of this chapter introduces some recently developed techniques for the control and tuning of accelerators with a focus on a feedback based extremum seeking method for automatic tuning and optimization.

9.1.1 Beam Dynamics

The typical coordinate system for discussing particle accelerator beam dynamics is shown in Fig. 9.1. The Lorentz force equation:

$$\frac{d\mathbf{P}}{dt} = e\left(\mathbf{E} + \frac{\mathbf{v}}{c} \times \mathbf{B}\right), \tag{9.1}$$

describes charged particle dynamics. In (9.1) e is electron charge, $\mathbf{v}$ is velocity, $v = |\mathbf{v}|$, $\mathbf{P} = \gamma m \mathbf{v}$ the relativistic momentum, $\gamma = 1/\sqrt{1 - \frac{v^2}{c^2}}$ the Lorentz factor, c the speed of light, $\mathbf{E}$ the electric field and $\mathbf{B}$ the magnetic field. In a particle accelerator $\mathbf{E}$ and $\mathbf{B}$ sources include electromagnetic accelerating fields, other charged particles, and magnets used for steering and focusing of the beams. While electric fields are used to accelerate particles, magnetic fields guide the particles along a design trajectory and keep them from diverging transversely. We start by reviewing betatron oscillations, a form of oscillatory motion which is common to all particle accelerators [19–24].

Betatron oscillations are a general phenomenon occurring in all particle accelerators and are of particular importance in circular machines. For a particle traveling at the designed beam energy, $p = p_0$, the transverse equations are given by Hill's equation

$$x'' = K_x(s)x, \qquad y'' = K_y(s)y, \tag{9.2}$$

with (x, y) being the transverse particle locations relative to the accelerator axis (see Fig. 9.1), s (or z) is a parametrization of particle location along the axis of the accelerator, and $x'(s) = dx(s)/ds$. In a ring, the function $K_{x,y}(s + L) = K_{x,y}(s)$ is L-periodic, where L is the circumference of the accelerator, and depends on magnetic field strengths. Equation (9.2) resembles a simple harmonic oscillator with position-

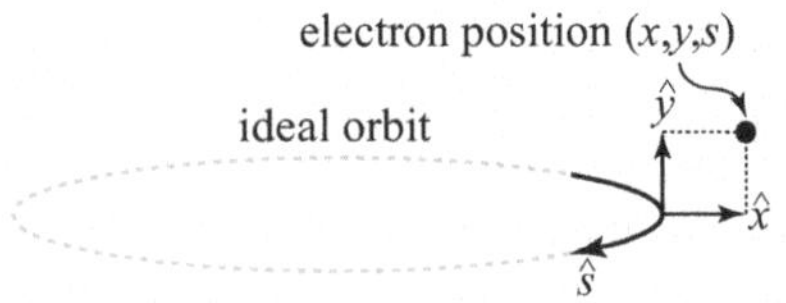

Fig. 9.1 A coordinate system centered on the ideal particle orbit. Distance along the orbit is parametrized by s. Transverse offset from the axis of the orbit is given by x and y

dependent $K_{x,y}(s)$. The solution of (9.2) is of the form

$$p_{x,y}(s) = A\sqrt{\beta_{x,y}(s)}\cos\left(\psi_{x,y}(s) + \delta\right), \quad \psi_{x,y}(s) = \int_0^s \frac{d\sigma}{\beta_{x,y}(\sigma)}, \tag{9.3}$$

where $\beta_{x,y}(s)$ are the periodic solutions of the system of equations

$$\beta'''_{x,y}(s) + 4K_{x,y}(s)\beta'_{x,y}(s) + 2K'_{x,y}(s)\beta(s)_{x,y} = 0, \tag{9.4}$$

$$\frac{1}{2}\beta(s)_{x,y}\beta''_{x,y}(s) - \frac{1}{4}\left(\beta'_{x,y}(s)\right)^2 + K_{x,y}(s)\beta^2_{x,y}(s) = 1. \tag{9.5}$$

The solutions of (9.3) are known as betatron oscillations and are periodic functions of s with varying amplitude and frequency [20].

In general, betatron motion is governed by equations of the form:

$$x''(s) = -K_x(x, y, s, \mathbf{P}, t)x(s) + F_x(x, x', y, y', s, \mathbf{P}, t), \tag{9.6}$$

$$y''(s) = -K_y(x, y, s, \mathbf{P}, t)y(s) + F_y(x, x', y, y', s, \mathbf{P}, t). \tag{9.7}$$

The nonlinear coupling between x and y depends not only on particle position, trajectory, energy deviation, and time.

Typically, quadrupole magnets focus the beam transversally, maintaining a tight bunch along the accelerator axis, and dipole magnets having only a non-zero y component of magnetic field direct the particles in a circular orbit in the (x, s) plane. The linear quadrupole and dipole magnetic field components give (9.6), (9.7) of the form

$$x'' = -\frac{p_0}{p}\left(\frac{1}{\rho^2} - K_1(s)\right)x + \frac{p - p_0}{p}\frac{1}{\rho}, \tag{9.8}$$

$$y'' = -\frac{p_0 K_1(s)}{p}y. \tag{9.9}$$

$K_1(s)$ is periodic and proportional to quadrupole field strength. The value $p = \sqrt{E^2/c^2 - m^2c^2}$ is the total kinetic momentum. p_0 is the designed kinetic momentum. The value ρ is the local radius of curvature [20].

Sources of nonlinearity and coupling in the functions F_x and F_y in (9.6), (9.7) are nonlinear magnetic field components, misaligned magnets, solenoid fields, magnetic field errors, and skew components of magnetic field gradients. Furthermore, all manufactured magnets are non-ideal and introduce nonlinear field components, higher order coupling terms given by [23]:

$$\Delta B_y + j\Delta B_x = B_0\sum_{n=0}^{\infty}(b_n + ja_n)(x + jy)^n. \tag{9.10}$$

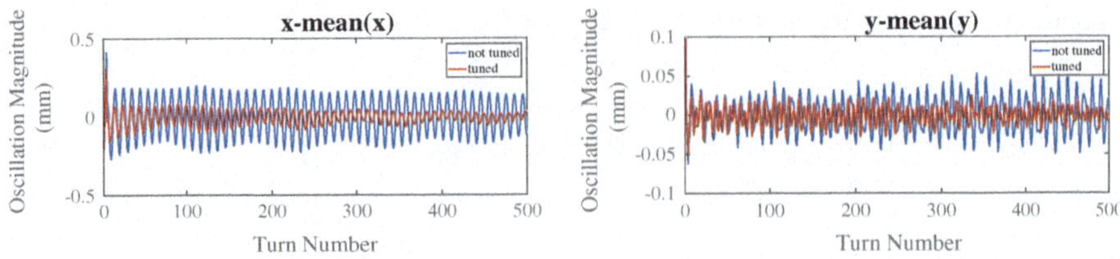

Fig. 9.2 BPM readings of x and y beam displacement over 500 turns, before and during tuning

Sometimes nonlinear magnets are purposely introduced into the accelerator lattice. For example, sextuple magnets are placed in regions of high dispersion to mitigate the fact that particles with various momentums experience non-equal forces from the same magnetic fields and their trajectories diverge (chromatic effects). Such magnets result in nonlinear coupling terms such as $(x^2 - y^2)$ and $(1 - \Delta)xy$, where $\Delta = (p - p_0)/p$ [20].

Betatron motion occurs in all accelerators, magnetic lattices are designed to minimize betatron oscillations. However, some regions of accelerators require large amplitude transverse particle motion. If this motion is not carefully, precisely controlled, excessive betatron oscillations are generated. One such section is a group of pulsed kicker magnets used to horizontally kick the beam out and then inject back into a machine. During injection kicks an imperfect match of parameters of the magnets results in the extremely large betatron oscillations, as shown in Fig. 9.2.

9.1.2 *RF Acceleration*

Particle acceleration in an RF field. For a particle passing through an RF cavity gap of length L, the energy gain due to an electromagnetic standing wave along the axis is given by

$$\Delta W = q \int_{-L/2}^{L/2} E(z) \cos(\omega t(z) + \phi) dz, \quad t(z) = \int_0^z \frac{dz}{v(z)}, \tag{9.11}$$

where $t(z)$ has been chosen such that the particle is at the center of the accelerating gap at $t = 0$, $\phi = 0$ if the particle arrives at the origin when the field is at a crest, and $v(z)$ is the velocity of the particle. This energy gain can be expanded as

$$\Delta W = q \int_{-L/2}^{L/2} E(z) \left[\cos(\omega t(z)) \cos(\phi) - \sin(\omega t(z)) \sin(\phi)\right] dz \tag{9.12}$$

and rewritten in the form

$$\Delta W = qV_0T\cos(\phi) = qE_0TL\cos(\phi), \quad E_0 = \frac{V_0}{L}, \tag{9.13}$$

where

$$V_0 = \int_{-L/2}^{L/2} E(Z)dz, \quad T = \frac{\int_{-L/2}^{L/2} E(z)\cos(\omega t(z))dz}{V_0} - \tan(\phi)\frac{\int_{-L/2}^{L/2} E(z)\sin(\omega t(z))dz}{V_0}, \tag{9.14}$$

and T known as the transit-time factor. For typical RF accelerating cavities, the electric field is symmetric relative to the center of the gap and the velocity change within an accelerating gap for a relativistic particle is negligible so $\omega t(z) \approx \omega z/v = 2\pi z/\beta\lambda$, where $\beta = v/c$ and $\beta\lambda$ is the distance a particle travels in one RF period. We can then rewrite the transit-time factor as

$$T = \frac{\int_{-L/2}^{L/2} E(z)\cos(2\pi z/\beta\lambda)\,dz}{V_0}. \tag{9.15}$$

Assuming that the electric field is constant $E(z) \equiv E_0$ within the gap, we get

$$T = \frac{\sin(\pi L/\beta\lambda)}{\pi L/\beta\lambda}, \tag{9.16}$$

and plugging back into (9.13) we get

$$\Delta W = \frac{qE_0\beta\lambda}{\pi}\cos(\phi)\sin\left(\frac{\pi L}{\beta\lambda}\right), \tag{9.17}$$

which is, as expected, maximized for $\phi = 0$ and $L = \beta\lambda/2$, that is for a particle that spends the maximal half of an RF period being accelerated through the cavity. This however would not be an efficient form of acceleration as most of the time the particle would see a much smaller than maximal RF field. For a given voltage gain V_0, we get a maximum $T = 1$ with $L = 0$, which is not realizable. Actual design values of T depend on individual cavity geometries and desired efficiency.

9.1.3 Bunch Compression

For maximal acceleration, we typically choose $\phi = 0$, especially for highly relativistic electrons. However, sometimes a nonzero ϕ is chosen either for longitudinal bunching or to purposely introduce an energy gradient along the electron bunch which can then be utilized for bunch compression. We define ϕ as the relative phase between a particle and the zero crossing of the RF field, such that earlier particles, with $\phi < 0$ will receive a higher energy gain than later particles with $\phi > 0$. The energy offset of a particle at phase ϕ at the exit of the RF compressor cavity, relative to the reference particle, is given by

$$\Delta E_1 = \Delta E_0 - \frac{qV_{\text{rf}}}{E}\sin(\phi), \tag{9.18}$$

where V_{rf} is the compressor voltage, E is beam energy, ΔE_0 is the initial energy offset. Next the beam is transported through a dispersive section with non-zero R_{56}, where

$$R_{56}(s) = \int_{s_0}^{s} \frac{R_{16}(s')}{\rho(s')} ds', \tag{9.19}$$

where R_{16} is the transverse displacement resulting from an energy error in a dispersive region of the accelerator. The energy offset is then translated to a longitudinal position offset according to

$$\Delta z_1 = \Delta z_0 + R_{56}\Delta E_1 = \Delta z_0 + R_{56}\left(\Delta E_0 - \frac{qV_{\text{rf}}}{E}\sin(\phi)\right). \tag{9.20}$$

For an RF field of frequency ω_{rf}, the phase ϕ relative to the RF at position offset Δz_0 is given by $\phi = -\omega_{\text{rf}}\Delta z_0/c$. If this phase is small, we can expand sine and rewrite both the energy and position change as

$$\Delta z_1 \approx \left(1 + R_{56}\frac{\omega_{\text{rf}}V_{\text{rf}}}{cE}\right)\Delta z_0 + R_{56}\Delta E_0, \tag{9.21}$$

$$\Delta E_1 = \Delta E_0 - \frac{eV_{\text{rf}}\omega_{\text{rf}}}{cE}\Delta z_0. \tag{9.22}$$

Therefore the final bunch length can be approximated as

$$\sigma_{zf} = \sqrt{\left(1 + R_{56}\frac{eV_{\text{rf}}\omega_{\text{rf}}}{Ec}\right)^2 \sigma_{z0}^2 + R_{56}^2\sigma_{\Delta E_0}^2}, \tag{9.23}$$

where σ_{z0} is the initial bunch length and $\sigma_{\Delta E_0}$ is the initial beam energy spread [26], with maximal compression for an RF system adjusted such that $R_{56}\frac{eV_{\text{rf}}\omega_{\text{rf}}}{Ec} \approx -1$.

9.1.4 RF Systems

For a right-cylindrical conducting cavity of radius R_c, as shown in Fig. 9.3, the 010 transverse-magnetic resonant mode, referred to as $\mathbf{TM_{010}}$, is used for acceleration

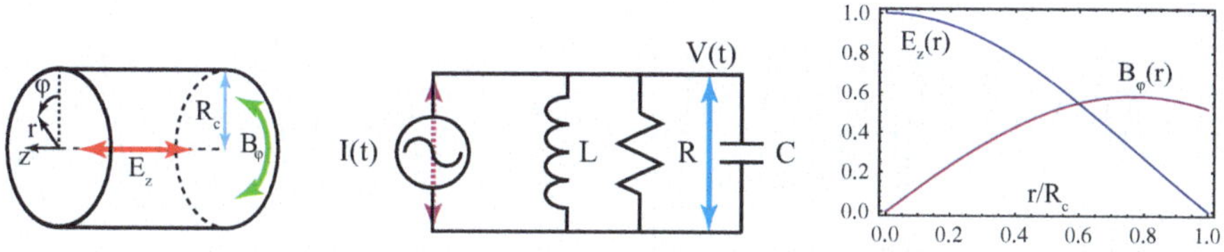

Fig. 9.3 **Left**: Electromagnetic field orientations for $\mathbf{TM}_{010}$ accelerating mode of a right cylindrical RF cavity. **Center**: RLC circuit approximation of the dynamics of a single RF mode. **Right**: The axial electric field is maximal on axis and zero at the walls of the cavity and the opposite is true of the azimuthal magnetic field

because along the axis this mode has a large oscillating electric field and no magnetic field, as shown in Fig. 9.3. The electromagnetic fields of the $\mathbf{TM_{010}}$ mode are:

$$\mathbf{E}(r,t) = E_0 J_0 \left(\frac{2.405r}{R_c}\right) e^{i\omega_0 t}\hat{\mathbf{z}} = \mathbf{E}_z(r) e^{i\omega_0 t}\hat{\mathbf{z}}, \tag{9.24}$$

$$\mathbf{B}(r,t) = -iE_0\sqrt{\frac{\epsilon}{\mu}} J_1 \left(\frac{2.405r}{R_c}\right) e^{i\omega_0 t}\hat{\varphi} = \mathbf{B}_\varphi(r) e^{i\omega_0 t}\hat{\varphi}, \tag{9.25}$$

where J_0 and J_1 are Bessel functions of the first kind with zero and first order, respectively, and the resonant frequency is given by

$$\omega_0 = \frac{2.405c}{R_c}, \qquad c = \text{speed of light}. \tag{9.26}$$

The dynamics of such a single mode of an RF cavity with resonant frequency f_0 can be approximated as

$$\ddot{V}_{\text{cav}} + \frac{\omega_0}{Q_L}\dot{V}_{\text{cav}} + \omega_0^2 V_{\text{cav}} = \frac{1}{C}\dot{I}, \tag{9.27}$$

where $\dot{V} = \frac{dV}{dt}$, $\ddot{V} = \frac{d^2V}{dt^2}$, $\omega_0 = 2\pi f_0$, Q_L is the loaded quality factor of the resonant cavity, L and C are the inductance and capacitance of the cavity structure, respectively, such that $\sqrt{LC} = \frac{1}{\omega_0}$, and $I = I_c + I_b$ is the input current driving the RF fields, the sources of which are both the RF generator, I_c, and the beam itself I_b [19, 27, 28].

For a driving current of the form

$$I_u(t) = I_0\cos(\omega_0 t), \tag{9.28}$$

after the fast decay of some transient terms, the cavity response is of the form

$$V_{\text{cav}}(t) = RI_0\left(1 - e^{-t/\tau}\right)\cos(\omega_0 t), \quad \tau = \frac{2Q_L}{\omega_0}. \tag{9.29}$$

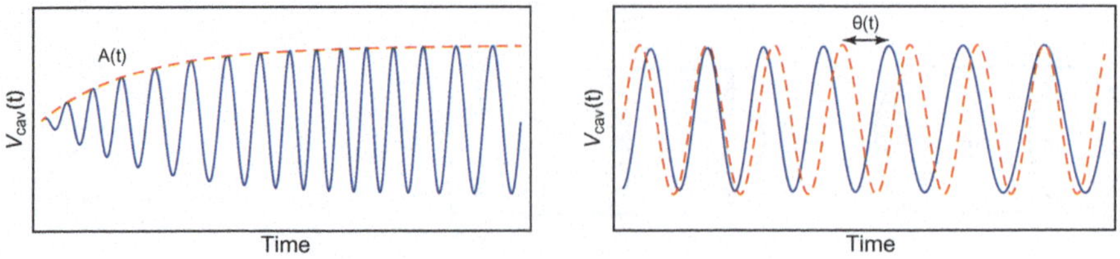

Fig. 9.4 Amplitude of the cavity field and its phase relative to a reference signal

9.1.5 Need for Feedback Control

Although (9.29) implies that for a desired accelerating gradient one must simply choose the correct input power level and drive the cavity, as shown in Fig. 9.4. However, in the real world simply choosing set points for an RF drive signal does not work because of un-modeled time varying disturbances which perturb cavity fields from their desired set points. These disturbances include:

1. Temperature variation-induced resonance frequency drifts on the time scales of minutes to hours.
2. Mechanical vibrations which alter the cavity resonance frequency on the times scale of milliseconds.
3. RF source voltage and current fluctuations on the time scale of microseconds.
4. RF source voltage droop on the time scale of microseconds.

Furthermore, even if a desired accelerating voltage could be reached within a desired rise time, when the beam that is to be accelerated shows up, it itself perturbs the fields both by interacting with the oscillating electrons in the cavity walls and by drawing energy out of the cavity via the electric field which accelerates it, causing both amplitude and phase changes on the time scales of nanoseconds which must be compensated for in order to maintain proper acceleration of subsequent beam bunches.

Therefore real time active feedback control is always necessary, both to bring cavity voltage amplitudes and phases to their required set points before beam can be properly accelerated and during beam acceleration in order to maintain tight bounds on beam-induced cavity field errors, known as beam loading.

From the above discussions it is clear that all of the disturbances experienced by the RF systems immediately couple into the transverse and longitudinal beam dynamics. Similarly, many of the beam dynamics, including the effects of space charge forces, magnet misalignments, and energy deviations alter a particle's position within a bunch and therefore the phase of the RF system relative to the particle's arrival time and therefore the entire accelerator is a completely coupled system in terms of the final beam phase space distribution relative to the RF systems, magnet systems, and the forces due to the particles in the beam itself.

9.1.6 Standart Proportional Integral (PI) Control for RF Cavity

The vast majority of accelerator systems, such as RF feedback and power converters are typically controlled at fixed set points with simple, classical, proportional integral (PI) controllers. Therefore we start with a detailed overview of RF cavity phase and amplitude PI control. To develop feedback controllers we must consider the coupled beam-cavity-RF source system. We consider only the ω_0 frequency component of the beam, $A_b(t)\cos(\omega_0 t + \theta_b(t))$, an RF driving current of the form $I_c(t) = A_c(t)\cos(\omega t + \theta_c(t))$, and a cavity field of the form $V_{\text{cav}}(t) = A_{\text{cav}}(t)\cos(\omega t + \theta_{\text{cav}}(t))$. The single second order differential equations describing the cavity dynamics, (9.27), can then be simplified to two coupled, linear, first order differential equations:

$$\dot{I} = -\omega_{\frac{1}{2}} I - \Delta\omega Q + \beta_{I,c} I_c + \beta_{I,b} I_b, \tag{9.30}$$

$$\dot{Q} = \Delta\omega I - \omega_{\frac{1}{2}} Q + \beta_{Q,c} Q_c + \beta_{Q,b} Q_b, \tag{9.31}$$

where $\Delta\omega = \omega - \omega_0$ is the difference between RF generator and cavity resonance frequencies, $\omega_{\frac{1}{2}} = \omega_0/2Q_L$, and the I and Q quantities represent

$$I(t) = A_{\text{cav}}(t)\cos(\theta_{\text{cav}}(t)), \quad I_c(t) = A_c(t)\cos(\theta_c(t)), \quad I_b(t) = A_b(t)\cos(\theta_b(t)), \tag{9.32}$$

$$Q(t) = A_{\text{cav}}(t)\sin(\theta_{\text{cav}}(t)), \quad Q_c(t) = A_c(t)\sin(\theta_c(t)), \quad Q_b(t) = A_b(t)\sin(\theta_b(t)), \tag{9.33}$$

from which amplitudes and phases can be calculated according to

$$A_\bullet(t) = \sqrt{I_\bullet^2(t) + Q_\bullet^2(t)}, \qquad \theta_\bullet(t) = \arctan\left(\frac{Q_\bullet(t)}{I_\bullet(t)}\right). \tag{9.34}$$

Equations (9.30), (9.31) can be written in the compact linear form

$$\dot{\mathbf{x}} = A\mathbf{x} + B_c\mathbf{u} + B_b\mathbf{d}, \quad \mathbf{x} = \begin{bmatrix} I \\ Q \end{bmatrix}, \quad A = \begin{bmatrix} -\omega_{\frac{1}{2}} & -\Delta\omega \\ \Delta\omega & -\omega_{\frac{1}{2}} \end{bmatrix}, \quad \mathbf{u} = \begin{bmatrix} I_c \\ Q_c \end{bmatrix}, \quad \mathbf{d} = \begin{bmatrix} I_b \\ Q_b \end{bmatrix}, \tag{9.35}$$

where $\mathbf{u}$ refers to the control vector, and the beam itself, $\mathbf{d}$, is thought of as a disturbance. The goal of RF feedback control is typically to maintain the cavity field as given by $\mathbf{x}$ at a desired set point thereby ensuring proper acceleration of the beam. In addition to providing a simple, linear approximation of the dynamics of the beam, cavity, and RF generator system, (9.35) is very useful because a typical digital RF system does not have access to the raw cavity voltage signal $V_{\text{cav}}(t)$, but rather to $I_{\text{cav}}(t)$ and $Q_{\text{cav}}(t)$, which are provided by down sampling the cavity field signal. For example, at the Los Alamos Neutron Science Center (LANSCE) linear accelerator, $f_{RF} = 201.25$ MHz RF signals of the form $V_{\text{cav}}(t) = A_{\text{cav}}(t)\cos(2\pi f_{RF} t + \theta_{\text{cav}}(t))$ are first mixed down via local oscillators to signals at an intermediate frequency

$f_{IF} = 25\,\text{MHz}$, of the form $A_{\text{cav}}(t)\cos(2\pi f_{IF}t + \theta_{\text{cav}}(t))$, which can be expanded in the I, Q formalism as:

$$\begin{aligned}
&A_{\text{cav}}(t)\cos(2\pi f_{IF}t + \theta_{\text{cav}}(t)) \\
&= \underbrace{A_{\text{cav}}(t)\cos(\theta_{\text{cav}}(t))}_{I_{\text{cav}}(t)}\cos(2\pi f_{IF}t) - \underbrace{A_{\text{cav}}(t)\sin(\theta_{\text{cav}}(t))}_{Q_{\text{cav}}(t)}\cos(2\pi f_{IF}t) \\
&= I_{\text{cav}}(t)\cos(2\pi f_{IF}t) - Q_{\text{cav}}(t)\cos(2\pi f_{IF}t)\sin(2\pi f_{IF}t).
\end{aligned} \tag{9.36}$$

Then, by oversampling the signal (9.36) at the rate $f_s = 4 \times f_{IF}$, the analog to digital converter (ADC) collects samples at time steps $nt = \frac{n}{fs}$:

$$V_{\text{cav}}\left(\frac{n}{4f_{IF}}\right) = I_{\text{cav}}\left(\frac{n}{4f_{IF}}\right)\cos\left(\frac{n\pi}{2}\right) - Q_{\text{cav}}\left(\frac{n}{4f_{IF}}\right)\sin\left(\frac{n\pi}{2}\right), \tag{9.37}$$

directly receiving the samples:

$$\{I_{\text{cav}}(0), -Q_{\text{cav}}(t_s), -I_{\text{cav}}(2t_s), Q_{\text{cav}}(3t_s), \ldots\}. \tag{9.38}$$

The job of the RF control system is to maintain the cavity fields at amplitude and phase set points, $A_s(t)$ and $\theta_s(t)$, respectively, which translate into I and Q set points: $I_s(t) = A_s(t)\cos(\theta_s(t))$, $Q_s(t) = A_s(t)\sin(\theta_s(t))$. The most simple typical RF feedback control system first compares the cavity I and Q signals to their set points and calculates error signals $I_e(t) = I_{\text{cav}}(t) - I_s(t)$, $Q_e(t) = Q_{\text{cav}}(t) - Q_s(t)$, and then performs proportional-integral feedback control of the form

$$I_c(t) = -k_p I_e(t) - k_i \int_0^t I_e(\tau)d\tau, \qquad Q_c(t) = -k_p Q_e(t) - k_i \int_0^t Q_e(\tau)d\tau. \tag{9.39}$$

Typically particle accelerators are pulsed at rates of tens to hundreds of Hz. For example, in the LANSCE accelerator, the RF drive power is turned on for 1 ms at a rate of 120 Hz. Once RF is turned on, cavity fields build up and reach steady state within a few hundred microseconds, after which the cavities are ready to accelerate the beam, whose sudden arrival perturbs the cavity fields, as shown in Fig. 9.5.

Although the initial I and Q set points are in the forms of smooth ramps, as seen from the shape of the cavity field amplitude in Fig. 9.5, once the field has reached steady state and before the beam has arrived, the set points are fixed in order to maintain a precise field amplitude and phase offset of the bunches relative to the RF zero crossing. Therefore, in what follows we consider the cavity set points only after steady state has been reached and they are therefore constants of the form:

$$I_s(t \geq T_{\text{rise}}) \equiv I_s(T_{\text{rise}}) = I_r, \quad Q_s(t \geq T_{\text{rise}}) \equiv Q_s(T_{\text{rise}}) = Q_r. \tag{9.40}$$

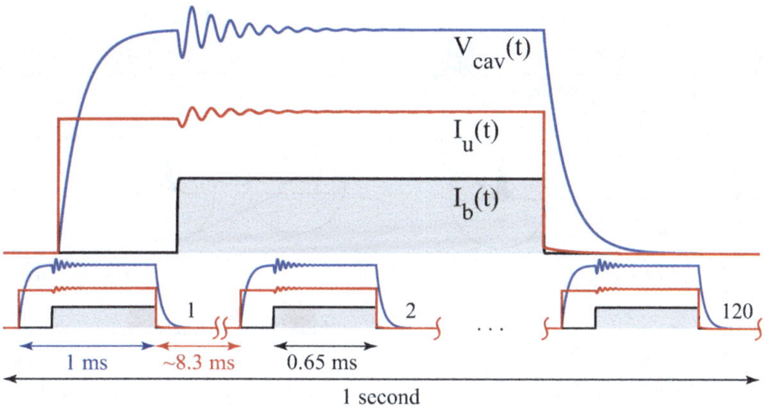

Fig. 9.5 The RF source, $I_c(t)$, is turned on at a rate of 120 Hz, for ~1 ms per pulse. The beam, $I_b(t)$, arrives around ~350 μs into the pulse after the cavity field, $V_{\mathrm{cav}}(t)$, has had time to settle. The beam's arrival disrupts the cavity field's steady state

Plugging the feedback (9.39) into the cavity dynamics (9.35) and rewriting the dynamics in terms of the error variables, we are then left with the closed loop system

$$\dot{\mathbf{x}}_e = A\mathbf{x}_e + A\mathbf{x}_r - k_p B_c \mathbf{x}_e - k_i B_c \int_0^t \mathbf{x}_e(\tau)d\tau + B_b\mathbf{d}, \quad \mathbf{x}_e = \begin{bmatrix} I_e \\ Q_e \end{bmatrix}, \quad \mathbf{x}_r = \begin{bmatrix} I_r \\ Q_r \end{bmatrix}. \tag{9.41}$$

Taking the Laplace transform of both sides of (9.41), assuming that we are at steady state so that $\mathbf{x}_e(T_{\mathrm{rise}}) = 0$, we get

$$\begin{aligned} s\mathbf{X}_e(s) &= A\mathbf{X}_e(s) + \frac{1}{s}A\mathbf{x}_r - k_p B_c \mathbf{X}_e(s) - \frac{1}{s}k_i B_c \mathbf{X}_e(s) + B_d \mathbf{D}(s) \\ &\Longrightarrow \\ \mathbf{X}_e(s) &= \left(s^2 I - s\left(k_p B_c - A\right) + k_i B_c\right)^{-1}\left(A\mathbf{x}_r + sB_d\mathbf{D}(s)\right). \end{aligned} \tag{9.42}$$

The gains, k_i and k_p of the simple PI feedback control loop are then tuned in order to maintain minimal error despite the disturbances $A\mathbf{x}_r$ and $sB_d\mathbf{D}(s)$. The constant term $A\mathbf{x}_r$ is due to the natural damping of the RF cavity and is easily compensated for. The more important and more difficult to deal with term is $sB_d\mathbf{D}(s)$, which, in the time domain is proportional to the derivative of the beam current $B_d\dot{\mathbf{d}}(t)$. Because the beam is typically ramped up to an intense current very quickly (tens of microseconds) or consists of an extremely short pulse, the derivative term is extremely disruptive to the cavity field phase and amplitude. Some typical beam current and bunch timing profiles are shown in Fig. 9.6. Currently LCLS is able to accelerate 1 nC during extremely powerful ~3 μs RF pulses, with a separation of 8.3 ms between bunches. The European XFEL is pushing orders of magnitude beyond the LCLS bunch timing with 1 nC pulses separated by only 220 ns. This is extremely challenging for an

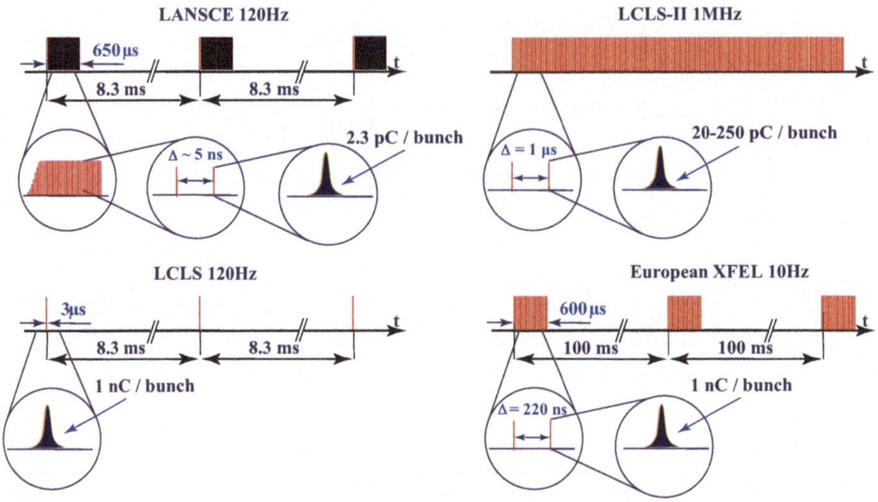

Fig. 9.6 Beam current time profiles of several accelerators are shown

RF system which must maintain field amplitude and phase set points and recover between bunches. The proposed MaRIE accelerator will push this problem another order of magnitude in attempting to accelerate high charge pulses with only ~2.5 ns of separation.

Although the PI controller used in (9.41) can theoretically hold the error $\mathbf{x}_e$ arbitrarily close to zero arbitrarily fast by choosing large enough gains k_i and k_p relative to the magnitude of the beam disturbance $\left\| B_d \dot{\mathbf{d}}(t) \right\|$, in practice all control gains are limited by actuator saturation, response time, and most importantly, delay in the feedback loop. A typical RF feedback loop is shown in Fig. 9.7 and may experience as much as 5 μs of round trip delay, which is an large delay relative to beam transient times.

Consider for example the following scalar, delay system, where the goal is to quickly drive $x(t)$ to zero from an arbitrary initial condition, but only being able to do so based on a controller which uses a delayed measurement of $x(t)$, $x(t-D)$. Considering a simply proportional feedback control, $u = -kx$, for the system

$$\dot{x}(t) = u(x(t-D)) \quad \implies \quad \dot{x}(t) = -kx(t-D), \tag{9.43}$$

taking Laplace transforms we get

$$sX(s) - x(0) = -ke^{-Ds}X(s) \quad \implies \quad X(s) = \frac{x(0)}{s + ke^{-Ds}}. \tag{9.44}$$

If we assume the delay is small, $D \ll 1$, we can approximate $e^{-Ds} \approx 1 - Ds$, invert the Laplace transform and get the solution

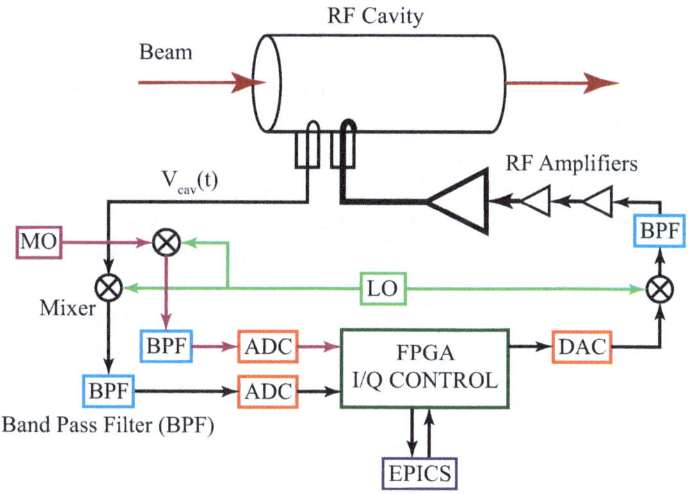

Fig. 9.7 Typical digital RF control setup with signals coming from the cavity into the digital FPGA-based controller and then back out through a chain of amplifiers

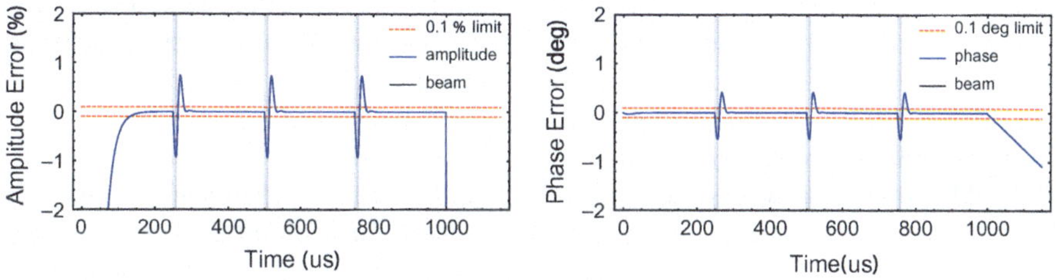

Fig. 9.8 Cavity field errors with frequency shift, RF power droop, beam loading, and simple proportional-integral feedback control

$$x(t) = x(0)e^{\gamma t}, \quad \gamma = \frac{-k}{1 - kD}, \tag{9.45}$$

which exponentially converges to 0 for $\gamma < 0$, requiring that k satisfy $\frac{1}{D} > k > 0$, a limit on possible stabilizing values of the feedback control gain. If our system (9.43) had an external disturbance, $d(t)$ the gain limit would be a major limitation in terms of compensating for large or fast $d(t)$.

Because of such limitations, a feedback only LLRF system's response to beam loading would typically look like the results shown in Fig. 9.8, where each intense beam pulse causes a large deviation of the accelerating field's voltage from the design phase and amplitude, which must be restored before the next bunch can be properly accelerated.

9.2 Advanced Control and Tuning Topics

For problems which can be accurately modeled, such as systems that do not vary with time and for which extensive, detailed diagnostics exist, there are many powerful optimization methods such as genetic algorithms (GA), which can be used during the design of an accelerator by performing extremely large searches over parameter space [29]. Such multi-objective genetic algorithms (MOGA) have been applied for the design of radio frequency cavities [30], photoinjectors [31], damping rings [32], storage ring dynamics [33], lattice design [34], neutrino factory design [35], simultaneous optimization of beam emittance and dynamic aperture [36], free electron laser linac drivers [37] and various other accelerator physics applications [38]. One extension of MOGA is multi-objective particle swarm optimization, has been used for emittance reduction [39]. Brute force approaches such as GA and MOGA search over the entire parameter space of interest and therefore result in global optimization, however, such model-based approaches are only optimal relative to the specific model which they are using, which in practice rarely exactly matches the actual machine when it is built. Differences are due to imperfect models, uncertainty, and finite precision of construction. Therefore, actual machines settings undergo extensive tuning and tweaking in order to reach optimal performance. Recently efforts have been made to implement a GA method on-line for the minimization of beam size at SPEAR3 [40]. Robust conjugate direction search (RCDS) is another optimization method. RCDS is model independent, but at the start of optimization in must learn the conjugate directions of the given system, and therefore is not applicable to quickly time-varying systems [41, 42]. Optimization of nonlinear storage ring dynamics via RCDS and particle swarm has been performed online [43].

Although many modern, well behaved machines can possibly be optimized with any of the methods mentioned above, and once at steady state, the operation may not require the fast re-tuning future light sources will require algorithms with an ability to quickly switch between various operating conditions and to handle quickly time-varying systems, based only on scalar measurements, rather than a detailed knowledge of the system dynamics, when compensating for complex collective effects. If any of the methods above were used, they would have to be repeated every time component settings were significantly changed and it is highly unlikely that they would converge or be well behaved during un-modeled, fast time-variation of components. Therefore, a model-independent feedback-based control and tuning procedure is required which can function on nonlinear and time varying systems with many coupled components.

The type of tuning problems that we are interested in have recently been approached with powerful machine learning methods [15, 44], which are showing very promising results. However, these methods require large training sets in order to learn how to reach specific machine set points, and interpolate in between. For example, if a user requests a combination of beam energy, pulse charge, and bunch length, which was not a member of a neural network-based controller's learning set, the achieved machine performance is not predictable. Furthermore, machine components slowly drift with time and un-modeled disturbances are present and limit any learning-based algorithm's abilities. Extremum seeking (ES) is a simple, local, model-independent algorithm for accelerator tuning, whose speed of convergence allows for the optimization and real-time tracking of many coupled parameters for time-varying nonlinear systems. Because ES is model independent, robust to noise, and has analytically guaranteed parameter bounds and update rates, it is useful for real time feedback in actual machines. One of the limitations of ES is that it is a local optimizer which can possible be trapped in local minima.

It is our belief that the combination of ES and machine learning methods will be a powerful method for quickly tuning FELs between drastically different user desired beam and light properties. For example, once a deep neural network (NN) has learned a mapping of machine settings to light properties for a given accelerator based on collected machine data, it can be used to quickly bring the machine within a local proximity of the required settings for a given user experiment. However, the performance will be limited by the fact that the machine changes with time, that the desired experiment settings were not in the training data, and un-modeled disturbances. Therefore, once brought within a small neighborhood of the required settings via NN, ES can be used to achieve local optimal tuning, which can also continuously re-tune to compensate for un-modeled disturbances and time variation of components. In the remainder of this chapter we will focus on the ES method, giving a general overview of the procedure and several simulation and in-hardware demonstrations of applications of the method. Further details on machine learning approaches can be found in [15, 44] and the references within.

9.3 Introduction to Extremum Seeking Control

The Extremum seeking method described in this chapter is a recently developed general approach for the stabilization of noisy, uncertain, open-loop unstable, time-varying systems [6, 7]. The main benefits of this approach are:

1. The method can tune many parameters of unknown, nonlinear, open-loop unstable systems, simultaneously.
2. The method is robust to measurement noise and external disturbances and can track quickly time-varying parameters.
3. Although operating on noisy and analytically unknown systems, the parameter updates have analytically guaranteed constraints, which make it safe for in-hardware implementation.

This method has been implemented in simulation to automatically tune large systems of magnets and RF set points to optimize beam parameters [11], it has been utilized in hardware at the proton linear accelerator at the Los Alamos Neutron Science Center to automatically tune two RF buncher cavities to maximize the RF system's beam acceptance, based only on a noisy measurement of beam current [12], it has been utilized at the Facility for Advanced Accelerator Experimental Tests, to non-destructively predict electron bunch properties via a coupling of simulation and machine data [13], it has been utilized for bunch compressor design [45], and has been used for the automated tuning of magnets in a time-varying lattice to continuously minimize betatron oscillations at SPEAR3 [8]. Furthermore, analytic proofs of convergence for the method are available for constrained systems with general, non-differentiable controllers [9, 10].

9.3.1 Physical Motivation

It has been shown that unexpected stability properties can be achieved in dynamic systems by introducing fast, small oscillations. One example is the stabilization of the vertical equilibrium point of an inverted pendulum by quickly oscillating the pendulum's pivot point. Kapitza first analyzed these dynamics in the 1950s [46]. The ES approach is in some ways related to such vibrational stabilization as high frequency oscillations are used to stabilize desired points of a system's state space and to force trajectories to converge to these points. This is done by creating cost functions whose minima correspond to the points of interest, allowing us to tune a large family of systems without relying on any models or system knowledge. The method even works for unknown functions, where we do not choose which point of the state space to stabilize, but rather are minimizing an analytically unknown function whose noisy measurements we are able to sample.

To give an intuitive 2D overview of this method, we consider finding the minimum of an unknown function $C(x, y)$. We propose the following scheme:

$$\frac{dx}{dt} = \sqrt{\alpha\omega}\cos\left(\omega t + kC(x, y)\right) \tag{9.46}$$

$$\frac{dy}{dt} = \sqrt{\alpha\omega}\sin\left(\omega t + kC(x, y)\right). \tag{9.47}$$

Note that although $C(x, y)$ enters the argument of the adaptive scheme, we do not rely on any knowledge of the analytic form of $C(x, y)$, we simply assume that it's value is available for measurement at different locations (x, y).

The velocity vector,

$$\mathbf{v} = \left(\frac{dx}{dt}, \frac{dy}{dt}\right) = \sqrt{\alpha\omega}\left[\cos\left(\theta(t)\right), \sin\left(\theta(t)\right)\right], \tag{9.48}$$

$$\theta(t) = \omega t + kC(x(t), y(t)), \tag{9.49}$$

has constant magnitude, $\|\mathbf{v}\| = \sqrt{\alpha\omega}$, and therefore the trajectory $(x(t), y(t))$ moves at a constant speed. However, the rate at which the direction of the trajectories' heading changes is a function of ω, k, and $C(x(t), y(t))$ expressed as:

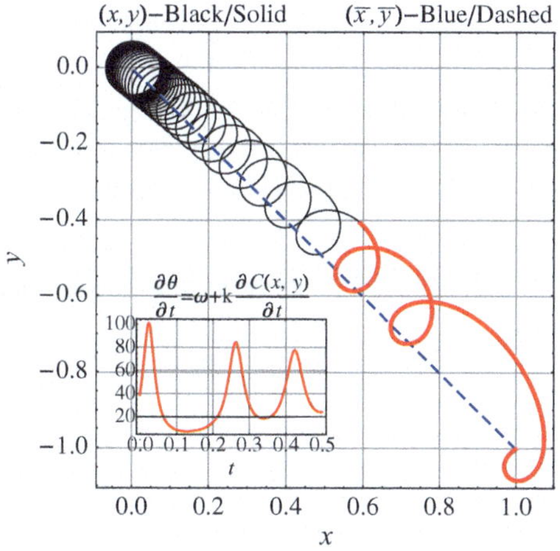

Fig. 9.9 The subfigure in the bottom left shows the rotation rate, $\frac{\partial\theta}{\partial t} = \omega + \frac{\partial C(x,y)}{\partial t}$, for the part of the trajectory that is bold red, which takes place during the first 0.5 s of simulation. The rotation of the parameters' velocity vector $\mathbf{v}(t)$ slows down when heading towards the minimum of $C(x, y) = x^2 + y^2$, at which time $k\frac{\partial C}{\partial t} < 0$, and speeds up when heading away from the minimum, when $k\frac{\partial C}{\partial t} > 0$. The system ends up spending more time heading towards and approaches the minimum of $C(x, y)$

$$\frac{d\theta}{dt} = \omega + k\left(\frac{\partial C}{\partial x}\frac{dx}{dt} + \frac{\partial C}{\partial y}\frac{dy}{dt}\right). \tag{9.50}$$

Therefore, when the trajectory is heading in the correct direction, towards a decreasing value of $C(x(t), y(t))$, the term $k\frac{\partial C}{\partial t}$ is negative so the overall turning rate $\frac{\partial\theta}{\partial t}$ (9.50), is decreased. On the other hand, when the trajectory is heading in the wrong direction, towards an increasing value of $C(x(t), y(t))$, the term $k\frac{\partial C}{\partial t}$ is positive, and the turning rate is increased. On average, the system ends up approaching the minimizing location of $C(x(t), y(t))$ because it spends more time moving towards it than away.

The ability of this direction-dependent turning rate scheme is apparent in the simulation of system (9.46), (9.47), in Fig. 9.9. The system, starting at initial location $x(0) = 1, y(0) = -1$, is simulated for 5 s with update parameters $\omega = 50, k = 5, \alpha = 0.5$, and $C(x, y) = x^2 + y^2$. We compare the actual system's (9.46), (9.47) dynamics with those of a system performing gradient descent:

$$\frac{d\bar{x}}{dt} \approx -\frac{k\alpha}{2}\frac{\partial C(\bar{x}, \bar{y})}{\partial \bar{x}} = -k\alpha\bar{x} \tag{9.51}$$

$$\frac{d\bar{y}}{dt} \approx -\frac{k\alpha}{2}\frac{\partial C(\bar{x}, \bar{y})}{\partial \bar{y}} = -k\alpha\bar{y}, \tag{9.52}$$

whose behavior our system mimics on average, with the difference

$$\max_{t \in [0,T]} \|(x(t), y(t)) - (\bar{x}(t), \bar{y}(t))\| \tag{9.53}$$

made arbitrarily small for any value of T, by choosing arbitrarily large values of ω.

Towards the end of the simulation, when the system's trajectory is near the origin, $C(x, y) \approx 0$, and the dynamics of (9.46), (9.47) are approximately

$$\frac{\partial x}{\partial t} \approx \sqrt{\alpha\omega} \cos(\omega t) \Longrightarrow x(t) \approx \sqrt{\frac{\alpha}{\omega}} \sin(\omega t) \tag{9.54}$$

$$\frac{\partial y}{\partial t} \approx \sqrt{\alpha\omega} \sin(\omega t) \Longrightarrow y(t) \approx -\sqrt{\frac{\alpha}{\omega}} \cos(\omega t), \tag{9.55}$$

a circle of radius $\sqrt{\frac{\alpha}{\omega}}$, which is made arbitrarily small by choosing arbitrarily large values of ω. Convergence towards a maximum, rather than a minimum is achieved by replacing k with $-k$.

9.3.2 General ES Scheme

For general tuning, we consider the problem of locating an extremum point of the function $C(\mathbf{p}, t) : \mathbb{R}^n \times \mathbb{R}^+ \to \mathbb{R}$, for $\mathbf{p} = (p_1, \ldots, p_n) \in \mathbb{R}^n$, when only a noise-corrupted measurement $y(t) = C(\mathbf{p}, t) + n(t)$ is available, with the analytic form of C unknown. For notational convenience, in what follows we sometimes write $C(\mathbf{p})$ or just C instead of $C(\mathbf{p}(t), t)$.

The explanation presented in the previous section used $\sin(\cdot)$ and $\cos(\cdot)$ functions for the x and y dynamics to give circular trajectories. The actual requirement for convergence is for an independence, in the frequency domain, of the functions used to perturb different parameters. In what follows, replacing $\cos(\cdot)$ with $\sin(\cdot)$ throughout makes no difference.

Theorem 1 *Consider the setup shown in Fig. 9.10 (for maximum seeking we replace k with $-k$):*

$$\dot{p}_i = \sqrt{\alpha\omega_i} \cos(\omega_i t + ky), \qquad y = C(\mathbf{p}, t) + n(t) \tag{9.56}$$

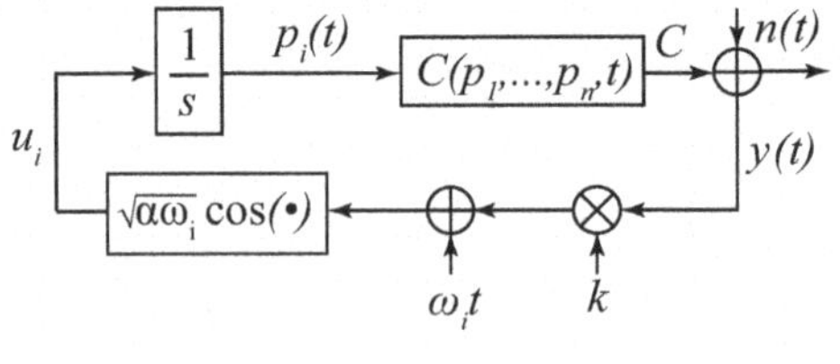

Fig. 9.10 Tuning of the ith component p_i of $\mathbf{p} = (p_1, \ldots, p_n) \in \mathbb{R}^n$. The symbol $\frac{1}{s}$ denotes the Laplace Transform of an integrator, so that in the above diagram $p_i(t) = p_i(0) + \int_0^t u_i(\tau) d\tau$

where $\omega_i = \omega_0 r_i$ *such that* $r_i \neq r_j$ $\forall i \neq j$ *and* $n(t)$ *is additive noise. The trajectory of system (9.56) approaches the minimum of* $C(\mathbf{p}, t)$, *with its trajectory arbitrarily close to that of*

$$\dot{\bar{\mathbf{p}}} = -\frac{k\alpha}{2}\nabla\mathbf{C}, \quad \bar{\mathbf{p}}(0) = \mathbf{p}(0) \tag{9.57}$$

with the distance between the two decreasing as a function of increasing ω_0. *Namely, for any given* $T \in [0, \infty)$, *any compact set of allowable parameters* $\mathbf{p} \in K \subset \mathbb{R}^m$, *and any desired accuracy* δ, *there exists* $\omega_0^\star$ *such that for all* $\omega_0 > \omega_0^\star$, *the distance between the trajectory* $\mathbf{p}(t)$ *of (9.56) and* $\bar{\mathbf{p}}(t)$ *of (9.57) satisfies the bound*

$$\max_{\mathbf{p}, \bar{\mathbf{p}} \in K, t \in [0,T]} \|\mathbf{p}(t) - \bar{\mathbf{p}}(t)\| < \delta. \tag{9.58}$$

Remark 1 One of the most important features of this scheme is that on average the system performs a gradient descent of the actual, unknown function C despite feedback being based only on its noise corrupted measurement $y = C(\mathbf{p}, t) + n(t)$.

Remark 2 The stability of this scheme is verified by the fact that an addition of an un-modeled, possibly destabilizing perturbation of the form $\mathbf{f}(\mathbf{p}, t)$ to the dynamics of $\dot{\mathbf{p}}$ results in the averaged system:

$$\dot{\bar{\mathbf{p}}} = \mathbf{f}(\bar{\mathbf{p}}, t) - \frac{k\alpha}{2}\nabla\mathbf{C}, \tag{9.59}$$

which may be made to approach the minimum of C, by choosing $k\alpha$ large enough relative to the values of $\left\|(\nabla\mathbf{C})^T\right\|$ and $\|\mathbf{f}(\bar{\mathbf{p}}, t)\|$.

Remark 3 In the case of a time-varying max/min location $\mathbf{p}^\star(t)$ of $C(\mathbf{p}, t)$, there will be terms of the form:

$$\frac{1}{\sqrt{\omega}}\left|\frac{\partial C(\mathbf{p}, t)}{\partial t}\right|, \tag{9.60}$$

which are made to approach zero by increasing ω. Furthermore, in the analysis of the convergence of the error $\mathbf{p}_e(t) = \mathbf{p}(t) - \mathbf{p}^\star(t)$ there will be terms of the form:

$$\frac{1}{k\alpha}\left|\frac{\partial C(\mathbf{p}, t)}{\partial t}\right|. \tag{9.61}$$

Together, (9.60) and (9.61) imply the intuitively obvious fact that for systems whose time-variation is fast, in which the minimum towards which we are descending is quickly varying, both the value of ω and of the product $k\alpha$ must be larger than for the time-invariant case.

Remark 4 In the case of different parameters having vastly different response characteristics and sensitivities (such as when tuning both RF and magnet settings in the same scheme), the choices of k and α may be specified differently for each component p_i, as k_i and α_i, without change to the above analysis.

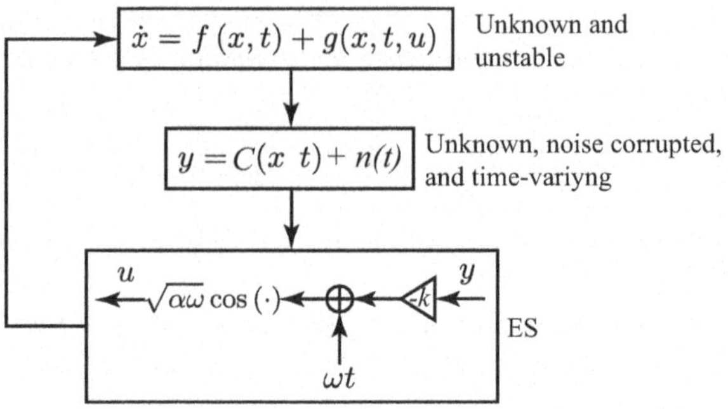

Fig. 9.11 ES for simultaneous stabilization and optimization of an unknown, open-loop unstable system based on a noise corrupted scalar measurement

A more general form of the scheme for simultaneous stabilization and optimization of an n-dimensional open-loop unstable system with analytically unknown noise-corrupted output function $C(\mathbf{x}, t)$ is shown in Fig. 9.11, but will not be discussed in detail here.

9.3.3 ES for RF Beam Loading Compensation

The ES method described above has been used both in simulation and optimization studies and has been implemented in hardware in accelerators. We now return to the RF problem described in Sect. 9.1.6, where we discussed the fact that due to delay-limited gains and power limitations, the sudden transient caused by beam loading greatly disturbs the RF fields of accelerating cavities which must be re-settled to within prescribed bounds before the next bunches can be brought in for acceleration. ES has been applied to this beam loading problem in the LANSCE accelerator via high speed field programmable gate array (FPGA).

In order to control the amplitude and phase of the RF cavity accelerating field, the $I(t) = A(t)\cos(\theta(t))$ and $Q(t) = A(t)\sin(\theta(t))$ components of the cavity voltage signal were sampled as described in Sect. 9.1.6, at a rate of 100 MS/s during a 1000 μs RF pulse. The detected RF signal was then broken down into 10 μs long sections and feed forward $I_{\text{ff},j}(n)$ and $Q_{\text{ff},j}(n)$ control outputs were generated for each 10 μs long section, as shown in Fig. 9.12.

Remark 5 In the discussion and figures that follow, we refer to $I_{\text{cav}}(t)$ and $Q_{\text{cav}}(t)$ simply as $I(t)$ and $Q(t)$.

The iterative extremum seeking was performed via finite difference approximation of the ES dynamics:

$$\frac{x(t+dt) - x(t)}{dt} \approx \frac{dx}{dt} = \sqrt{\alpha\omega}\cos(\omega t + kC(x,t)), \tag{9.62}$$

by updating the feedforward signals according to

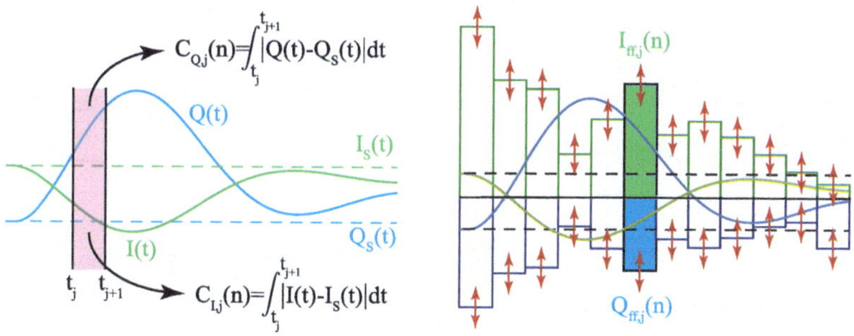

Fig. 9.12 **Top**: Iterative scheme for determining I and Q costs during 1–10 μs intervals. **Bottom:** ES-based feedforward outputs for beam loading transient compensation

$$I_{\mathrm{ff},j}(n+1) = I_{\mathrm{ff},j}(n) + \Delta\sqrt{\alpha\omega}\cos\left(\omega n\Delta + kC_{I,j}(n)\right), \tag{9.63}$$

and

$$Q_{\mathrm{ff},j}(n+1) = Q_{\mathrm{ff},j}(n) + \Delta\sqrt{\alpha\omega}\sin\left(\omega n\Delta + kC_{Q,j}(n)\right), \tag{9.64}$$

where the individual I and Q costs were calculated as

$$C_{I,j}(n) = \int_{t_j}^{t_{j+1}} |I(t) - I_s(t)|\, dt, \tag{9.65}$$

$$C_{Q,j}(n) = \int_{t_j}^{t_{j+1}} |Q(t) - Q_s(t)|\, dt. \tag{9.66}$$

Note that although the I_j and Q_j parameters were updated on separate costs, they were still dithered with different functions, $\sin(\cdot)$ and $\cos(\cdot)$, to help maintain orthogonality in the frequency domain. The feed forward signals were then added to the PI and static feed forward controller outputs. Running at a repetition rate of 120 Hz, the feedback converges within several hundred iterations or a few seconds.

These preliminary experimental results are shown in Fig. 9.13 and summarized in Table 9.1. The maximum, rms, and average values are all calculated during a 150 μs window which includes the beam turn on transient to capture the worst case scenario. The ES-based scheme is a >2× improvement over static feed-forward in terms of maximum errors and a >3× improvement in terms of rms error. With the currently used FPGA, the ES window lengths can be further reduced from 10 μs to 10 ns and with the latest FPGAs down to 1 ns, which will greatly improve the ES performance.

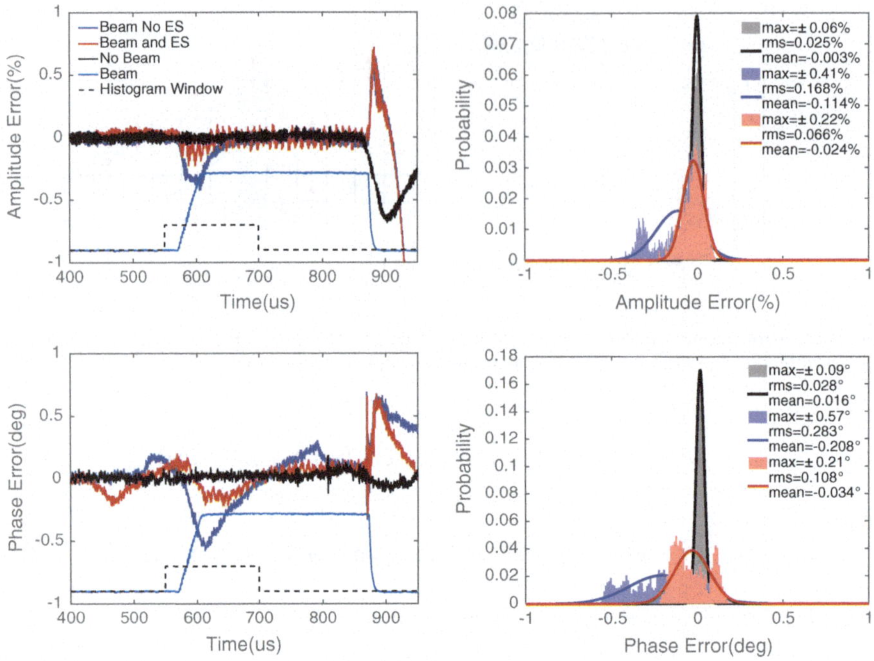

Fig. 9.13 Phase and amplitude errors shown before, during, and after beam turn-on transient. The histogram data shown is collected during the dashed histogram window, and cleaned up via 100 point moving average after raw data was sampled at 100 MS/s. **Black**: Beam OFF. **Blue**: Beam ON, feedback, and static feed-forward only. **Red**: Beam ON, feedback, static feed-forward, and iterative ES feed-forward

Table 9.1 ES performance during beam turn on transient

	No Beam	Beam, No ES	Beam and ES
max A error (%)	±0.06	±0.41	±0.22
rms A error (%)	0.025	0.168	0.066
mean A error (%)	−0.003	−0.114	−0.024
max θ error (%)	±0.09	±0.57	±0.21
rms θ error (%)	0.028	0.283	0.108
mean θ error (%)	0.016	−0.208	−0.034

9.3.4 ES for Magnet Tuning

ES has also been tested in hardware for magnet-based beam dynamics tuning, as described in Sect. 9.1.1. At the SPEAR3 synchrotron at LCLS, ES was used for continuous re-tuning of the eight parameter system shown in Fig. 9.14, in which the delay, pulse width, and voltage of two injection kickers, K_1 and K_2, as well as

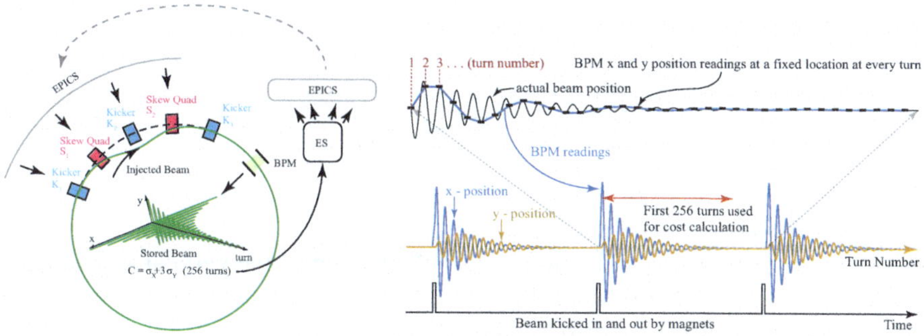

Fig. 9.14 Kicker magnets and skew quadrupole magnets. When the beam is kicked in and out of orbit, because of imperfect magnet matching, betatron oscillations occur, which are sampled at the BPM every time the beam completes a turn around the machine

the current of two skew quadrupoles S_1 and S_2, were tuned in order to optimize the injection kicker bump match, minimizing betatron oscillations. At SPEAR3, we simultaneously tuned 8 parameters: (1). $p_1 = K_1$ delay. (2). $p_2 = K_1$ pulse width. (3). $p_3 = K_1$ voltage. (4). $p_4 = K_2$ delay. (5). $p_5 = K_2$ pulse width. (6). $p_6 = K_2$ voltage. (7). $p_7 = S_1$ current. (8). $p_8 = S_2$ current. The parameters are illustrated in Figs. 9.14, 9.15. While controlling the voltage for the kicker magnets K_1, K_2, and the current for the skew quadrupole magnets S_1, S_2, in each case a change in the setting resulted in a change in magnetic field strength.

The cost function used for tuning was a combination of the horizontal, σ_x, and vertical, σ_y, variance of beam position monitor readings over 256 turns, the

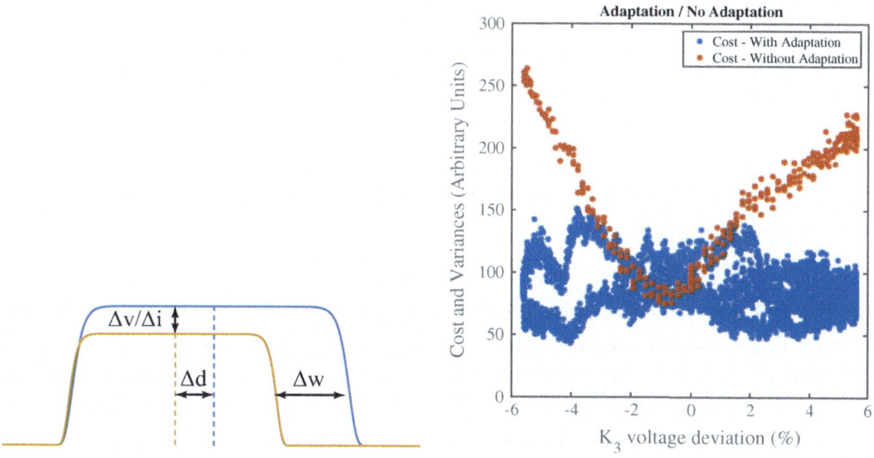

Fig. 9.15 **Left**: Kicker magnet delay (d), pulse width (w), and voltage (v) were adaptively adjusted, as well as the skew quadrupole magnet currents (i). **Right**: Comparison of beam quality with and without adaptation

minimization of which resulted in decreased betatron oscillations,

$$C = \sqrt{\frac{1}{256}\sum_{i=1}^{256}(x(i)-\bar{x})^2} + \sqrt{\frac{9}{256}\sum_{i=1}^{256}(y(i)-\bar{y})^2}$$
$$= \sigma_x + 3\sigma_y, \tag{9.67}$$

where the factor of 3 was added to increase the weight of the vertical oscillations, which require tighter control since the vertical beam size is much smaller and therefore users are more sensitive to vertical oscillations.

The cost was based on beam position monitor (BPM) measurements in the SPEAR3 ring based on a centroid x and y position of the beam recorded at each revolution, as shown in Fig. 9.14. Variances σ_x and σ_y were calculated based on this data, as in (9.67). Feedback was implemented via the experimental physics and industrial control system (EPICS) [47].

To demonstrate the scheme's ability to compensate for an uncertain, time-varying perturbation of the system, we purposely varied the voltage (and therefore resulting magnetic field strength) of the third kicker magnet, $K_3(t)$. The kicker voltage was varied sinusoidally over a range of $\pm 6\%$ over the course of 1.5 h, which is a very dramatic and fast change relative to actual machine parameter drift rates and magnitudes. The ES scheme was implemented by setting parameter values, kicking an electron beam out and back into the ring, and recording beam position monitor data for a few thousand turns. Based on this data the cost was calculated as in (9.67), based on a measurement of the horizontal and vertical variance of beam position monitor readings. The magnet settings were then adjusted, the beam was kicked again, and a new cost was calculated. This process was repeated and the cost was iteratively, continuously minimized.

Figure 9.14 shows the cost, which is a function of betatron oscillation, versus magnet setting $K_3(t)$, with and without ES feedback. For large magnetic field deviations, the improvement is roughly a factor of 2.5.

9.3.5 ES for Electron Bunch Longitudinal Phase Space Prediction

The Facility for Advanced Accelerator Experimental Tests (FACET) at SLAC National Accelerator Laboratory produces high energy electron beams for plasma wakefield acceleration [48]. For these experiments, precise control of the longitudinal beam profile is very important. FACET uses an x-band transverse deflecting cavity (TCAV) to streak the beam and measure the bunch profile (Fig. 9.16a). Although the TCAV provides an accurate measure of the bunch profile, it is a destructive measurement; the beam cannot be used for plasma wakefield acceleration (PWFA) once it has been streaked. In addition, using the TCAV to measure the bunch profile requires

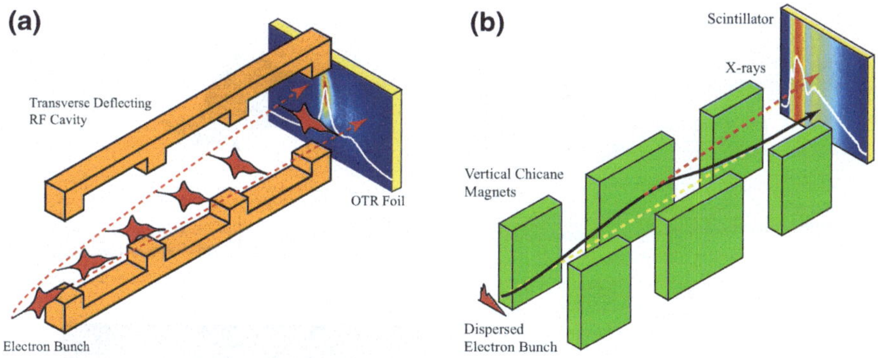

Fig. 9.16 The energy spectrum is recorded as the electron bunch passes through a series of magnets and radiates x-rays. The intensity distribution of the X-rays is correlated to the energy spectrum of the electron beam (**a**). This non-destructive measurement is available at all times, and used as the input to the ES scheme, which is then matched by adaptively tuning machine parameters in the simulation. For the TCAV measurement, the electron bunch is passed through a high frequency (11.4 GHz) RF cavity with a transverse mode, in which it is streaked and passes through a metallic foil (**b**). The intensity of the optical transition radiation (OTR) is proportional to the longitudinal charge density distribution. This high accuracy longitudinal bunch profile measurement is a destructive technique

adjusting the optics of the final focus system to optimize the resolution and accuracy of measurement. This makes it a time consuming process and prevents on-the-fly measurements of the bunch profile during plasma experiments.

There are two diagnostics that are used as an alternative to the TCAV that provide information about the longitudinal phase space in a non-destructive manner. The first is a pyrometer that captures optical diffraction radiation (ODR) produced by the electron beam as it passes through a hole in a metal foil. The spectral content of the ODR changes with bunch length. The pyrometer is sensitive to the spectral content and the signal it collects is proportional to $1/\sigma_z$, where σ_z is the bunch length. The pyrometer is an excellent device for measuring variation in the shot-to-shot bunch profile but provides no information about the shape of the bunch profile or specific changes to shape. The second device is a non-destructive energy spectrometer consisting of a half-period vertical wiggler located in a region of large horizontal dispersion. The wiggler produces a streak of X-rays with an intensity profile that is correlated with the dispersed beam profile. There X-rays are intercepted by a scintillating YAG crystal and imaged by a CCD camera (Fig. 9.16b). The horizontal profile of the x-ray streak is interpreted as the energy spectrum of the beam [49].

The measured energy spectrum is observed to correlate with the longitudinal bunch profile in a one-to-one manner if certain machine parameters, such as chicane optics, are fixed. To calculate the beam properties based on an energy spectrum measurement, the detected spectrum is compared to a simulated spectrum created with the 2D longitudinal particle tracking code, LiTrack [50]. The energy spread of short electron bunches desirable for plasma wakefield acceleration can be uniquely

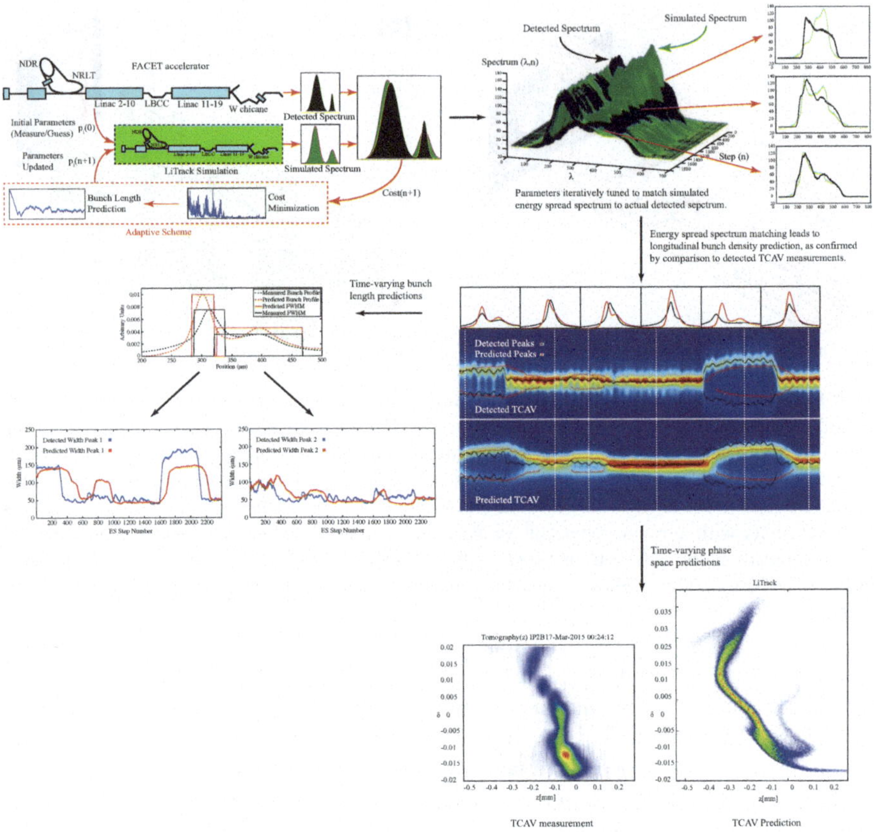

Fig. 9.17 ES scheme at FACET

correlated to the beam profile if all of the various accelerator parameters which influence the bunch profile and energy spread are accounted for accurately. Unfortunately, throughout the 2 km facility, there exist systematic phase drifts of various high frequency devices, mis-calibrations, and time-varying uncertainties due to thermal drifts. Therefore, in order to effectively and accurately relate an energy spectrum to a bunch profile, a very large parameter space must be searched and fit by LiTrack, which effectively limits and prevents the use of the energy spectrum measurement as a real time measurement of bunch profile.

Figures 9.16 and 9.17 show the overall setup of the tuning procedure at FACET. A simulation of the accelerator, LiTrack is run in parallel to the machines operation. The simulation was initialized with guesses and any available measurements of actual machine settings, $\mathbf{p} = (p_1, \ldots, p_n)$. We emphasize that these are only guesses because even measured values are noisy and have arbitrary phase shift errors. The electron beam in the actual machine was accelerated and then passed through a series of deflecting magnets, as shown in Figs. 9.16b and 9.17, which created X-rays, whose intensity distribution can be correlated to the electron bunch density via LiTrack. This non-destructive measurement is available at all times, and used as the input to the ES scheme, which is then matched by adaptively tuning machine parameters in the simulation. Once the simulated and actual spectrum were matched, certain beam properties could be predicted by the simulation.

Each parameter setting has its own influence on electron beam dynamics, which in turn influenced the separation, charge, length, etc, of the leading and trailing electron bunches.

The cost that our adaptive scheme was attempting to minimize was then the difference between the actual, detected spectrum, and that predicted by LiTrack:

$$C(\mathbf{x}, \hat{\mathbf{x}}, \mathbf{p}, \hat{\mathbf{p}}, t) = \int \left|\tilde{\psi}(\mathbf{x}, \mathbf{p}, t, \nu) - \hat{\psi}(\hat{\mathbf{x}}, \hat{\mathbf{p}}, t, \nu)\right|^2 d\nu, \tag{9.68}$$

in which $\tilde{\psi}(\mathbf{x}, \mathbf{p}, t, \nu)$ was a noisy measurement of the actual, time-varying (due to phase drift, thermal cycling...) energy spectrum, and $\hat{\psi}(\hat{\mathbf{x}}, \hat{\mathbf{p}}, t, \nu)$ was the LiTrack, simulated spectrum, $\mathbf{x}(t) = (x_1(t), \ldots, x_n(t))$ represents various aspects of the beam, such as bunch length, beam energy, bunch charge, etc. at certain locations throughout the accelerator, $\mathbf{p}(t) = (p_1(t), \ldots, p_n(t))$ represents various time-varying uncertain parameters of the accelerator itself, such as RF system phase drifts and RF field amplitudes throughout the machine, $\mathbf{x}(t)$ are approximated by their simulated estimates $\hat{\mathbf{x}}(t) = \left(\hat{x}_1(t), \ldots, \hat{x}_n(t)\right)$ and actual system parameters, $\mathbf{p}(t)$, are approximated by virtual parameters $\hat{\mathbf{p}}(t) = \left(\hat{p}_1(t), \ldots, \hat{p}_n(t)\right)$.

The problem was then to minimize the measurable, but analytically unknown function C, by adaptively tuning the simulation parameters $\hat{\mathbf{p}}$. The hope was that, by finding simulation machine settings which resulted in matched spectrums, we would also match other properties of the real and simulated beams, something we could not simply do by setting the simulation parameters to the exact machine settings, due to unknowns, such as time-varying, arbitrary phase shifts.

LiTrackES simulates large components of FACET as single elements. The critical elements of the simulation are the North Damping Ring (NDR) which sets the initial bunch parameters including the bunch length and energy spread, the North Ring to Linac (NRTL) which is the first of three bunch compressors, Linac Sectors 2–10 where the beam is accelerated and chirped, the second bunch compressor in Sector 10 (LBCC), Linac Sectors 11–19 where the beam is again accelerated and chirped, and finally the FACET W-chicane which is the third and final bunch compressor.

We calibrated the LiTrackES algorithm using simultaneous measurements of the energy spectrum and bunch profile while allowing a set of unknown parameters to converge. After convergence we left a subset of these calibrated parameters fixed, as they are known to vary slowly or not at all and performed our tuning on a much smaller subset of the parameters:

- p_1: NDR bunch length
- p_2: NRTL energy offset
- p_3: NRTL compressor amplitude

- p_4: NRTL chicane T_{566}
- p_5: Phase Ramp

"Phase ramp" refers to a net phase of the NDR and NRTL RF systems with respect to the main linac RF. Changing the phase ramp parameter results in a phase set offset in the linac relative to some desired phase.

LiTrackES, the combination of ES and LiTrack, as demonstrated, is able to provide a quasi real time estimate of many machine and electron beam properties which are either inaccessible or require destructive measurements. We plan to improve the convergence rate of LiTrackES by fine tuning the adaptive scheme's parameters, such as the gains k_i, perturbing amplitudes α_i and dithering frequencies ω_i. Furthermore, we plan on taking advantage of several simultaneously running LiTrackES schemes, which can communicate with each other in an intelligent way, and each of which has slightly different adaptive parameters/initial parameter guesses, which we believe can greatly increase both the rate and accuracy of the convergence. Another major goal is the extension of this algorithm from monitoring to tuning. We hope to one day utilize LiTrackES as an actual feedback to the machine settings in order to tune for desired electron beam properties.

9.3.6 ES for Phase Space Tuning

For the work described here, a measured XTCAV image was utilized and compared to the simulated energy and position spread of an electron bunch at the end of the LCLS as simulated by LiTrack. The electron bunch distribution is given by a function $\rho(\Delta E, \Delta z)$ where $\Delta E = E - E_0$ is energy offset from the mean or design energy of the bunch and $\Delta z = z - z_0$ is position offset from the center of the bunch. We worked with two distributions:

$$\begin{aligned} \text{XTCAV measured} &: \rho_{\text{TCAV}}(\Delta E, \Delta z), \\ \text{LiTrack simulated} &: \rho_{\text{LiTrack}}(\Delta E, \Delta z). \end{aligned}$$

These distributions were then integrated along the E and z projections in order to calculate 1D energy and charge distributions:

$$\begin{aligned} &\rho_{E,\text{TCAV}}(\Delta E), \quad \rho_{z,\text{TCAV}}(\Delta z), \\ &\rho_{E,\text{LiTrack}}(\Delta E), \quad \rho_{z,\text{LiTrack}}(\Delta z). \end{aligned}$$

Finally, the energy and charge spread distributions were compared to create cost values:

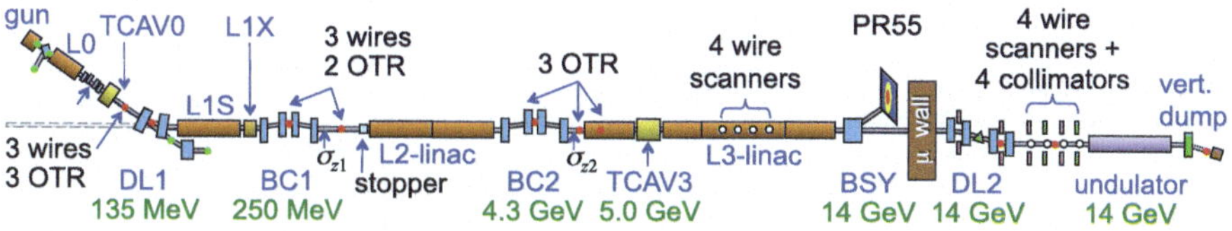

Fig. 9.18 Components of the LCLS beamline

$$C_E = \int \left[\rho_{E,\text{TCAV}}(\Delta E) - \rho_{E,\text{LiTrack}}(\Delta E)\right]^2 d\Delta E, \tag{9.69}$$

$$C_z = \int \left[\rho_{z,\text{TCAV}}(\Delta z) - \rho_{z,\text{LiTrack}}(\Delta z)\right]^2 d\Delta z, \tag{9.70}$$

whose weighted sum was combined into a single final cost:

$$C = w_E C_E + w_z C_z. \tag{9.71}$$

Iterative extremum seeking was then performed via finite difference approximation of the ES dynamics (Fig. 9.18):

$$\frac{\mathbf{p}(t + dt) - \mathbf{p}(t)}{dt} \approx \frac{d\mathbf{p}}{dt} = \sqrt{\alpha\omega}\cos(\omega t + kC(\mathbf{p}, t)), \tag{9.72}$$

by updating LiTrack model parameters, $\mathbf{p} = (p_1, \ldots, p_m)$, according to

$$p_j(n + 1) = p_j(n) + \Delta\sqrt{\alpha\omega_j}\cos\left(\omega_j n\Delta + kC(n)\right), \tag{9.73}$$

where the previous step's cost is based on the previous simulation's parameter settings,

$$C(n) = C(\mathbf{p}(n)). \tag{9.74}$$

The parameters being tuned were:

1. **L1S phase**: typically drifts continuously and is repeatedly corrected via an invasive phase scan. Within some limited range a correct bunch length can be maintained by the existing feedback system. This parameter is used for optimizing machine settings and FEL pulse intensity. When the charge off the cathode is changed, L1S phase must be adjusted manually.
2. **L1X phase**: must be changed if L1S phase is changed significantly. This linearizes the curvature of the beam.
3. **BC1 energy**: control bunch length and provides feedback to L1S amplitude.
4. **L2 phase**: drifts continuously with temperature, is a set of multiple Klystrons, all of which cycle in amplitude and phase. Feedback is required to introduce the correct energy chirp required for BC2 peak current/bunch length set point. Tuned to maximize FEL intensity and minimize jitter.

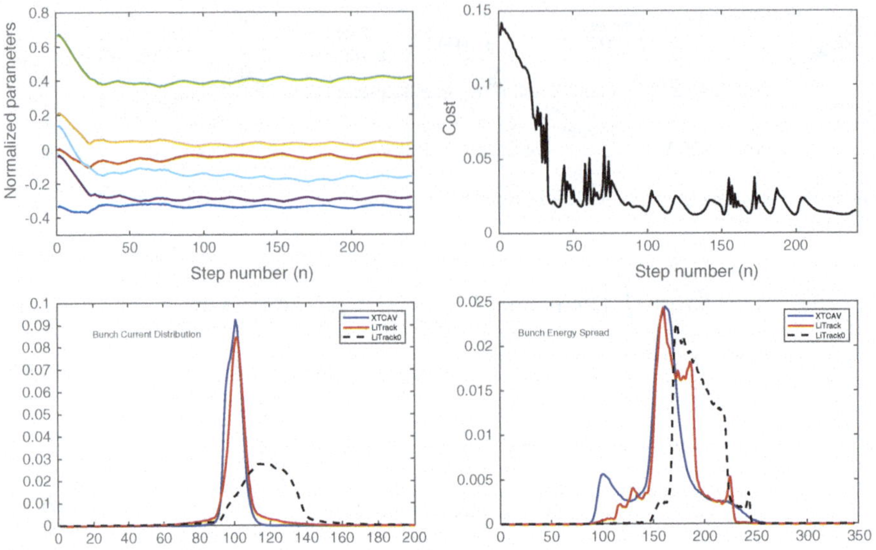

Fig. 9.19 Parameter convergence and cost minimization for matching desired bunch length and energy spread profiles

5. **BC2 energy**: drifts due to Klystron fluctuations, must be changed to optimize FEL pulse intensity for exotic setups.
6. **L3 phase**: drifts continuously with temperature, based on a coupled system of many Klystrons.

Machine tuning work has begun with general analytic studies as well as simulation-based algorithm development focused on the LCLS beam line, using SLACs LiTrack software, a code which captures most aspects of the electron beams phase space evolution and incorporates noise representative of operating conditions. The initial effort focused on developing ES-based auto tuning of the electron beam's bunch length and energy spread by varying LiTrack parameters in order to match LiTrack's output to an actual TCAV measurement taken from the accelerator by tuning bunch compressor energies and RF phases. The results are shown in Figs. 9.19 and 9.20. Running at a repetition rate of 120 Hz, the simulated feedback would have converged within 2 s on the actual LCLS machine.

Preliminary results have demonstrated that ES is a powerful tool with the potential to automatically tune an FEL between various bunch properties such as energy spread and bunch length requirements by simultaneously tuning multiple coupled parameters, based only on a TCAV measurement at the end of the machine. Although the simulation results are promising, It remains to be seen what the limitations of the method are in the actual machine in terms of getting stuck in local minima and time of convergence. We plan on exploring the extent of parameter and phase space through which we can automatically move.

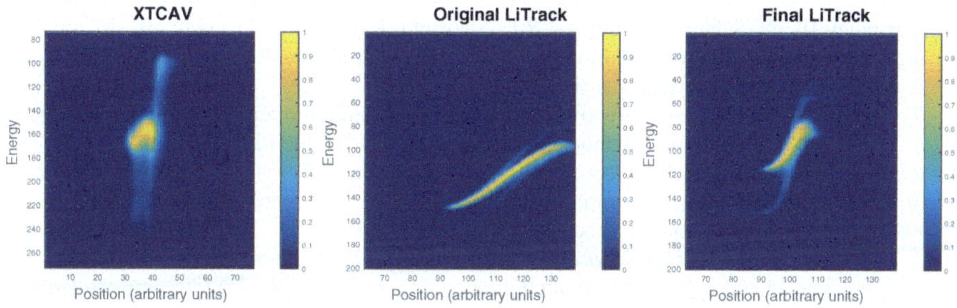

Fig. 9.20 Measured XTCAV, original LiTrack and final, converged LiTrack energy versus position phases space of the electron bunch shown

9.4 Conclusions

The intense bunch charges, extremely short bunch lengths, and extremely high energies of next generation FEL beams result in complex collective effects which couple transverse and longitudinal dynamics and therefore all of the RF and magnet systems and their influence on the quality of the light being produced. These future light sources, especially 4th generation FELs, face major challenges both in achieving extremely tight constraints on beam quality and in quickly tuning between various, exotic experimental setups. We have presented a very brief and simple introduction to some of the beam dynamics important to accelerators and have introduced some methods for achieving better beam quality and faster tuning. Based on preliminary results it is our belief is that a combination of machine learning and advanced feedback methods such as ES have great potential towards meeting the requirements of future light sources. Such a combination of ES and machine learning has recently been demonstrated in a proof of principle experiment at the Linac-Coherent Light Source FEL [51]. During this experiment we quickly trained a simple neural network to obtain an estimate of a complex and time-varying parameter space, mapping longitudinal electron beam phase space (energy vs time) to machine parameter settings. For a target longitudinal phase space, we used the neural network to give us an initial guess of the required parameter settings which brought us to within a neighborhood of the correct parameter settings, but did not give a perfect match. We then used ES-based feedback to zoom in on and track the actual optimal time-varying parameters settings.

References

1. T.O. Raubenheimer, Technical challenges of the LCLS-II CW X-RAY FEL, in *Proceedings of the International Particle Accelerator Conference*, Richmond, VA, USA (2015)
2. C. Schmidt et al., Recent developments of the European XFEL LLRF system, in *Proceedings of the International Particle Accelerator Conference*, Shanghai, China (2013)

3. J. Bradley III, A. Scheinker, D. Rees, R.L. Sheffield, High power RF requirements for driving discontinuous bunch trains in the MaRIE LINAC, in *Proceedings of the Linear Particle Accelerator Conference*, East Lansing, MI, USA (2016)
4. R. Sheffield, Enabling cost-effective high-current burst-mode operation in superconducting accelerators. Nucl. Instrum. Methods Phys. Res. A **758**, 197–200 (2015)
5. R. Akre, A. Brachmann, F.J. Decker, Y.T. Ding, P. Emma, A.S. Fisher, R.H. Iverson, Tuning of the LCLS Linac for user operation, in *Conf. Proc. C110328: 2462-2464*, 2011 (No. SLAC-PUB-16643) (SLAC National Accelerator Laboratory, 2016)
6. A. Scheinker, Ph.D. thesis, University of California, San Diego, Nov 2012
7. A. Scheinker, Model independent beam tuning, in *Proceedings of the 4th International Particle Accelerator Conference*, Beijing, China (2012)
8. A. Scheinker, X. Huang, J. Wu, Minimization of betatron oscillations of electron beam injected into a time-varying lattice via extremum seeking. IEEE Trans. Control Syst. Technol. (2017). https://doi.org/10.1109/TCST.2017.2664728
9. A. Scheinker, D. Scheinker, Bounded extremum seeking with discontinuous dithers. Automatica **69**, 250–257 (2016)
10. A. Scheinker, D. Scheinker, Constrained extremum seeking stabilization of systems not affine in control. Int. J. Robust Nonlinear Control (to appear) (2017). https://doi.org/10.1002/rnc.3886
11. A. Scheinker, X. Pang, L. Rybarcyk, Model-independent particle accelerator tuning. Phys. Rev. Accel. Beams **16**(10), 102803 (2013)
12. A. Scheinker, S. Baily, D. Young, J. Kolski, M. Prokop, In-hardware demonstration of model-independent adaptive tuning of noisy systems with arbitrary phase drift. Nucl. Instrum. Methods Phys. Res. Sect. A **756**, 30–38 (2014)
13. A. Scheinker, S. Gessner, Adaptive method for electron bunch profile prediction. Phys. Rev. Accel. Beams **18**(10), 102801 (2015)
14. S.G. Biedron, A. Edelen, S. Milton, Advanced controls for accelerators, in *Compact EUV & X-ray Light Sources* (Optical Society of America, 2016), p. EM9A-3
15. A.L. Edelen, S.G. Biedron, B.E. Chase, D. Edstrom, S.V. Milton, P. Stabile, Neural networks for modeling and control of particle accelerators. IEEE Trans. Nucl. Sci. **63**(2), 878–897 (2016)
16. Y.B. Kong, M.G. Hur, E.J. Lee, J.H. Park, Y.D. Park, S.D. Yang, Predictive ion source control using artificial neural network for RFT-30 cyclotron. Nucl. Instrum. Methods Phys. Res. Sect. A: Accel. Spectrom. Detect. Assoc. Equip. **806**, 55–60 (2016)
17. M. Buchanan, Depths of learning. Nat. Phys. **11**(10), 798–798 (2015)
18. X. Huang, J. Corbett, J. Safranek, J. Wu, An algorithm for online optimization of accelerators. Nucl. Instrum. Methods Phys. Res. Sect. A: Accel. Spectrom. Detect. Assoc. Equip. **726**, 77–83 (2013)
19. T.P. Wangler, *RF Linear Accelerators* (Wiley, 2008)
20. R. Ruth, Single particle dynamics in circular accelerators, in *AIP Conference Proceedings*, vol. 153, No. SLAC-PUB-4103 (1986)
21. H. Wiedemann, *Particle Accelerator Physics* (Springer, New York, 1993)
22. D.A. Edwards, M.J. Syphers, *An Introduction to the Physics of High Energy Accelerators* (Wiley-VCH, 2004)
23. S.Y. Lee, *Accelerator Physics* (World Scientific Publishing, 2004)
24. M. Reiser, *Theory and Design of Charged Particle Beams* (Wiley-VCH, 2008)
25. C.X. Wang, A. Chao, Transfer matrices of superimposed magnets and RF cavity, No. SLAC-AP-106 (1996)
26. M.G. Minty, F. Zimmermann, *Measurement and Control of Charged Particle Beams* (Springer, 2003)
27. J.C. Slater, Microwave electronics. Rev. Modern Phys. **18**(4) (1946)
28. J. Jackson, *Classical Electrodynamics* (Wiley, NJ, 1999)
29. M. Borland, Report No. APS LS-287 (2000)
30. R. Hajima, N. Taked, H. Ohashi, M. Akiyama, Optimization of wiggler magnets ordering using a genetic algorithm. Nucl. Instrum. Methods Phys. Res. Sect. A **318**, 822 (1992)

31. I. Bazarov, C. Sinclair, Multivariate optimization of a high brightness dc gun photo injector. Phys. Rev. ST Accel. Beams **8**, 034202 (2005)
32. L. Emery, in *Proceedings of the 21st Particle Accelerator Conference*, Knoxville, 2005 (IEEE, Piscataway, NJ, 2005)
33. M. Borland, V. Sajaev, L. Emery, A. Xiao, in *Proceedings of the 23rd Particle Accelerator Conference*, Vancouver, Canada, 2009 (IEEE, Piscataway, NJ, 2009)
34. L. Yang, D. Robin, F. Sannibale, C. Steier, W. Wan, Global optimization of an accelerator lattice using multiobjective genetic algorithms. Nucl. Instrum. Methods Phys. Res. Sect. A **609**, 50 (2009)
35. A. Poklonskiy, D. Neuffer, Evolutionary algorithm for the neutrino factory front end design. Int. J. Mod. Phys. A **24**, 5 (2009)
36. W. Gao, L. Wang, W. Li, Simultaneous optimization of beam emittance and dynamic aperture for electron storage ring using genetic algorithm. Phys. Rev. ST Accel. Beams **14**, 094001 (2011)
37. R. Bartolini, M. Apollonio, I.P.S. Martin, Multiobjective genetic algorithm optimization of the beam dynamics in linac drivers for free electron lasers. Phys. Rev. ST Accel. Beams **15**, 030701 (2012)
38. A. Hofler, B. Terzic, M. Kramer, A. Zvezdin, V. Morozov, Y. Roblin, F. Lin, C. Jarvis, Innovative applications of genetic algorithms to problems in accelerator physics. Phys. Rev. ST Accel. Beams **16**, 010101 (2013)
39. X. Huang, J. Safranek, Nonlinear dynamics optimization with particle swarm and genetic algorithms for SPEAR3 emittance upgrade. Nucl. Instrum. Methods Phys. Res. Sect. A **757**, 48–53 (2014)
40. K. Tian, J. Safranek, Y. Yan, Machine based optimization using genetic algorithms in a storage ring. Phys. Rev. Accel. Beams **17**, 020703 (2014)
41. X. Huang, J. Corbett, J. Safranek, J. Wu, An algorithm for online optimization of accelerators. Nucl. Instrum. Methods Phys. Res. A **726**, 77–83 (2013)
42. H. Ji, S. Wang, Y. Jiao, D. Ji, C. Yu, Y. Zhang, X. Huang, Discussion on the problems of the online optimization of the luminosity of BEPCII with the robust conjugate direction search method, in *Proceedings of the International Particle Accelerator Conference*, Shanghai, China (2015)
43. X. Huang, J. Safranek, Online optimization of storage ring nonlinear beam dynamics. Phys. Rev. ST Accel. Beams **18**(8), 084001 (2015)
44. A.L. Edelen et al., Neural network model of the PXIE RFQ cooling system and resonant frequency response (2016). arXiv:1612.07237
45. B.E. Carlsten, K.A. Bishofberger, S.J. Russell, N.A. Yampolsky, Using an emittance exchanger as a bunch compressor. Phys. Rev. Spec. Top.-Accel. Beams **14**(8), 084403 (2011)
46. P.L. Kapitza, Dynamic stability of a pendulum when its point of suspension vibrates. Sov. Phys. JETP **21**, 588–592 (1951)
47. R.L. Dalesio, J.O. Hill, M. Kraimer, S. Lewis, D. Murray, S. Hunt, W. Watson, M. Clausen, J. Dalesio, The experimental physics and industrial control system architecture: past, present, and future. Nucl. Instrum. Methods Phys. Res. Sect. A **352**(1), 179–184 (1994)
48. M.J. Hogan, T.O. Raubenheimer, A. Seryi, P. Muggli, T. Katsouleas, C. Huang, W. Lu, W. An, K.A. Marsh, W.B. Mori, C.E. Clayton, C. Joshi, Plasma wakefield acceleration experiments at FACET. New J. Phys. **12**, 055030 (2010)
49. J. Seeman, W. Brunk, R. Early, M. Ross, E. Tillman, D. Walz, SLC energy spectrum monitor using synchrotron radiation. SLAC-PUB-3495 (1986)
50. K. Bane, P. Emma, *LiTrack*: a fast longitudinal phase space tracking code. SLAC-PUB-11035 (2005)
51. A. Scheinker, A. Edelen, D. Bohler, C. Emma, and A. Lutman, Demonstration of model-independent control of the longitudinal phase space of electron beams in the Linac-coherent light source with Femtosecond resolution. Phys. Rev. Lett. **121**(4), 044801 (2018)

Index

T. Lookman et al. (eds.), *Materials Discovery and Design*, Springer Series in Materials Science 280, https://doi.org/10.1007/978-3-319-99465-9

The manufacturer's authorised representative in the EU is Springer Nature Customer Service Centre GmbH, Europaplatz 3, 69115 Heidelberg, Germany. If you have any concerns regarding our products, please contact ProductSafety@springernature.com

Printed and bound by CPI Group (UK) Ltd, Croydon, CR0 4YY
05/10/2026
02228138-0011